Tourism and Development in the Developing World

Tourism is widely considered to be an effective tool for economic development, particularly in less-developed countries. However, despite almost universal recognition of tourism's development potential, the extent to which economic and social progress is linked to the growth of a country's tourism sector remains the subject of intense debate. *Tourism and Development in the Developing World* offers a thorough overview of the tourism–development relationship. Focusing specifically on the less-developed world and drawing on contemporary case studies, this updated second edition questions widely held assumptions on the role of tourism in development and seeks to highlight the challenges faced by destinations seeking to achieve development through tourism.

The introductory chapter establishes the foundation for the book, exploring the meaning and objectives of development, reviewing theoretical perspectives on the developmental process, and assessing the reasons why less-developed countries are attracted to tourism as a development option. The concept of sustainable development, as the most widely adopted contemporary model of development, is then introduced and its links with tourism critically assessed. Subsequent chapters explore the key issues associated with tourism and development, including the rise of globalization; the tourism planning and development process; the relationship between tourism and communities within which it is developed; the management implications of trends in the demand for and uptake of tourism; and an analysis of the consequences of tourism development for destination environments, economies and societies. A new chapter considers the challenges of climate change, sustainability of resource supply (oil, water and food), global economic instability, political instability and changing demographics. Finally, the issues raised throughout the book are drawn together in a concluding chapter that assesses the tourism and development 'dilemma'.

Combining an overview of essential concepts, theories and knowledge with an analysis of contemporary issues and debates in tourism and development, this new edition will be an invaluable resource for those investigating tourism issues in developing countries. The book will be of interest to students of tourism, development, geography and area studies, international relations and politics, and sociology.

David J. Telfer is Associate Professor in the Department of Tourism Management at Brock University, Canada.

Richard Sharpley is Professor of Tourism and Development, University of Central Lancashire, UK.

Routledge Perspectives on Development

Series Editor: Professor Tony Binns, University of Otago

Since it was established in 2000, the same year as the Millennium Development Goals were set by the United Nations, the *Routledge Perspectives on Development* series has become the pre-eminent international textbook series on key development issues. Written by leading authors in their fields, the books have been popular with academics and students working in disciplines such as anthropology, economics, geography, international relations, politics and sociology. The series has also proved to be of particular interest to those working in interdisciplinary fields, such as area studies (African, Asian and Latin American studies), development studies, environmental studies, peace and conflict studies, rural and urban studies, travel and tourism.

If you would like to submit a book proposal for the series, please contact the Series Editor, Tony Binns, on: jab@geography.otago.ac.nz

Published:

Third World Cities, 2nd Edition
David W. Drakakis-Smith

Rural–Urban Interactions in the Developing World
Kenneth Lynch

Environmental Management and Development
Chris Barrow

Tourism and Development in the Developing World
Richard Sharpley and David J. Telfer

Southeast Asian Development
Andrew McGregor

Postcolonialism and Development
Cheryl McEwan

Conflict and Development
Andrew Williams and Roger MacGinty

Disaster and Development
Andrew Collins

Non-Governmental Organisations and Development
David Lewis and Nazneen Kanji

Cities and Development
Jo Beall

Gender and Development, 2nd Edition
Janet Momsen

Economics and Development Studies
Michael Tribe, Frederick Nixson and Andrew Sumner

Water Resources and Development
Clive Agnew and Philip Woodhouse

Theories and Practices of Development, 2nd Edition
Katie Willis

Food and Development
E. M. Young

An Introduction to Sustainable Development, 4th Edition
Jennifer Elliott

Latin American Development
Julie Cupples

Religion and Development
Emma Tomalin

Development Organizations
Rebecca Schaaf

Climate Change and Development
Thomas Tanner and Leo Horn-Phathanothai

Global Finance and Development
David Hudson

Population and Development, 2nd Edition
W. T. S. Gould

Conservation and Development
Andrew Newsham and Shonil Bhagwat

Tourism and Development in the Developing World, 2nd Edition
David J. Telfer and Richard Sharpley

Forthcoming:

Cities and Development, 2nd Edition
Sean Fox and Tom Goodfellow

Conflict and Development, 2nd Edition
Andrew Williams and Roger MacGinty

Children, Youth and Development, 2nd Edition
Nicola Ansell

South Asian Development
Trevor Birkenholtz

Gender and Development, 3rd Edition
Janet Momsen

Natural Resource Extraction and Development
Roy Maconachie and Gavin M. Hilson

Tourism and Development in the Developing World

SECOND EDITION

David J. Telfer and Richard Sharpley

LONDON AND NEW YORK

Second edition published 2016
by Routledge
2 Park Square, Milton Park, Abingdon, Oxon OX14 4RN

and by Routledge
711 Third Avenue, New York, NY 10017

Routledge is an imprint of the Taylor & Francis Group, an informa business

First edition published by Routledge 2008

British Library Cataloguing in Publication Data
A catalogue record for this book is available from the British Library

Library of Congress Cataloging in Publication Data
Telfer, David J.
Tourism and development in the developing world/
David J. Telfer and Richard Sharpley. – Second edition.
pages cm. – (Routledge perspectives on development)
Includes bibliographical references and index.
1. Tourism – Developing countries. 2. Sustainable development –
Developing countries. I. Sharpley, Richard, 1956–
II. Title. G155.A1S476 2015
338.4'791091724 – dc23
2015025859

ISBN: 978-1-138-92173-3 (hbk)
ISBN: 978-1-138-92174-0 (pbk)
ISBN: 978-1-315-68619-6 (ebk)

Typeset in Times New Roman and Franklin Gothic
by Florence Production Ltd, Stoodleigh, Devon, UK

Loughborough
COLLEGE est 1909

To Olivia, Rosie, Kyoko and Sakura

Contents

Plates

Figures

Tables

Boxes

Preface to the second edition

In 2008, the year that the first edition of this book was published, global international tourist arrivals totalled 917 million. Also in that year, the world faced a financial crisis that represented a threat both to the future growth of tourism and to destinations that depend upon tourism for their development. However, despite an expected decline the following year, by 2010 international arrivals had recovered and increased to 940 million and, since then, have continued to grow. Indeed, a major milestone was reached in 2012 when for the first time annual international arrivals exceeded one billion. Moreover, much of that growth has occurred in the developing world, with significant implications for the role of tourism in development.

At the same time, since the first edition was published, significant advances have been made in knowledge and understanding of the relationship between tourism and development in general and tourism development processes in particular. Approaches to tourism development such as community-based tourism, pro-poor tourism, ecotourism and volunteer tourism are more widely critiqued, while the concept of responsibility in tourism supply and demand has gained wider currency. Equally, major transformations have occurred in perceptions and knowledge of, and approaches to, development. Grand meta-theories of development, including sustainable development, are increasingly criticized as attention turns to specific developmental challenges such as poverty reduction, equity and human rights, and issues such as the politics of failing states, all of which are of direct relevance to the study of tourism and development.

It must also be recognized that the political, economic and sociocultural environments within which tourism occurs have undergone, and continue to experience, significant changes. While some such changes, such as increased deregulation of markets, have proved to be beneficial, the potential contribution of tourism as an agent of development faces a number of emerging challenges, not least climate change. However, other challenges also exist, including the sustainability of resource supply, global economic instability, political instability and changing demographics.

The purpose of the second edition is to explore what we refer to as the tourism development dilemma taking into account these transformations in both tourism itself and in knowledge and understanding of tourism and development practices and processes. Each chapter has been revised and fully updated, drawing wherever possible on the most recent data, concepts, policies and academic debates. In particular, it advances the discussion of development goals and processes beyond what has become the 'impasse' of sustainable development, exploring in greater detail approaches to tourism development that reflect the more 'human-centred' nature of contemporary development, as well as the more recent shift towards what is referred to as 'global development'. It also includes a new chapter exploring contemporary and future challenges to tourism and development. In so doing, it builds on the first edition to present a contemporary, critical yet succinct exploration of the relationship between tourism and development.

Preface to the first edition

Tourism is increasingly being seen as an attractive development option for many parts of the developing world. In some developing nations, it may in fact be the only viable means of stimulating development. However, as developing countries opt into this industry, they face what is referred to in this book as a tourism development dilemma.

Developing nations are seeking the potential benefits of tourism, such as increased income, foreign exchange, employment and economic diversification; nevertheless, these developmental benefits may in fact fail to materialize. In entering this global competitive industry, developing countries may find tourism only benefits the local elite or multinational corporations or is achieved at significant economic, social or environmental costs. The challenge in this dilemma is then accepting or managing the negative outcomes of the tourism development process for the potential long-term benefits offered by tourism.

The purpose of this book is to explore the nature of the tourism development dilemma by investigating the challenges and opportunities facing developing countries pursuing tourism as a development option. The book begins with an examination of the nature of developing countries along with why they are attracted to such a volatile industry as a preferred development tool. It is important to consider to what extent tourism can contribute to overall development broadly defined, and so Chapter 1 also examines the evolution of development thought whereby development is no longer tied solely to economic criteria. The second chapter examines the nature of sustainable development and its relationship to tourism, which has become a much-contested concept focusing not only on the physical environment, but also on the economic, social and cultural environment. This chapter sets the stage for the remainder of the text by raising key issues, including the influence of globalization on tourism (Chapter 3), the tourism planning and development process (Chapter 4), community responses to tourism (Chapter 5), consumption of tourism (Chapter 6) and an analysis of tourism impacts (Chapter 7). The concluding chapter draws together the main issues in the book, presenting a tourism development dilemma framework that illustrates the complexity of often-interconnected forces at work in using tourism as a development tool. While it is argued that there is a development imperative and a sustainable development imperative, it is important to recognize the challenges of implementing the ideals of sustainability in the context of the realities in the tourism industry in developing countries.

The focus of the book is to present an introductory-level text that explores the relationship between tourism and development, and it is designed in part to be a successor to John Lea's 1988 *Tourism and Development in the Third World*, originally published in the Routledge *Introductions to Development* series.

 # Acknowledgements

The authors would like to thank Andrew Mould, Sarah Gilkes and all their colleagues at Routledge for their patience and assistance with this second edition. We would also like to thank Hui Di Wang and Tom and Hazel Telfer for their photographs. Finally, we would also like to thank, as always, Julia Sharpley and Atsuko Hashimoto for their support during the writing of the second edition of the book.

The cover photo (beach tourism in Zanzibar) is by Richard Sharpley.

1 Introduction: tourism in developing countries

Learning objectives

When you have finished reading this chapter, you should be able to:

- appreciate the characteristics of underdevelopment in developing countries
- understand why tourism is selected as a development option for developing countries
- identify global tourism market shares and the changing nature of tourism
- be familiar with the different approaches to tourism and development

Over the last 60 years, tourism has evolved into one of the world's most powerful, yet controversial, socio-economic forces. As ever-greater numbers of people have achieved the ability, means and freedom to travel, not only has tourism become increasingly democratized (Urry and Larsen 2012), but also both the scale and scope of tourism have grown remarkably. In 1950, for example, just over 25 million international tourist arrivals were recorded worldwide. By 2012, that figure had surpassed the one billion mark (UNWTO 2013a), or, putting it another way, 2012 was the first year in the history of tourism that more than 1,000 million international cross-border movements (to be precise, 1,035 million) were made by people classified as tourists. Since then, international tourism has continued its inexorable growth, with international arrivals expected to have exceeded 1.1 billion in 2014. Moreover, if domestic tourism (that is, people visiting destinations within their own country) is included, the total global volume of tourist trips is estimated to be between six and 10 times higher than that figure. For example, in China alone, an estimated 2.61 billion domestic tourism trips, or more than double the total number of worldwide international arrivals, were made in 2011 (National Bureau of Statistics 2013). Little wonder, then, that tourism has been described as the 'largest peaceful movement of people across cultural boundaries in the history of the world' (Lett 1989: 265).

As participation in tourism has grown, so too has the number of countries that play host to tourists. Although just 12 nations accounted for almost half of all international arrivals in 2013, including traditionally popular destinations such as France, Spain, the USA and the UK, not only have a number of newer destinations, such as Turkey, Thailand and Malaysia, joined the list of top international tourist destinations, but also many others have claimed a place on the international tourism map. At the same time, numerous more distant, exotic places have, in recent years, enjoyed a rapid increase in tourism. Indeed, throughout the last decade, the Asia and Pacific region in particular has witnessed the highest and most sustained growth in arrivals globally, with a number of nations in that region, such as Cambodia, Myanmar, Thailand and Vietnam, having experienced higher than world average growth in tourism (UNWTO 2014a). Such is the global scale of contemporary tourism that the WTO currently publishes annual tourism statistics for around 200 countries.

Reflecting this dramatic growth in scale and scope, tourism's global economic contribution has also become increasingly significant. International tourism alone generated over US$1,159 billion in 2013 (UNWTO 2014a), and, if current forecasts prove to be correct, this figure could rise to US$2 trillion by 2020 (WTO 1998). By the end of the last century, tourism also represented the world's most valuable export category, and although it subsequently fell back to fourth place, it is now, according to the World Travel and Tourism Council, the world's fifth largest 'industry' in terms of direct GDP after financial services, mining, communication services and education. If indirect and induced economic contribution is included, then tourism overall generates more than $6.3 trillion annually, accounting for roughly 10 per cent of global GDP and employment (WTTC 2012).

Given this remarkable growth and economic significance, it is not surprising that tourism has long been considered an effective means of achieving economic and social development in destination areas. Indeed, the most common justification for the promotion of tourism is its potential contribution to development, particularly in the context of developing countries. That is, although it is an important economic sector, and frequently a vehicle of both rural and urban economic regeneration in many industrialized nations, it is within the developing world that attention is most frequently focused upon tourism as a developmental catalyst. In many such countries, not only has tourism long been an integral element of national development strategies (Jenkins 1991) – though often, given a lack of viable alternatives, an option of 'last resort' (Lea 1988) – but also it has become an increasingly significant sector of the economy, providing a vital source of employment, income and foreign exchange, as well as a potential means of redistributing wealth from the richer nations of the world. As the 2001 UN Conference on Trade and Development noted, for example, 'tourism development appears to be one of the most valuable avenues for reducing the marginalization of LDCs from the global economy' (UNCTAD 2001: 1).

Importantly, though, the introduction of tourism does not inevitably set a nation on the path to development. In other words, many developing countries are, at first sight, benefiting from an increase in tourist arrivals and consequential foreign exchange earnings. However, the unique characteristics of tourism as a social and economic activity, the complex relationships between the various elements of the international tourism system and transformations in the global political economy of which tourism is a part all serve to reduce its potential developmental contribution. Not only is tourism highly susceptible to external forces and events, such as political upheaval (e.g. Egypt, where continuing political instability following the 2011 Arab Spring revolution resulted in tourism revenues in 2013 remaining approximately half the 2010 figure), natural disasters (e.g. the Indian Ocean tsunami in December 2004) or health scares (e.g. Mexico in 2009), but many countries have become increasingly dependent upon tourism as an economic sector, which by and large remains dominated by wealthier, industrialized nations (Reid 2003). Moreover, the political, economic and social structures within developing countries frequently restrict the extent to which the benefits of tourism development are realized. The factors that influence tourism's potential developmental contribution are summarized in Figure 1.1.

Figure 1.1 *Influences on tourism's development*

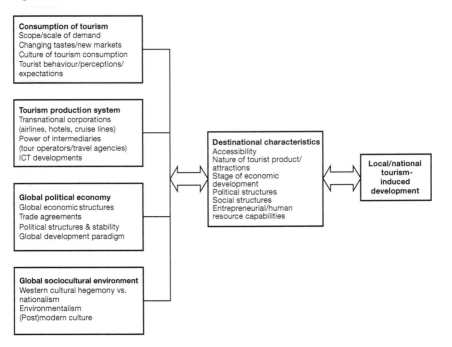

Many of these issues will be addressed throughout this book. However, the fundamental point is that there exists what may be referred to as a tourism development dilemma. That is, tourism undoubtedly represents a potentially attractive (and frequently the only viable) means of stimulating social and economic development in destination areas and nations, yet frequently that development fails to materialize, benefits only local elites or is achieved at significant economic, social or environmental cost to local communities. The dilemma for many developing countries, therefore, lies in the challenge of accepting or managing such negative consequences for the potential longer-term benefits offered by tourism development.

The purpose of this book is to explore these challenges and opportunities facing developing countries that pursue tourism as a development option. In so doing, it will critically appraise contemporary perspectives on tourism and development, and, in particular, sustainable tourism development, which, despite increasing doubts with respect to its legitimacy and viability, remains the dominant tourism development paradigm. However, the first task is to consider the concepts of underdevelopment/development and the relevance of tourism as a development option. It is with this that the rest of this introductory chapter is concerned.

Focus and definitions

As noted above, this book is primarily concerned with tourism in developing countries. The term 'developing country' is, of course, subject to wide interpretation and is often used interchangeably with other terminology, such as 'Third World' or 'less-developed country' or, more generally, 'the South'. However, it usefully contrasts a country or group of countries (the 'developing world') with those that are 'developed', although, similarly, there is no established convention for defining a nation as 'developed'. Nevertheless, the developed countries of the world, those that are technologically and economically advanced, that enjoy a relatively high standard of living and have modern social and political structures and institutions, are generally considered to include Japan, Australia and New Zealand in Oceania, Canada and the USA in North America, and the countries that formerly comprised Western Europe. Some commentators also include Israel, Singapore, Hong Kong and South Korea as developed countries.

Of course, categorizing the countries of the world as either 'developed' or 'developing' oversimplifies a complex global political economy. Not only does the developing world include countries that vary enormously in terms of economic and social development, with some, such as the BRIC group – Brazil, Russia, India and China – as well as the South East Asian 'tiger' economies, assuming an ever-increasingly important position in the global economy, but new trade or political alignments cut across the developed–developing dichotomy. For example, the so-called G20, or Group of Twenty, established in 1999, promotes dialogue between industrialized and those emerging-market countries not considered to be adequately included in global economic discussion and governance (see www.g20.org). The group comprises Argentina, Australia, Brazil, Canada, China, France, Germany, India, Indonesia, Italy, Japan, Mexico, Russia, Saudi Arabia, South Africa, South Korea, Turkey, the UK and the USA, countries that account for 90 per cent of global GDP and 80 per cent of world trade.

Nevertheless, for the purposes of this book, the term 'developing country' embraces all nation states that are not generally recognized as being developed, including the transitional economies of the former 'Second World' and contemporary, centrally planned economies. Though covering an enormous diversity of countries that may demand subcategorization, this focus nevertheless reflects Britton's (1982) metropolitan/periphery political-economic model, which, arguably, still defines the structure of international tourism, though less so than in the past. That is, despite the increasing numbers of developing countries with significant tourism sectors relative to both their national economies and global tourism more generally, recent figures reflect the continuing domination of the developed world in terms of both international arrivals and receipts; in 2012, the 32 developed countries collectively accounted for approximately 49 per cent of global arrivals and 54 per cent of global receipts, albeit a reduction on the 2002 figures of 54 per cent and 61 per cent, respectively. However, according to the UNWTO (2014b: 13), 'By 2030, the majority of all international tourist arrivals (57 per cent) will be in emerging economy destinations', pointing to a shift in the balance of world tourism and, indeed, the importance of considering tourism and development in these countries as a whole.

The terms 'tourism' and 'development' also require definition. Regarding tourism, most introductory tourism texts consider the issue in some depth (for example, Sharpley 2002; Fletcher *et al.* 2013), while, generally, many definitions have been proposed. However, these may be classified under two principal headings:

Technical definitions attempt to identify different categories of tourist for statistical or legislative purposes. Various parameters have been established to define a tourist, such as minimum (one day) and maximum (one year) lengths of stay, minimum distance from home travelled (160 km) and purpose, such as 'holiday' or 'business' (WTO/UNSTAT 1994), though a useful overall definition has been proposed by the UK's Tourism Society:

> Tourism is the temporary short-term movement of people to destinations outside the places where they normally live and work, and their activities during their stay at these places; it includes movement for all purposes, as well as day visits or excursions.

Conceptual definitions attempt to convey the meaning or function of tourism as a particular social institution (see Burns and Holden 1995; Sharpley 2008). Typically, these emphasize the nature of tourism as a leisure activity that contrasts with normal, everyday life (perhaps the most commonly held perception of what tourism is) and provide a basis for assessing tourist behaviour and attitudes. Interestingly, some commentators suggest that tourism has become enmeshed in contemporary social life to such an extent that it can no longer be seen in isolation from other social practices. In other words, tourism is 'increasingly being interpreted as but one, albeit highly significant, dimension of social mobility' (Hall 2005: 21), and hence should be studied within the framework of the mobilities paradigm. For the purposes of this book, however, tourism is considered to be an identifiable and quantifiable social activity.

Development is a more complex concept and one that 'seems to defy definition' (Cowen and Shenton 1996: 3); despite many attempts to clarify its meaning, it remains a contested concept, reflecting the fact, perhaps, that development can or should be thought of only in relation to the needs or aims of particular societies and the ways in which those societies seek to address their societal challenges (Hettne 2009). Moreover, according to the post-development school, it is also a global concept that, over the last half-century, has failed in its objectives and should, therefore, be abandoned (Rahnema and Bawtree 1997). Nevertheless, it remains a term in common usage, and one that refers to both a process that societies undergo and the goal or outcome of that process – that is, the development process in a society may result in it achieving the state or condition of development. It is also a term that, although most usually considered in the context of the developing countries, relates to every nation in the world. In other words, a society that is 'developed' does not cease to change or progress (or even regress); the nature of that change may, however, be different to changes in less-developed societies.

Traditionally, development was measured in economic terms, typically GNP or per capita GDP. Indeed, during the 1950s and 1960s, development and economic growth were considered to be synonymous (Mabogunje 1980). However, as Seers (1969) argued, this revealed nothing about improvements (or lack of) in the distribution of wealth, the reduction of poverty, employment and other factors such as education, housing, healthcare and so on. Thus, development became seen as a much broader concept, embracing at least five dimensions (see Goulet 1992):

● *an economic component* – wealth creation and equitable access to resources;

- *a social component* – improvements in health, housing, education and employment;
- *a political dimension* – assertion of human rights, appropriate political systems;
- *a cultural dimension* – protection or affirmation of cultural identity and self-esteem; and
- *the full-life paradigm* – preservation and strengthening of a society's symbols, beliefs and meaning systems.

To these should be added an ecological component, not only reflecting the emergence of environmental sustainability as a fundamental parameter of contemporary approaches to development, but, as discussed in Chapter 2, the basis of the concept of sustainable tourism development. Collectively, these dimensions are broadly reflected in the UN Millennium Project's goals and targets, as well as in the UN Post-2015 Development Agenda (see Box 1.1). The most widely accepted measure of development is the annual UNDP Human Development Index (HDI), which ranks countries according to a variety of economic and social indicators measuring health, education and living standards (see also Dasgupta and Weale 1992). However, the UNDP stresses that the concept of human development is far broader than what is presented in the HDI or other composite indices in the Human Development Reports, including the Inequality-Adjusted HDI, the Gender Inequality Index and the Multidimensional Poverty Index (UNDP 2015). A more recent attempt at measuring development, first published in 2013, is the Social Progress Index (SPI) developed by the Social Progress Imperative, a non-profit organization, which incorporates basic human needs, foundations of well-being and opportunity. The index tracks social and environmental outcomes (*The Economist* 2013).

Overall, then, development, as currently understood, is essentially a social phenomenon focusing on the betterment of the human condition, or what a recently devised development index refers to as 'social progress':

> Social progress is the capacity of a society to meet the basic human needs of its citizens, establish the building blocks that allow citizens and communities to enhance and sustain the quality of their lives, and create the conditions for all individuals to reach their full potential.

(Porter *et al.* 2013: 7)

Box 1.1

The UN Millennium Project and beyond

The United Nations Millennium Project was an independent advisory body that, drawing on research undertaken by 10 task forces collectively comprising more than 260 development experts, advised the UN on appropriate strategies to achieve an internationally agreed set of global development targets, including reducing poverty, hunger, disease and environmental degradation, by 2015. If these goals are met, it was claimed that half a billion people would be lifted out of poverty and a further 250 million would no longer suffer from hunger. The Project had eight Millennium Development Goals (MDGs), each of which has specific targets to be met by 2015.

Goal 1: Eradicate extreme poverty and hunger

> *Target 1:* Reduce by half the proportion of people whose income is less than $1 a day
> *Target 2:* Reduce by half the proportion of people who suffer from hunger

Goal 2: Achieve universal primary education

> *Target 3:* Ensure that children everywhere are able to complete full primary schooling

Goal 3: Promote gender equality and empower women

> *Target 4:* Eliminate gender disparity in all levels of education

Goal 4: Reduce child mortality

> *Target 5:* Reduce the under-5 mortality rate by two-thirds

Goal 5: Improve maternal health

> *Target 6:* Reduce the maternal mortality rate by three-quarters

Goal 6: Combat HIV/AIDS, malaria and other diseases

> *Target 7:* Halt/reverse the spread of HIV/AIDS
> *Target 8:* Halt/reverse the incidence of malaria and other serious diseases

Goal 7: Ensure environmental sustainability

> *Target 9:* Integrate the principles of sustainable development into national development policies
> *Target 10:* Halve the proportion of people without access to basic sanitation and drinking water
> *Target 11:* Achieve a significant improvement in the lives of 100 million slum dwellers

Goal 8: Develop a global partnership for development

> *Target 12:* Develop an open, non-discriminatory trading and financial system
> *Target 13:* Address the special needs of least-developed countries
> *Target 14:* Address the special needs of landlocked developing countries and small island developing states
> *Target 15:* Deal comprehensively with the developing counties' debt problems

The UN has posted fact sheets on their website (www.un.org/millenniumgoals/bkgd.shtml) on the various goals and targets, illustrating what has been achieved and what further work needs to be done. For example, under Goal 1, the target on the global proportion of people living in extreme poverty was met five years ahead of schedule. However, the UN also notes that the numbers of people going hungry remains disturbingly high. In addition to evaluating goal progress, the UN is actively drafting a post-2015 sustainable development agenda. The process has incorporated a wide-ranging consultation including panels of experts, summits, national and global thematic consultations and 'My World: The United Nations Global Survey for a Better World'. Efforts at creating the post-2015 sustainable development agenda are further documented in Box 2.1.

Source: Adapted from www.unmillenniumproject.org

For Sen (1999), the goal of such social development is simply 'freedom' or, more precisely, human capability, 'capability [being] treated as the freedom to promote or achieve combinations of valuable functionings' (McGillivray 2008: 34). Similarly, the UNDP, though recognizing that there remains no consensus over the term, most recently defines human development as:

> the expansion of people's freedoms to live long, healthy and creative lives; to advance other goals they have reason to value; and to engage actively in shaping development equitably and sustainably on a shared planet. People are both the beneficiaries and drivers of human development, as individuals and in groups.
>
> (UNDP 2010: 22)

Such an approach to development embraces three elements: well-being, empowerment and agency, and justice, the latter including the expansion of equity, sustaining outcomes over time and respecting human rights and other goals of society (UNDP 2010).

Development, then, is a complex, multidimensional concept that may be defined as the continuous and positive change in the economic, social, political and cultural dimensions of the human condition, guided by the principle of freedom of choice and limited by the capacity of the environment to sustain such change. The question to be addressed now, however, is: What are the particular characteristics of any society or country that define it as underdeveloped?

Underdevelopment and development

Many of the problems facing developing countries, such as poverty, inequality, vulnerability, poor healthcare and a lack of educational opportunities, are widely recognized; they are also reflected in the goals of international development programmes, such as the UN Millennium Project. The UNDP Human Development Report for 2014, for example, was titled *Sustaining Human Progress: Reducing Vulnerabilities and Building Resilience* and stressed that vulnerability threatens human development. Moreover, development is arguably regaining a dominant position within international politics, embodied in the international fight against poverty that has built upon the foundations established by the *Make Poverty History* campaign in the UK in 2005. However, the specific characteristics of underdevelopment are less clear. That is, many of the problems facing developing countries are the outcome, not the cause, of underdevelopment, and, as a consequence, it is also unclear to what extent particular developmental vehicles, such as tourism, are effective means of addressing these problems and challenges.

It is also important to point out that there is, of course, an enormous diversity of countries that comprise the developing world as defined for the purposes of this book. Geographical, political, historical, economic and sociocultural characteristics and structures all influence a country's level or rate of development (Todaro and Smith 2011), as well as its tourism development potential. However, as noted above, developing countries are typically classified according to national and/or per capita income, non-economic development indicators, such as life expectancy, literacy or environmental factors, or a combination of the two. The World Bank, for example, classifies all countries according to per capita Gross National Income (see Table 1.1), accepting that level of income does not necessarily reflect development status. Consequently, a number of the 75 countries classified as high income are not generally considered to be among the group of developed nations.

Table 1.1 *Per capita GNI country classifications for (2005) and 2015*

	Low-income economies	Lower-middle-income economies	Upper-middle-income economies	High-income economies
Per capita GNI	$1,045 or less ($735 or less)	$1,046–$4,125 ($736–2,935)	$4,126–$12,745 ($2,936–$9,075)	$12,746 or above ($9,076 or above)
Number of countries in group	34 (64)	50 (54)	55 (34)	75 (56)

Source: World Bank (2005, 2015a)

In Table 1.1, 2005 figures are included (in brackets) as a reference point for the most recent (2015) classifications. A comparison suggests that over the last decade, significant progress has been made in raising income levels across the developing world, although care must be taken in making such an assertion. For example, the term least-developed countries (LDCs) is also increasingly used to distinguish the world's poorest nations from the total of approximately 160 developing countries. To be added to the UN's official list of LDCs (which, following the 2012 review, comprised a total of 48 states, significantly more than the number of 'low-income economies' indicated in Table 1.1), a country must have a per capita income below $992 (to be raised to $1,035 in the forthcoming 2015 review), as well as satisfying complex 'economic vulnerability' and 'human resource weakness' criteria. Many of these countries are referred to by de Rivero (2001) as NNEs, or 'non-viable national economies', suggesting that they cannot be regarded as 'developing' countries in any sense of the word. Interestingly, a number of LDCs have either established or nascent tourism sectors, which, though small by international standards, are significant in terms of the local economy – for example, in the West African country of the Gambia, the tourism sector is relatively small in terms of arrivals (157,000), yet it contributes some 50 per cent of total exports. In cases such as this, however, tourism may be considered an economic survival strategy as opposed to a development strategy.

The characteristics of underdevelopment

Developing countries typically share a number of features that characterize the condition of underdevelopment.

(1) *Economic dependence on a large, traditional agricultural sector and the export of primary products*: Most developing country economies are dependent on agricultural production and exports for employment, income and foreign exchange earnings. Conversely, the industrial/manufacturing sector may be small and technologically deficient. Typically, over 60 per cent of the workforce is employed in agriculture in developing countries, compared with less than 5 per cent in developed countries. At the same time, low productivity and international price support mechanisms limit their ability to compete in global markets for primary products.

(2) *Low standards of living*: A variety of factors contribute collectively to low standards of living. Reference has already been made to low income levels (per capita income), although it is important to point out that average income gives no indication of income inequality within countries. It has been observed that few developing countries 'enjoy the luxury of having less than 20 per cent of their population below the poverty line' (de Rivero 2001: 64), and in some countries the contrast is stark. In India, some 30 per cent of the population are officially considered to live in poverty (an income of less than $1.25 a day), although one study in 2010 suggested that when non-economic measures of poverty were also taken into account, 55 per cent of the country's then population of 1.1 billion were living in poverty (Kumar 2010). Conversely, in China, the world's fastest-growing economy, remarkable progress has been made in reducing poverty levels, latest figures suggesting that just 6.3 per cent of the population live below the $1.25 threshold (World Bank 2015b). Indeed, between 1990 and 2010, the total number of people globally living in extreme poverty halved, falling to 22 per cent of the world's population, although in some regions slower progress has been made. In sub-Saharan Africa, for example, the proportion of people living on less than $1.25 a day fell from 56 per cent to 48 per cent during this 20-year period (UN 2014a). Care should also be taken in interpreting the data; although the overall incidence of poverty in sub-Saharan Africa has fallen, population growth has meant that the actual number of people in that region living in poverty almost doubled between 1981 and 2011, from 210 million to 415 million (World Bank 2015a).

(3) *High population growth and high unemployment/ underemployment*: Over 80 per cent of the world's population live in developing countries, a proportion that will continue to grow given higher birth rates on average (around 2 per cent annually) than in

developed countries (about 0.5 per cent). In the period between 1995 and 2025, the population of many developing countries will double. Consequently, under- and unemployment in developing countries, averaging between 8 and 15 per cent of the workforce, though often double this figure among the 15–24 age group, will increase significantly.

(4) *Economic fragility*: The economies of many developing countries are weak, characterized by low financial reserves, severe balance of payment deficits and high levels of international debt. Limited natural resources and industrial production necessitates high levels of imports to meet basic needs, yet exports typically cover around only two-thirds of developing countries' import bills. The resultant levels of international debt and interest payments have resulted in many developing countries becoming ensnared in the debt trap, hence the frequent calls for their debt to be written off by Western creditors.

(5) *Limited or unstable sociopolitical structures*: While underdevelopment is frequently claimed to result from inequalities in the global distribution of economic and political power (with international tourism widely seen as a manifestation of such inequality), the political and social structures within developing countries may also determine the extent to which development occurs. Although the last quarter of the twentieth century witnessed the dramatic spread of democratization (Potter 2000) – though not necessarily with a corresponding increase in development – the distribution of power in developing countries tends to favour a small, powerful elite whose position is frequently strengthened and legitimized by the democratization process. Consequently, the extent to which development occurs in any particular country is widely considered to reflect the extent of effective governance, with a lack of development being linked to the concept of the 'failing state' (Di John 2010). In other words, the ability or willingness on the part of the state to fulfil its obligations to its citizens may determine the nature of development both generally and in the specific context of tourism (Torres and Anderson 2004; Sharpley and Ussi 2014).

Inevitably, these characteristics of underdevelopment are not equally evident in all developing countries, while other indicators, such as gender-related issues (Momsen 2004), the ability to exercise human rights, or safety and security, must also be included as measures of development. Moreover, many developed nations also have 'less-developed' regions and face a number of developmental challenges, whether environmental, social (crime, inequality, education, health) or economic (poverty, unemployment). Nevertheless, within the developing world, tourism is increasingly viewed as a means of addressing underdevelopment, which, by implication, suggests that tourism may also impact positively on some or all of these specific challenges. The extent to which this occurs in practice is, of course, the subject of this book.

Development paradigms

Having introduced the characteristics of underdevelopment, it is also important to review briefly how development theory (that is, a combination of the ideological *ends* of development and the strategic *means* of achieving them) has evolved over time. A full consideration of development theory can be found elsewhere, both within the development literature (for example, Todaro and Smith 2011; Peet and Hartwick 2015) and the tourism literature (for example, Telfer 2015a). Importantly, however, just as the meaning of development has changed over the last half-century, so too have the dominant perspectives, or paradigms, on how development may be encouraged or achieved. To a great extent, these have been reflected in the evolution of tourism development theories in particular, although, as discussed later in this chapter, the extent to which a causal relationship exists between development theory and tourism development (both theory and practice) is not always clear.

It should also be noted that a number of what can be thought of as 'sub-theories' of development also exist. These are, in effect, specific development policies that are normally followed at a national level as opposed to more overarching or 'grand' theories of development. For example, import-substitution policies were dominant in Latin America in the 1950s and 1960s while, more recently, state-led or 'statist' approaches to development have enjoyed a resurgence in many countries (Wade 2004; see also Clancy 1999). Here, however, we are concerned with broader development paradigms and their relationship with tourism development in particular.

In general, the 'story' of development theory is one of a shift from traditional, top-down economic growth-based models to a more broad-based approach with an emphasis on bottom-up planning, satisfaction of people's basic needs, sustainability and human development. An emerging paradigm of global development is evolving in response to deal with issues such as climate change. More specifically, seven identifiable development paradigms have emerged chronologically since the end of the Second World War; these are summarized in Table 1.2. Importantly, emerging paradigms did not replace preceding ones; despite criticisms, elements of each paradigm remain relevant today. The timelines are only guides as to when paradigms gained prominence.

(1) *Modernization*: Modernization theory is based on the concept that all societies follow an inevitable evolutionary path from traditional to modern, characterized by a transformation from

Table 1.2 *The evolution of development theory*

Period	Development paradigm	Theoretical perspectives
1950s–1960s	**Modernization**	Stages of growth
		Diffusion: growth impulses/trickle-down effect
1950s–1970s	**Dependency**	Neocolonialism: underdevelopment caused by exploitation by developed countries
		Dualism: poverty functional to global economic growth
		Structuralism: domestic markets, state involvement
Mid-1970s–1980s	**Economic neo-liberalism**	Free market: free competitive markets/privatization
		Structural adjustment: competitive exports/market forces
		One world: new world financial systems
1970s–early 1980s	**Alternative development**	Basic needs: focus on food, housing, education, health
		Grass roots: people-centred development
		Gender: gender relations/empowerment
		Sustainable development: environmental management
Late 1980s–early 1990s	**The impasse and post-development**	Postmodern critique of metanarratives of development discourse; pluralistic approaches that value local knowledge and solutions
1990s–2000s	**Human development**	Human development: human rights, freedom, democracy, poverty reduction and pro-poor growth, good governance
		State-led development
		Focus on civil society and social capital
		Transnational social movements: environment, peace, indigenous peoples, feminists, etc.
		Culture: different world views are accommodated
		Human security; challenging the 'failed state'
2000s and 2010s	**Global development**	Focus on enhancing global international relations and governance through yet to be built supranational political institutions

Source: Adapted from Telfer (2015a: 36–7)

agriculture to industry, from rural to urban and from traditional to modern (that is, Western) values and social institutions (Harrison 1988). Progress along this evolutionary path is considered to be dependent upon economic growth as the basis of development and, according to Rostow (1967), only when a country has reached the 'take-off stage', manifested in the emergence of one or more significant industries that induce growth in associated sectors, can it begin to modernize or develop.

A variety of theories and strategies are embraced by the modernization paradigm, although most commonly the focus is usually on the introduction of a 'growth pole' (an industry or economic sector) from which 'growth impulses' diffuse throughout the region, thereby stimulating modernization. The development of a tourist destination can be considered as one such growth pole.

(2) *Dependency*: Dependency theory, sometimes referred to as underdevelopment theory, arose in the 1960s as a critique of the modernization paradigm. Essentially, it proposes that the condition of underdevelopment results not from the particular socio-economic characteristics of less-developed countries (as suggested by modernization theory), but from the external and internal political, economic and institutional structures that keep them in a dependent position relative to developed countries. In other words, global political and economic relations are such that wealthier, more powerful Western nations are able to exploit weaker, peripheral nations (often mirroring earlier colonial ties), thereby limiting developmental opportunities within less-developed countries. Thus, underdevelopment can be explained by an unequal international capitalistic system within which developing countries are unable to 'break out of a state of economic dependency and advance to an economic position beside the major capitalist industrial powers' (Palma 1995: 162). Various theoretical perspectives on dependency theory exist, although tourism, as a global industry that remains largely dominated by Western businesses and tourists, has long been considered a manifestation of the paradigm (Britton 1982; Bastin 1984; Nash 1989).

(3) *Economic neo-liberalism*: As a reaction or 'counter-revolution' to interventionist, Keynsian economic policy (Brohman 1996), economic liberalism, which became popular during the Reagan-Thatcher era of the 1980s, espoused the role of international trade in export-led economic development. Proposing that the problems facing developing countries arose from excessive state intervention,

its supporters argued that the path to development lay in promoting market liberalization, the privatization of state enterprises and the overall reduction of state intervention. As a result, international loan programmes administered by the World Bank and IMF to promote development were conditional upon adjustments to economic structures and political policies in recipient nations, hence the term 'Structural Adjustment Lending Programme' (SALP) (Mosley and Toye 1988). Tourism development in many countries has benefited from international structural funding (Inskeep and Kallenberger 1992), although, generally, SALPs have since been widely discredited (Harrigan and Mosley 1991).

(4) *Alternative development*: Signifying a departure from (or an alternative to) the preceding Western-centric, economic growth-based development paradigms, alternative development adopts a resource-based, bottom-up approach that focuses primarily on human and environmental concerns. Recognizing that development is a complex, multilayered process embracing not only economic growth, but broader social, cultural, political and environmental factors, its fundamental tenet is that development should be endogenous. That is, it is a process that starts within, and is guided by the needs of, each society; it is not something that should be implemented or imposed by other societies. It also emphasizes the importance of satisfying basic needs and of encouraging self-reliance (Galtung 1986), while environmental management is also a key element. Since the late 1980s, the alternative development paradigm has become more widely adopted as sustainable development, a concept that, though highly contested, continues to dominate global development policy. Not coincidentally, the 1980s concept of 'alternative tourism' (Smith and Eadington 1992) also provided the basis for what became, and has remained, the dominant tourism development paradigm, namely sustainable tourism development. This is considered in more detail in Chapter 2.

(5) *The impasse and post-development*: The perceived failure of preceding development policies (including sustainable development) led to claims of an 'impasse' in development studies. That is, a variety of factors, including the realization that the gap between rich and poor nations was not diminishing, the collapse of socialism as a political system, the inherently oxymoronic nature of sustainable development, the diminishing role of the nation state in an increasingly globalized world and recognition that developing countries are not homogenous, and hence not responsive to

development 'blueprints' or metanarratives (Schuurman 1993), led many to suggest that not only was the concept of development as a global project obsolete, but also that development theory had reached an impasse. Essentially, following what has been referred to as the lost decade of development of the 1980s, there were calls for the idea of development to be rejected. From the ensuing vacuum in development thinking, there emerged the 'post-development' school, an eclectic collection of approaches to development that broadly favoured traditional, non-modern/Western philosophies and cultures, and emphasized local engagement, community involvement and autonomy from the state in development processes. However, post-development offered no real solutions to the development impasse while, in the specific context of tourism, development policies continued to be framed by the sustainable development paradigm at the same time as embracing approaches to tourism development that reflected what has come to be termed human development.

(6) *Human development*: In comparison to earlier development paradigms based upon identifiable philosophical and political foundations, human development refers to what may be described as a movement within development policy that, from the early 1990s, embraced a variety of approaches and practices unified by a focus on improving the human condition. For example, the UNDP adopted the Human Development Index, incorporating indicators such as life expectancy, education and income as measures of human development. Issues including human rights and security, democracy and good governance, debt cancellation and poverty reduction were also gaining both political legitimacy and public appeal while Sen's (1999) concept of development as freedom based upon the expansion of human capabilities neatly captured the growing emphasis on people as the focus of development. There was a move away from market liberalization, questioning its role in driving development, towards a recognition of the regulatory role of the state to support the market and citizens (Bately, cited in Telfer 2015a). At the same time, attention turned to the diminishing role of the state in a globalizing world and, in particular, the necessary contribution of civil society (manifested in, for example, the activities of NGOs) and social capital in supporting human development, as well as the increasing influence of transnational organizations that sought, for example, to highlight what some see as the destructive power of capitalism on the individual human condition. Culture emerged as key factor in development thinking, owing in part to the threat of globalization to cultural diversity (Racliffe 2006, cited in Telfer

2015a). SALPs from economic neo-liberalism were replaced by Poverty Reduction Strategy Papers with the shift in focus to poverty reduction. However, these too were heavily criticized. In short, human development represents a diversity of approaches that collectively seek to enhance human capabilities and security and address human deprivations, not least poverty reduction, which, since the late 1990s, has become either directly or indirectly the focus of newer approaches to and forms of tourism, such as pro-poor tourism, responsible tourism and volunteer tourism (see Chapters 5 and 6). However, it should be noted that in the context of tourism development globally, these effectively remain on the margins.

(7) *Global development*: It has long been recognized that development on a global scale is dependent upon effective international (regional and global) collaboration and cooperation in the spheres of trade and economy, politics and global security, the environment and climate change, and so on. As Held (2010: 220) suggests:

> today, there is a newfound recognition that global problems cannot be solved by any one nation state acting alone, nor by states just fighting their corner in regional blocs. What is required is collective and collaborative action – something the states of the world have not been good at, and which they need to reconsider and advance if the most pressing issues are to be adequately tackled.

This, in turn, requires the establishment of new supranational political organizations, which can work towards achieving consensus to support global development (Hettne 2009) (for example, UN Millennium Development Goals, Intergovernmental Panel on Climate Change). To some extent, such supranational organizations exist, yet, while many processes and challenges are global (for example, the economy, the environment, climate change), politics remains by and large national. Hence, global development is an aspirational yet necessary paradigm.

As this brief chronology of development theory demonstrates, both the process and objectives of development have evolved over time, from relatively simplistic economic growth models through the more complex notion of sustainable development to a more specific focus on human development along with an emerging paradigm of global development. As is discussed shortly, approaches to tourism development have also evolved from its role as a vehicle of economic development (Diamond 1977) to the contemporary focus on sustainable tourism, while there is evidence of increasing concern for human development through, for example, community tourism development, pro-poor tourism and volunteer tourism. Under global development, there are also increasing concerns related to tourism and climate change. Why is it, however, that tourism has been adopted so widely as a developmental option?

Why tourism?

As previously noted, few if any nations of the world have not become tourist destinations and, for many, it has become an integral element of national development policy. In China, for example, tourism has long been a fundamental strand of the Western Region Development Strategy, which aims to promote the socio-economic development of the country's western provinces, covering some 70 per cent of China's total land area (Zhang *et al.* 1999). Moreover, even for countries with a limited history of tourism, or those with a less apparent need to turn to tourism, such as some Middle East oil states, it has become a development option of choice (see Box 1.2).

In some instances, of course, tourism may represent the only realistic development path; that is, for some developing countries, there is simply no other choice (Brown 1998: 59). More positively, however, the most compelling reason for adopting tourism as a development strategy is its potential contribution to the local or national economy, in particular to the balance of payments (Mihalič 2015). Many developing countries suffer severe balance of payment deficits and, as an export, tourism may represent a significant source of foreign exchange earnings. It is also widely considered to be a labour-intensive industry and, hence, an effective source of employment in destination areas, whether direct employment in hotels, restaurants and so on, or indirect/informal employment (Farver 1984; Cukier and Wall 1994).

Beyond these basic economic drivers, however, a number of factors underpin the attraction of tourism as a development option.

Tourism is a growth industry. As noted above, international tourism has demonstrated remarkable and consistent growth over the last 60 years, averaging 6.2 per cent annual growth in arrivals and over 10 per cent annual growth in receipts since 1950. However, the rate of growth has been steadily declining. During the 1990s, for example, the average annual growth in tourist arrivals worldwide

Box 1.2

Tourism and development in Qatar

Once a relatively poor Gulf state known primarily for its pearling industry, Qatar is now one of the richest countries in the world in terms of per capita income; according to World Bank data, its per capita GDP in 2013 amounted to almost $94,000 (World Bank 2015c), more than double the average for high-income countries and placing it third in the world rankings after Luxembourg and Norway. The country is also ranked 31 (out of 187) in the most recent Human Development Index (HDI), although it is not included in the list of recognized advanced/developed economies, despite being ranked higher in the HDI than some countries on that list. Nevertheless, Qatar has experienced rapid social and economic development over the last half-century, its two million inhabitants enjoy a high standard of living and the achievement of advanced economy status is a key objective of its current development policy. And yet, for the last decade, the country has been focusing on developing a tourism sector.

Qatar's wealth is, of course, based almost entirely on its production and export of oil and gas, which together account for more than half of GDP and government revenues and around 85 per cent of export earnings. Currently, the country has the third largest proven reserves of natural gas in the world, sufficient to maintain current production levels for up to 200 years; however, proven oil reserves, though substantial, amount to 25 billion barrels, or roughly 57 years of output at current levels. This, in part, explains the fact that Qatar is now following the path of its Arabian Gulf neighbours, notably Dubai, Abu Dhabi and Oman, in seeking to diversify its economy away from dependence on finite oil and gas reserves into tourism and other activities. At the same time, however, volatility in global oil prices has proved to be the principal driving force behind economic diversification. For example, downward pressure on oil prices in the latter half of the 1990s was followed by rising prices throughout the 2000s, while in late 2014 and early 2015, the price of oil fell dramatically in the face of lower demand and higher supply resulting from new sources (particularly shale oil in the US). Political uncertainties in many oil-producing nations also represent a further source of price instability, hence the desire on the part of many oil-dependent economies, including Qatar, to seek greater economic stability and sustainability though diversification into new sectors, including tourism.

In 2000, Qatar attracted 378,000 international tourist arrivals; by 2013, this figure had risen to 1.3 million, pointing to successful growth in the sector underpinned by substantial investment in attractions and facilities. For example, as part of a tourism master plan launched in 2004, $15 billion was earmarked for investment in luxury hotels, museums and other attractions, including Pearl Island off Doha, the aim of the plan to increase arrivals to 1.4 million by 2008. This was not achieved, yet the latest plan, the Qatar National Tourism Sector Strategy 2030, launched in 2014, has set some ambitious targets, including achieving seven million arrivals and boosting revenue from $1.3 billion in 2012 to $11 billion by 2030. Underpinning this growth in tourism, the Strategy focuses on developing cultural and urban tourism and MICE (meetings, incentives, conferences and exhibitions) tourism, as well as encouraging sports, health, beach and educational tourism. Key to success will be broadening the market base. Around 90 per cent of arrivals comprise mainly regional business visitors, and in 2013, fewer than 37,000 European tourists visited Qatar. Consequently, the Strategy seeks to increase the proportion of leisure tourists to 64 per cent of all international arrivals by 2030. The hosting of the FIFA World Cup in 2022 may help achieve this, but at the same time may prove to be something of a double-edged sword. That is, the controversy surrounding the World Cup bidding process may arguably have tarnished the country's image, while research has demonstrated that more generally, Qatar has a weak image among potential international markets and may struggle to differentiate itself from competitor destinations, specifically Dubai and Abu Dhabi, with established and recognized tourism sectors (Morakabati et al. 2014). Hence, even with substantial investment in attractions and facilities, developing a successful tourism sector in an increasingly competitive global market remains challenging.

Source: Sherwood (2006); *Gulf Times* (2014); Qatar Tourism Authority (2015)

Table 1.3 *International tourism arrivals and receipts growth rates, 1950–2000*

Decade	Arrivals (average annual increase %)	Receipts (average annual increase %)
1950–1960	10.6	12.6
1960–1970	9.1	10.1
1970–1980	5.6	19.4
1980–1990	4.8	9.8
1990–2000	4.2	6.5

Source: Adapted from WTO (2005)

was 4.2 per cent, the lowest rate since the 1950s, although 2004 witnessed a remarkable growth of 10 per cent over the previous year (Table 1.3).

Over the last decade, despite a number of challenges including the global economic crisis of 2008, the average annual increase in arrivals has continued at around 4.5 per cent, pointing to both the resilience of tourism to external forces and the likelihood of the UNWTO's long-standing forecast of 1.6 billion international arrivals and $2 trillion in international receipts by 2020 being easily met, if not exceeded (see Plates 1.1 and 1.2). Thus, tourism remains one of the world's fastest-growing industries and, as a consequence, it is seen essentially as a safe development option.

Nevertheless, it is important to point out that certain periods have witnessed low or even negative growth. The financial crisis of 2008, for example, contributed directly to a 3.8 per cent decline in global tourist arrivals in 2009, while, eight years earlier, a decline of 0.5 per cent in global arrivals was attributed to '9/11'. More typically, however, the effects of external influences are more locally or regionally defined. The Indian Ocean tsunami in December 2004, for example, had a devastating impact on the tourism industries of the Maldives, Sri Lanka and the Phuket area of Thailand, yet had little effect on total global arrivals (Sharpley 2005).

Tourism redistributes wealth. Tourism is, in principle, an effective means of transferring wealth, either through direct tourist expenditure or international investment in tourism infrastructure and facilities, from richer, developed countries. Through the promotion of domestic tourism, it also potentially redistributes wealth on a national scale – in India, for example, domestic tourism is significantly greater, in terms of tourist trips, than international tourism (Singh 2001), while

Plate 1.1 *Cuba, Varadero: hotel construction*

Source: Photo by D. Telfer

Plate 1.2 *Tunisia, near Monastir: hotel construction*

Source: Photo by D. Telfer

reference has already been made to the remarkable scale of domestic tourism in China. However, the gross value and net retention of tourist spending varies considerably from one destination to another – many destinations suffer 'leakages', whereby tourist expenditure finances the import of goods to meet tourists' needs – while overseas investment is conditioned by the global political economy of tourism (see Chapter 3).

Backward linkages. Given the variety of goods and services demanded by tourists in the destination, from accommodation to local transport and souvenirs, tourism potentially offers more opportunities than other industries for backward linkages throughout the local economy, whether directly meeting tourists' needs, such as the provision of food to hotels (Telfer 1996) (see Plate 1.3), or through indirect links with, for example, the construction industry. Again, the extent to which such linkages can develop depends upon a variety of factors, such as the availability of finance, the diversity and maturity of the local economy or the quality of locally produced goods (Rogerson 2012).

Plate 1.3 *Indonesia, Yogyakarta: a woman whose family owns and operates a small hotel in one of the tourist districts in the city is returning from a traditional market by three-wheel bicycle taxi after purchasing food for the hotel restaurant; she purchases most of the food in traditional markets circulating tourist expenditures into the local economy*

Source: Photo by D. Telfer

Tourism utilizes natural, 'free' infrastructure. The development of tourism is frequently based on existing natural or man-made attractions, such as beaches, wilderness areas or heritage sites (see Plate 1.4). Thus, tourism may be considered to have low 'start-up' costs when compared with other industries as such resources are, in a simplistic sense, 'free'. Increasingly, however, attempts are being made to place an economic value on the use of these basic resources (Mihalič 2015) while, inevitably, costs are incurred in the protection, upkeep and management of all tourism resources.

No tourism trade barriers. In many instances, individual countries or trading blocs, such as the European Union, impose restrictions of one form or another to protect their internal markets. In principle, international tourism faces no such trade barriers. That is, generating countries rarely place limitations on the right of their citizens to travel overseas, on where they visit and how much they spend (although travel advisories are one form of limitation on travel). Thus, destination countries have free and equal access to the international tourism market. However, the extent to which

Plate 1.4 *China: tourists on the Great Wall of China, a UNESCO World Heritage Site*

Source: Photo by H. Di Wang

destinations can take advantage of this 'barrier-free' market is, of course, determined by international competition in general and by the structure and control of the international tourism system in particular. Indeed, as the following section demonstrates, developing countries continue to enjoy a relatively limited share of global tourist arrivals and receipts.

Tourism demand

For present purposes, the demand for tourism can be considered from two perspectives:

1 Historical and contemporary patterns and flows of international tourism (that is, statistical data).
2 Transformations in the nature of the demand for tourism (that is, changes in styles of tourism).

International tourism trends and flows

Reference has already been made to the remarkable and continuing growth of international tourism since the 1950s. This growth has, at times, been restricted by a variety of events, including international conflicts, such as the 1991 Gulf War, the oil crises of the 1970s, global economic recession in the early 1980s and 1990s and late 2000s, the Arab Spring, and a variety of specific events such as health scares, natural disasters and, of course, terrorist activity. Only rarely, however, have global, as opposed to regional, arrivals experienced a downturn (Table 1.4).

Importantly, the global growth of international tourism has not been equitable. That is, not all parts of the world have experienced similar growth rates, and international tourism is still largely dominated by the industrialized world with major tourist flows being primarily between the more developed nations and, to a lesser extent, from developed to less-developed countries, although the more recent growth in overall international tourist arrivals has been primarily driven by an increase in tourism to and within the Asia-Pacific region. In particular, not only is China one of the world's most popular destinations, but in 2012 became the largest generator of international tourists, with more than 130 million Chinese people forecast to travel internationally in 2015. However, it should be

Table 1.4 *International tourist arrivals and receipts, 1950–2013*

Year	Arrivals (millions)	Receipts (US$bn)	Year	Arrivals (millions)	Receipts (US$bn)
1950	25.3	2.1	1999	639.6	464.5
1960	69.3	6.9	2000	687.0	481.6
1965	112.9	11.6	2001	686.7	469.9
1970	165.8	17.9	2002	707.0	488.2
1975	222.3	40.7	2003	694.6	534.6
1980	278.1	104.4	2004	765.1	634.7
1985	320.1	119.1	2005	806.6	682.7
1990	439.5	270.2	2006	847.0	742.0
1991	442.5	283.4	2007	903.0	856.0
1992	479.8	326.6	2008	917.0	939.0
1993	495.7	332.6	2009	882.0	851.0
1994	519.8	362.1	2010	940.0	927.0
1995	540.6	410.7	2011	995.0	1,042.0
1996	575.0	446.0	2012	1,035.0	1,075.0
1997	598.6	450.4	2013	1,087.0	1,159.0
1998	616.7	451.4			

Source: Adapted from UNWTO data

noted that more than half of the international visitors to China originate in Hong Kong, Macao and Taiwan, while the great majority of outbound tourism is to these and other countries in the region. As Table 1.5 shows, almost half of all international arrivals can be accounted for by just 12 nations, collectively attracting 46 per cent of total global arrivals in 2013.

Not surprisingly, a similar pattern is evident in terms of tourism receipts (Table 1.6). The US has long been the greatest beneficiary of international tourism, while in 2013 almost 49 per cent of global tourism receipts were earned by the top 10 countries. Of the top 10 most popular destinations for international tourism, seven are also ranked within the 10 largest tourism generators by expenditure (Table 1.7).

The pattern of international tourist flows is reflected in regional shares of tourist arrivals and receipts. It should be noted that Asia and Pacific figures are based on combined data from two regions previously referred to by the UNWTO as East Asia and Pacific (EAP) and South Asia.

Table 1.5 *World's top 12 international tourism destinations, 2013*

		Arrivals ('000)	Share of total (%)
1	France	84,700	7.8
2	US	69,768	6.4
3	Spain	60,661	5.6
4	China	55,686	5.1
5	Italy	47,704	4.4
6	Turkey	37,795	3.5
7	Germany	31,545	2.9
8	UK	31,169	2.9
9	Russian Federation	28,356	2.6
10	Thailand	26,547	2.4
11	Malaysia	25,715	2.4
12	Hong Kong	25,661	2.4

Source: Adapted from UNWTO (2014a)

Table 1.6 *World's top 10 international tourism earners, 2013*

		Receipts ($ billion)	Share of total (%)
1	US	139.6	12.0
2	Spain	60.4	5.2
3	France	56.1	4.8
4	China	51.7	4.5
5	Macao	51.6	4.5
6	Italy	43.9	3.8
7	Thailand	42.1	3.6
8	Germany	41.2	3.6
9	UK	40.6	3.5
10	Hong Kong	38.9	3.4

Source: Adapted from UNWTO (2014a)

Europe has long received the greatest proportion of international arrivals, although, as is evident from Table 1.8, despite continuing to enjoy an annual increase in international arrivals, its share of the global market has been steadily shrinking.

Conversely, the Asia-Pacific region has, in particular, enjoyed spectacular growth in tourist arrivals, overtaking the Americas in 2002 to become the world's second most visited region that year.

Indeed, during the 1990s, annual arrivals in the region doubled while receipts grew by 121 per cent, both figures being twice the global rate. The main destinations in EAP are China, Thailand, Hong Kong and Malaysia, although, interestingly, newer destinations in the region, including Vietnam, Cambodia, Myanmar, Laos and

Table 1.7 *World's top 10 international tourism generators by expenditure, 2013*

		Expenditure ($ billion)	Share of total (%)
1	China	128.6	11.1
2	US	86.2	7.4
3	Germany	85.9	7.4
4	Russian Federation	53.5	4.6
5	UK	52.6	4.5
6	France	42.4	3.7
7	Canada	35.2	3.0
8	Australia	28.4	2.4
9	Italy	27.0	2.3
10	Brazil	25.1	2.2

Source: Adapted from UNWTO (2014a)

Table 1.8 *Percentage share of international tourist arrivals by region, 1960–2013*

	Africa	Americas	Asia & Pacific	Europe	Middle East
1960	1.1	24.1	1.4	72.6	0.9
1970	1.5	25.5	3.8	68.2	1.1
1980	2.6	21.6	8.2	65.6	2.1
1990	3.3	20.4	12.7	61.6	2.2
1995	3.6	19.8	15.6	58.6	2.5
2000	4.0	18.6	16.8	57.1	3.5
2005	4.6	16.6	19.2	54.8	4.7
2006	4.9	16.0	19.7	54.6	4.8
2007	4.9	15.8	20.4	53.6	5.3
2008	4.9	16.1	20.1	52.9	6.0
2009	5.2	15.9	20.5	52.3	6.0
2010	5.2	15.9	21.7	50.7	6.4
2011	5.0	15.7	21.9	51.9	5.5
2012	5.1	15.8	22.6	51.6	5.0
2013	5.1	15.4	22.8	51.8	4.7

Source: Adapted from UNWTO data

Polynesia, have successfully developed their tourism sectors. As previously observed, much of this growth can be accounted for by intra-regional travel. Other regions of the world have also increased their share of the global tourism market. Annual international arrivals in the Middle East more than doubled during the 1990s, with Egypt, Bahrain, Jordan, and, in particular, Saudi Arabia and the UAE enjoying rapid growth, although since the 'Arab Spring' of 2010, this region as a whole has lost market share because of continuing political turmoil in some countries.

Thus, overall, there has been a gradual shift away from the traditional destinations of Europe and North America to other regions of the world. Indeed, according to the WTO (2005), during the period 1995–2002, over 30 nations enjoyed tourism growth at more than double the global rate. All of these were developing countries, although, collectively, the developing world, or what the UNWTO refers to as 'emerging economies', still attracts only 36 per cent of total international tourism receipts. Nevertheless, the arrivals/receipts data mask the importance of tourism to the economies of many developing countries. That is, although their arrivals figures tend to be relatively low, the contribution of tourism to the national economy is high. For example, data provided by the World Travel and Tourism Council (WTTC 2004) demonstrates that in 25 countries, all of them small island developing states, the tourism economy

Table 1.9 *Destinations with highest total contribution of tourism to GDP, 2013*

	Country	Tourism (direct and indirect) as % of GDP
1	Maldives	94.1
2	Macau	86.2
3	Aruba	84.1
4	British Virgin Islands	76.9
5	Vanuatu	64.8
6	Antigua and Barbuda	62.9
7	Anguilla	57.1
8	Seychelles	56.5
9	Former Dutch Antilles	46.7
10	Bahamas	46.0

Source: Adapted from WTTC (2014)

contributes more than 25 per cent of GDP, while, as can be seen from Table 1.9, the current top 10 counties ranked according to tourism's total contribution (direct and indirect) to GDP reveals a significant dependency on the sector. Therefore, a destination's share of global arrivals and receipts is less important, in developmental terms, than the relative importance of tourism to the local economy.

The nature of tourism demand

The dramatic growth and spread of international tourism over the last 60 years has been driven by a variety of factors. Typically, increases in wealth and free time, as well as technological advances in transport, are considered to be the principal influences on the development of tourism, although political change has, in more recent years, been a significant factor in the emergence of new destinations in the developing world, particularly in the former Soviet republics. At the same time, the emergence of a sophisticated travel industry, initially providing the 'package holiday' to mass markets inexperienced in international travel while, more recently, developing more varied products, from low-cost/no-frills flights to Internet-based 'dynamic packaging' (whereby tourists are able to construct their own package holiday by booking flights, accommodation, car hire and other services from different suppliers), has fuelled the growth of tourism.

However, of particular relevance to developing countries, the nature of tourism demand has also changed and evolved over the last 20 years. Although the standardized, sun-sea-sand package holiday remains the most popular form of tourism, at least among tourists from the developed, Western world, there has been a dramatic growth in the demand for more individualistic, active/participatory forms of tourism that provide a broader or more fulfilling experience, as evidenced by, for example, the growth in demand for cultural tourism, adventure tourism, heritage tourism, health and wellness (including medical) tourism, ecotourism and, more generally, an expansion of long-haul tourism. This, in turn, suggests that tourists have become more experienced, more discerning, more quality-conscious and more adventurous in the practice of consuming tourism experiences.

These issues are considered in more detail in Chapter 6, but the important point is that the demand for tourism has allegedly been characterized by the emergence of the so-called 'new' tourist and, more recently, the 'responsible' tourist. Tourists are now considered to be more flexible, more environmentally sensitive, more adventurous and more inclined to seek out meaningful experiences, and hence are increasingly travelling to different, more distant, untouched, exotic or new destinations. Whether or not this can be explained by the existence of the new, responsible tourist remains debatable, but there is no doubt that changing tastes in tourism demand represent a vital opportunity for developing countries; the challenge lies in the extent to which they are able to harness this potential means of development.

Tourism supply

The ability of tourists to enjoy travel or vacation experiences is largely dependent upon the multitude of organizations that collectively supply the goods and services required by tourists. Typically, these include transport both to and within a destination, accommodation, food and drink, entertainment (attractions or activities) and shopping, as well as associated services such as insurance and finance. At the same time, public sector organizations frequently support the supply of tourism through, for example, regional or national marketing or the provision of information services for visitors. In some cases, such as cruise holidays, the core elements of the tourism product (transport, accommodation, sustenance, entertainment) are supplied collectively, while the success of tour operators is based on their packaging of the different components into a single product – the package holiday. In other cases, of course, tourists independently organize their own holidays but, nevertheless, the key elements remain the same.

However, the manner in which tourism is supplied may have a significant impact on the contribution of tourism to development. In other words, although the tourism industry supplies similar components in the provision of tourist experiences, the characteristics of that supply may vary considerably in terms of scale, nature and control/ownership. At one extreme, for example, a destination may be characterized by standardized, large-scale, mass tourism development that is largely owned or controlled by overseas interests, potentially limiting tourism's developmental contribution and reducing the destination to a state of dependency (Britton 1982). At the other extreme, tourism development may be small-scale, appropriate to the local environment and locally owned and controlled, thereby potentially optimizing the developmental benefits to the local community. Indeed, such extremes may represent the two ends of a tourism development continuum. At one end lies, arguably, the least desirable situation, whereby a destination becomes, in a sense, an annex of the wealthy, developed countries, with tourism representing a modern form of colonial exploitation; at the other (ideal) end, tourism contributes effectively to local sustainable development.

In reality, most tourism developments fall somewhere between these two poles, while the relationship between the characteristics of tourism supply and its contribution to development is more complex. For example, over the last 30 years, the Dominican Republic has built an economically successful tourism sector, attracting some 4.5 million tourists annually (including about 500,000 Domincans domiciled overseas) and now generating around $4.5 billion in revenues and directly and indirectly contributing 15 per cent of the country's GDP and some 14 per cent of total employment. However, that success has resulted primarily from the development of what many would consider to be 'unsustainable' all-inclusive mass tourism resorts. Moreover, in many instances, the contribution of tourism to development is enhanced not by the scale and ownership of tourism resources, but by innovative schemes implemented by the international tourism industry or other relevant organizations. One such example is the Travel Foundation, a UK-based charity that works with the outbound travel industry to manage tourism more sustainably in destinations around the world. For instance, the Foundation has been working on sustainable tourism projects in Mexico for eight years, with one project that commenced in 2009 helping local people, through the economic benefits of tourism, to remain in their villages in the Yucatan Penninsula. Specifically, with support from Thomson Holidays and a local NGO, this project has helped a group of women in the community establish a 'jungle jams' business, the products initially being sold to Thomson's Sensatori Hotel in Cancún. The funds collected supported the project for four more years, and by 2012 some 5,200 kg of jam had been sold to 11 tourism businesses, generating an income of £32,448. Since then, the business has continued to thrive, with jam sales generating an average of over £16,000 per year compared to an average of £300 before the project started (see www.thetravelfoundation.org.uk).

The important point is that simplistic models of appropriate/ inappropriate tourism supply cannot meet the diversity of developmental contexts. In other words, local social, economic or political structures and developmental needs may dictate the nature of tourism supply. Thus, while the Himalayan kingdom of Bhutan has long pursued a restrictive policy on tourism development, strictly controlling visitor numbers to maintain the cultural integrity of the country (though recent years have witnessed some relaxation of this policy), other destinations, such as Cancún in Mexico, have followed a more expansionist tourism development policy based on foreign investment in mass tourism enterprises (see Clancy 1999; Telfer 2002a). Nevertheless, an integrated, community-focused approach to tourism supply is widely considered to be the most appropriate form of tourism development, reflecting the contemporary dominance of the sustainable tourism development paradigm (Mann 2000).

Tourism and development

Just as development theory in general has evolved over time, so too have approaches to tourism development in particular. Moreover, the evolution of tourism theory has also, by and large, reflected the evolution of development theory as described earlier in this chapter, although any relationship between the two is not always clear. Indeed, until recently (see, for example, Burns and Novelli 2008; Mowforth and Munt 2009; Holden 2013; Sharpley and Telfer 2015), there has been little interaction between the two fields of study, while successive approaches to tourism development have been linked primarily to concerns over the socio-environmental impacts of tourism (Dowling 1992) rather than an understanding of tourism and development processes. Nevertheless, based on an analysis of tourism literature, Jafari (1989) identifies four stages or 'platforms' of tourism theory that, to an extent, parallel development theory:

Advocacy: During the 1960s, tourism was viewed as a positive vehicle of national and international development. Reflecting modernization theory, its potential was considered to lie in its contribution to economic growth, this being measured by indicators such as income and employment generation and the multiplier effect. In short, tourism was seen as an effective developmental growth pole, as indeed it continues to be seen in some contexts.

Cautionary: From the late 1960s onwards, concern was increasingly expressed over the negative environmental and sociocultural consequences of tourism that were resulting both from the scope, scale and rapidity of tourism development and from the emerging political economy of tourism, which reflected the dependency paradigm. Thus, tourism theory at this stage was concerned with understanding tourism's impacts on destination environments and societies, with a particular focus upon centre-periphery dependency models (Høivik and Heiberg 1980).

Adaptancy: As a response to the preceding antithetical positions on tourism development, the 1980s witnessed the emergence of alternative, though idealistic, approaches to tourism. Variously referred to as green, appropriate, responsible, soft or alternative tourism, these attempted to transpose the principles of alternative development on to tourism, proposing appropriately scaled, locally owned and controlled development with the community as the primary instigators and beneficiaries of tourism. To an extent, this approach remains reflected in the contemporary (though contentious) concept of ecotourism.

Knowledge: As greater knowledge of tourism's developmental processes has emerged, the idealistic ambition of alternative tourism has been overtaken by a broader approach to tourism development that attempts to embrace the principles and objectives of sustainable development. Thus, tourism, as a specific developmental vehicle, has aligned itself with the contemporary development paradigm, although, as considered in detail in Chapter 2, sustainable tourism development has proved to be problematic both in its practical implementation and in its acceptance by many developing countries, which view the concept as evidence of continuing Western imperialism.

Approaches to tourism development have, then, evolved over time, from traditional, modernist economic growth models through to sustainable tourism development approaches that attempt to balance tourism as a profit-driven, resource-hungry activity with the developmental needs of destination environments and communities. The latter have collectively been referred to by the World Travel and Tourism Council as 'new tourism' (WTTC 2003), an approach that demands an effective, long-term partnership between the public and private sectors, though, interestingly, no explicit reference is made to community participation. In other words, the most recent approaches to tourism development, such as 'pro-poor tourism' (www.propoortourism.org.uk), have shifted the emphasis towards responsible activities on the part of the international tourism industry.

This, however, will do little to alleviate the dilemma facing tourism destinations in the developing world. That is, to many developing countries, tourism represents a potentially valuable development option, yet it is associated with a variety of costs or impacts, from environmental degradation to dependency on international corporations. At the same time, adopting a policy of larger-scale tourism development may provide greater economic benefits in terms of income and employment, though potentially greater impacts; conversely, the adoption of smaller-scale, appropriate tourism may lessen the impacts, but may also result in a reduced developmental contribution.

The challenge for developing countries is, therefore, to seek ways of resolving this tourism development dilemma, and it is with this that this book is concerned. The next chapter reviews sustainable tourism development, focusing in particular on contemporary debates that challenge or question the validity of the concept and highlighting issues that are then addressed in subsequent chapters. These include the relevance and influence of globalization on tourism in developing countries (Chapter 3) and an analysis of different forms of tourism development and their potential contribution to sustainable development in Chapter 4. Community involvement in tourism remains a central tenet of sustainable tourism development; means of achieving community-focused tourism development are explored in Chapter 5, while Chapter 6 considers tourism development from the perspective of the tourist, addressing in particular the implications of transformations in the consumption of tourism for sustainable tourism development. Chapter 7 highlights the impacts of tourism development and explores contemporary techniques for minimizing such impacts, while Chapter 8 introduces and explores a number of issues that represent challenges to tourism and development. Finally, Chapter 9 draws together the themes and debates raised throughout the book, comparing the realities of tourism planning and management in developing countries with the idealism of sustainable tourism development.

Discussion questions

1   Why is tourism selected as a development tool by so many developing nations?
2   What are the structural dimensions in developing countries that relate to underdevelopment, and what role can tourism play in addressing these problems?

3   Is there a new emerging tourism market?
4   How have development approaches to tourism changed over
     time?

Further reading

Mowforth, M. and Munt, I. (2009) *Tourism and Sustainability: Development and New Tourism in the Third World*, 3rd edn, London: Routledge.

This book is essential reading. It provides a comprehensive and detailed critique of contemporary approaches to tourism development within the framework of development, globalization, power relations and sustainability. In particular, it challenges the widespread optimism for new approaches to tourism, questioning the role of tourism within the global political economy.

Reid, D. (2003) *Tourism, Globalization and Development: Responsible Tourism Planning*, London: Pluto Press.

Questioning the potential contribution of corporate-dominated tourism development to the social and economic development in less-developed countries, this book provides a clear analysis of the relationship between tourism and development and, in particular, highlights the benefits of community-led tourism planning and development.

Sharpley, R. and Telfer, D. (2015) *Tourism and Development: Concepts and Issues*, 2nd edn, Bristol: Channel View Publications.

The first section of this book provides a detailed introduction to the relationship between tourism, development and development theories. This is followed by both a comprehensive analysis of key issues relevant to tourism's contribution to development and also a critique of contemporary challenges and barriers to the achievement of sustainable development through tourism.

Websites

www.propoortourism.org.uk
The ProPoor Tourism site provides access to up-to-date research reports and studies about the implementation and benefits of pro-poor tourism.

www.thetravelfoundation.org.uk
The Travel Foundation is a charitable organization that encourages the outbound travel industry to manage tourism more sustainably.

www.undp.org
This site should be accessed for information on development processes and progress in the developing world in general, and on the activities and programmes of the United Nations Development Programme in particular (including the UNDP's annual Human Development Report).

www.worldbank.org

The World Bank is a useful source of information regarding economic and social development goals, processes and statistical data in less-developed countries.

www2.unwto.org

This is the website of the World Tourism Organization (UNWTO), an essential information source for contemporary international tourism trends, statistics, global tourism policies and tourism development guidelines.

www.wttc.org

The World Travel and Tourism Council (WTTC) website is a valuable source of information and statistical data, in particular via its Document Resource Centre, which provides access to all published WTTC documents, research and policy statements.

2 Tourism and sustainable development

Learning objectives

When you have finished reading this chapter, you should be able to:

- understand the evolution, principles and objectives of the concept of sustainable development
- evaluate the key debates surrounding the definition, implementation and measurement of sustainable tourism development
- identify and assess contemporary approaches to sustainable tourism development
- understand the links between sustainable development, globalization and political economy

In 1987, the World Commission on Environment and Development (WCED) published its report, *Our Common Future* (WCED 1987). Better known, perhaps, as the Brundtland Report, this document argued that the most effective way to 'square the circle of competing demands for environmental protection and economic development' (Dresner 2002) was through the adoption of a new approach, namely sustainable development. Although the WCED did not in fact coin the term – the World Conservation Union had referred to sustainable development in its earlier *World Conservation Strategy* (IUCN 1980) – the report brought the concept to the attention of a much wider audience and, since the late 1980s, sustainable development has dominated global development policy. Not only has it provided the focus for a number of major international events, including the 1992 'Earth Summit' in Rio de Janiero, the World Summit on Sustainable Development ('Rio+10') in Johannesburg in 2002 and, more recently and again in Rio de Janeiro, the United Nations Conference on Sustainable Development ('Rio+20: The Future We Want'; see https://sustainabledevelopment.un.org/rio20), but also innumerable organizations in the public, private and voluntary sectors, and at the international, national and local levels, have embraced its principles and objectives.

Reflecting the emergence and growing acceptance of sustainable development in general, the concept of sustainable *tourism* development in particular also came to prominence in the early 1990s, subsequently achieving 'virtual global endorsement as the new [tourism] industry paradigm' (Godfrey 1996: 60). In both the public and private sectors, a plethora of policy documents, planning guidelines, statements of 'good practice', case studies, codes of conduct for tourists and other publications have since been produced, all broadly concerned with the issue of sustainable tourism development. Moreover, the importance of the concept continues to be highlighted within global development policy. The report of the 2002 World Summit on Sustainable Development, for example, proposed that sustainable tourism development should be promoted in order to 'increase the benefit from tourism resources for the population in host communities while maintaining the cultural and environmental integrity of the host communities and enhancing the protection of ecologically sensitive areas and natural heritages' (RWSSD 2002: 34), while the Rio+20 conference in 2012 called for 'enhanced support for sustainable tourism activities and relevant capacity-building in developing countries in order to contribute to the achievement of sustainable development' (UNCSD 2012a: 25). In short, for more than 20 years, tourism policy and planning has, by and large, been driven by the principles and objectives of sustainable tourism development, although, as we shall see, the extent to which those objectives have been achieved, as well as the viability of the concept more generally, remains debatable.

To a great extent, the widespread support for the concept of sustainable tourism development is not surprising. As noted in Chapter 1, the initial euphoria over the developmental potential of international tourism had, by the late 1960s, been replaced by a more cautionary approach as increasing numbers of commentators drew attention to the potentially destructive effects of tourism development. Initially, concerns were voiced by the 'limits to growth' school, which, reflecting the then prevailing criticism of unbridled economic growth (Schumacher 1974; Andersen 1991), called for restraint in the development of tourism (Mishan 1969; Young 1973). More specific studies of tourism's consequences followed in the late 1970s and early 1980s (Turner and Ash 1975; Smith 1977; de Kadt 1979; Mathieson and Wall 1982), and, by the 1990s, tourism (or, more specifically, mass tourism) was being described in almost apocalyptic terms as a 'spectre . . . haunting our planet' (Croall 1995: 1). As a result, the principles of sustainable tourism development, which address many (and often justifiable) concerns and criticisms of mass tourism, were widely adopted at national and destinational levels, as well as by certain sectors of the travel and tourism industry.

However, sustainable tourism development has always been and remains a highly controversial concept (Wheeller 1991), as indeed does its parental paradigm of sustainable development. Specifically, it is seen as divisive, polarizing the debate between sustainable or 'good' forms of tourism and unsustainable, mass (or 'bad') forms of tourism, as well as being an inflexible blueprint that cannot be adapted to different tourism developmental contexts. More generally, not only have many questioned the extent to which tourism can be mapped on to the broad principles and objectives of sustainable development (Sharpley 2000; Berno and Bricker 2001), but also it is accepted that there are few, if any, examples of 'true' sustainable tourism development in practice. Indeed, reflecting the claim discussed later in this chapter that sustainable development is inherently oxymoronic, Butler (2013: 223) observes that despite the inexorable and (in some quarters) celebrated growth of tourism, 'there has emerged the fantasy . . . of sustainable tourism'. Moreover, there is recent evidence to suggest that idealistic support for the concept is on the wane (Sharpley 2009c; Mundt 2011; Buckley 2013), while the adoption of a more pragmatic and 'responsible' approach to tourism development is being manifested in specific policies, such as pro-poor tourism, introduced later in this chapter.

Nevertheless, it is still claimed that 'tourism is in a special position in the contribution it can make to sustainable development' (UNEP/ WTO 2005: 9). In other words, for many countries, tourism represents the principal, if not only, route to development, and thus there is a need to ensure that it is developed in such a way that its contribution to a destination's sustainable development is optimized.

The purpose of this chapter, therefore, is to explore the principles and objectives of sustainable development before going on to consider the debates surrounding the concept of sustainable tourism development and the contemporary challenges it faces. Thus, the first question to be addressed is: What is sustainable development?

Sustainable development: towards a definition

Despite widespread support for and promotion of sustainable development, there remains a lack of consensus over the actual meaning of the term. That is, although it is relatively easy to define what sustainable development is *not*, saying what it *is* has proved to be more problematic (Washington 2015). It means different things to different people and is applied to innumerable contexts (including, of course, tourism). Indeed, some suggest that the strength of the concept lies is in its vagueness and ambiguity. For example, it is described by Giddings *et al.* (2002: 188) as a 'political fudge', purposefully ambiguous so as to be widely acceptable, while Kates *et al.* (2005) suggest it is sufficiently broad to act as a focus for disparate interest groups to work together. Either way, definitions abound – it has been observed, for example, that even by the early 1990s, over 70 definitions of sustainable development had been proposed (Steer and Wade-Gery 1993)! Nevertheless, the most popular and enduring remains the Brundtland Report's definition as 'development that meets the needs of the present without compromising the ability of future generations to meet their own needs' (WCED 1987: 48), though this is widely regarded as, at best, vague and, at worst, meaningless.

It is generally agreed, however, that sustainable development represents a 'meeting point for environmentalists and developers' (Dresner 2002: 64). In other words, sustainable development may be thought of as a combination of two factors, namely development and sustainability, although the latter term, somewhat confusingly, is sometimes used interchangeably with sustainable development. Such an approach does, perhaps, oversimplify matters. That is, both 'development' and 'sustainability' are open to interpretation – for example, it has long been recognized that environmental sustainability can be considered from either an ecocentric (strong sustainability) or techno-centric (weak sustainability) perspective. Nevertheless, for the purposes of this book, dividing sustainable development into its two constituent elements provides a useful basis for exploring the evolution of the concept.

'Development' in general and transformations in the meaning or interpretation of development in particular have already been addressed in Chapter 1. It is, then, necessary to consider the other half of the sustainable development 'equation', namely sustainability.

From conservation to sustainability

Just as development thinking has evolved from the narrow, classical economic growth perspective though the broader, alternative development paradigm to human development and global development, so too has the nature of environmental concern, or environmentalism, evolved through a number of stages, from conservation to sustainability – although it may be argued that the need for environmental sustainability has long been recognized. Mundt (2011: 20), for example, notes that the principle of sustainability was fundamental to the timber industry in the seventeenth century.

Concern about human impacts on the environment can be traced back almost to the beginning of civilization, while, throughout history, societies have suffered (and continue to suffer) from a variety of environmental problems such as overpopulation, resource depletion and pollution (McCormick 1995). However, it was not until around the 1850s that a formal and organized conservation movement began to emerge, signalling the advent of contemporary environmentalism. Influenced by a number of factors, including urban and industrial development, a greater interest in and knowledge of the natural world and an evolving amenity movement demanding greater access to natural spaces, a large number of organizations were established that sought either to protect particular species, such as Britain's Royal Society for the Protection of Birds, founded in 1889, or to preserve natural areas, such as the Sierra Club, founded by John Muir in 1892 in the US. All these organizations had a common aim, and one that was very much in opposition to the prevailing modernist techno-centric belief in the domination and exploitation of nature; that is, to promote the conservation (and public enjoyment) of natural resources.

This focus on conservation remained predominant until the mid-twentieth century. Moreover, despite the creation of international organizations, such as the International Union for the Conservation of Nature (IUCN), the scope of environmental concern was usually local and rarely transcended national boundaries or interests. From the 1960s, however, environmentalism became a popular ideology with a set of preoccupations that went far beyond the specific concerns of protecting threatened natural areas and species. Rather than focusing simply on resource depletion, the actual scientific, technological and economic processes upon which human progress was previously seen to depend were also questioned. For example, Rachel Carson's (1962) *Silent Spring* was specifically concerned with the misuse of synthetic pesticides, yet is seen by many as a landmark event in the history of modern environmentalism. Similarly, Hardin's (1968) 'The tragedy of the Commons' described in simple terms how individual overconsumption of a limited natural resource eventually brings ruin to all. More recently, movements such as 'Occupy' have come to represent a broader, political branch of environmentalism that considers capitalism to be the cause of global environmental and social problems.

At the same time, it was acknowledged that the by-products of industrialization, the so-called 'effluence of affluence', did not respect national boundaries; environmental problems, such as air and water pollution, frequently originated in one country but adversely affected another. Influenced by Boulding's (1992) notion of 'spaceship earth', environmentalism took on an international dimension. The earth became viewed as a closed system with finite resources and a limited capacity to absorb waste, and, as a result, the threat to the world's environment came to be seen as a global crisis. Thus, it was no coincidence that the motto of the United Nations Conference on the Human Environment (UNCHE) in 1972 was 'Only One Earth'.

Environmentalism, then, differed from earlier conservation in two respects. First, it addressed the entire human environment, embracing not only resource problems, such as acid rain, deforestation and whaling, but also the underlying technological, political and economic processes that led to such problems. International travel, particularly air travel, has of course come to be seen by many as one such process that makes a significant contribution to environmental degradation and what has become the dominant political-environmental issue of today, namely climate change (Hall *et al.* 2013). Second, it was more overtly political and activist –

environmentalists turned their attention to social and political issues and embraced other social movements, including the anti-war, anti-consumerism and civil rights movements, which flourished in the 1960s and 1970s. Nevertheless, their main concern was (and remains) the earth's capacity to support human existence, providing the foundation for what is now referred to as sustainability.

In essence, sustainability is based upon the notion that the human economic system of production and consumption is a subsystem of the global ecosystem. This, in turn, is the source of all inputs into the economic subsystem and the sink for all its wastes (Goodland 1992). The global ecosystem's source and sink functions have a finite capacity to supply respectively the needs of production/consumption and absorb the wastes resulting from production/consumption processes. Thus, the variables in the equation become:

(a) the rate at which the stock of natural (non-renewable) resources is depleted relative to the development of substitute, renewable resources;
(b) the rate at which waste is deposited back into the ecosystem relative to the assimilative capacity of the environment; and
(c) global population levels and per capita levels of consumption.

Sustainability is concerned with maintaining a balance between these variables, the potential for which is, to a greater extent, dependent on the effective management (or, perhaps, transformation) of the political, technological, economic and social institutions that determine the production and consumption processes. Each variable is of equal importance to the achievement of sustainability, although, in the two decades since the Kyoto Treaty of 1997, the greatest attention has focused on global warming and climate change. Indeed, since 1995, UN Climate Change Conferences have been held annually, although, with the exception of the conference in Montreal in 2005, where all but one member state (the US) agreed to speed up the measures agreed at Kyoto, relatively little progress has been made towards achieving a long-term agreement on action to combat climate change. In particular, the 2009 Copenhagen conference was seen as a major failure in this endeavour. Nevertheless, there remains continuing widespread concern for an issue that, as we shall see shortly, is of particular relevance to tourism.

Sustainable development: principles and objectives

During the 1970s, attention was primarily focused on limiting growth in order to reduce what were considered to be excessive demands upon the global ecosystem. The IUCN's *World Conservation Strategy* in 1980, referred to above, introduced the term 'sustainable development' and made some attempt to integrate development with conservation, though its emphasis was firmly on the latter. Thus, it was left to the Brundtland Report to combine development and environmental issues within a global strategy for sustainable development. This was subsequently followed by the lesser known, but no less important, *Caring for the Earth* document (IUCN 1991). Whereas the Brundtland Report's strategy for sustainable development was very much based on a return to neoclassical economic growth (Reid 1995) – that is, it espoused traditional economic growth-based development as the foundation for sustainable development – *Caring for the Earth* gave precedence to a strategy for 'sustainable living', the emphasis being on the adoption of sustainable lifestyles and, essentially, a transformation in people's attitudes towards consumption practices. Together, these reports provide a framework for identifying the key principles and objectives of sustainable development and, indeed, the prerequisites for its achievement.

Building on the basis that any form of development should occur within environmental limits, sustainable development should be guided by the following principles:

- *Holistic perspective*: development and sustainability are global challenges.
- *Futurity*: the emphasis should be on the long-term future.
- *Equity*: development should be fair and equitable both within and between generations.

As already observed, while Brundtland recognized the desirability of meeting basic needs, it argued for economic growth as a prerequisite for sustainable development. Conversely, *Caring for the Earth* argued for a fundamental transformation in people's approach to the environment in general and consumption in particular. However, both are important elements of sustainable development. A certain level of wealth is necessary to underpin development (hence the contemporary focus on poverty reduction), yet, globally, environmental pressures are not resource problems – they are human problems (Ludwig *et al.* 1993). Therefore, as summarized in Table 2.1, in order to meet the developmental and environmental objectives of sustainable development, a number of requirements must be satisfied.

The overall objectives of sustainable development, then, can be seen as:

- *environmental sustainability*: the conservation and effective management of resources;
- *economic sustainability*: longer-term prosperity as a foundation for continuing development;
- *social sustainability*: with a focus on alleviating poverty, the promotion of human rights, equal opportunity, political freedom and self-determination.

It is important to note, of course, that this discussion of the principles and objectives of sustainable development reflects a dominant operational/managerialist, and arguably Western, perspective. That is, it is concerned with principles and processes deemed necessary to preserve the world's resources for, implicitly, their subsequent exploitation. However, development is also an inherently political process; differing perspectives on development may reflect differing political ideologies. For example, neo-liberalism, which espouses both personal freedom and development based upon free-market economic growth, runs counter to the radical ecological movement that sees the resolution of environmental problems lying in intervention in and restrictions on individual freedoms (Martell 1994). In a similar sense, socialist political economy proposes that sustainability may be achieved by reducing the influence of capitalism and the drive for profit, which itself requires a transformation in relations of production towards collective ownership. Other political perspectives on development and sustainability include both the Marxist approach, which, some claim, contributes positively to environmentalism through emphasizing people's need to interact with their natural environment to overcome their sense of alienation and achieve personal freedom and fulfilment, and an eco-feminist perspective. The latter proposes that values and traits associated with femininity (such as an emphasis on interrelationships, balance and natural processes) may provide the foundation for sustainability (Martell 1994).

Table 2.1 *Sustainable development: principles and objectives*

Fundamental principles	• *Holistic approach*: development and environmental issues integrated within a global social, economic and ecological context.
	• *Futurity*: focus on long-term capacity for continuance of the global ecosystem, including the human subsystem.
	• *Equity*: development that is fair and equitable, and which provides opportunities for access to and use of resources for all members of all societies, both in the present and future.
Development objectives	• Millennium Development Goals and beyond.
	• Improvement of the quality of life for all people: education, life expectancy, opportunities to fulfil potential.
	• Satisfaction of basic needs: concentration on the nature of what is provided rather than income.
	• Self-reliance: political freedom and local decision-making for local needs.
	• Endogenous development.
Sustainability objectives	• Poverty reduction.
	• Sustainable population levels.
	• Minimal depletion of non-renewable natural resources.
	• Sustainable use of renewable resources.
	• Pollution emissions within the assimilative capacity of the environment.
Requirements for sustainable development	• Sustainable consumption: adoption of a new social paradigm relevant to sustainable living.
	• Sustainable production: biodiversity conservation; technological systems that can search continuously for new solutions to environmental problems.
	• Sustainable distribution systems: international and national political and economic systems dedicated to equitable development and resource use.
	• Technological systems that can search continuously for new solutions to environmental problems.
	• Global alliance facilitating integrated development policies at local, national and international levels.

Source: Adapted from Sharpley and Telfer (2015), based on Streeten (1977); WCED (1987); Pearce et al. (1989); IUCN (1991)

The UN has been working on a post-2015 development agenda to move beyond the 2015 UN Millennium Development Goals, as outlined in Box 1.1. A synthesis report on efforts to date by the UN Secretary General in December 2014 stressed the importance of continuing to work towards the Millennium Development Goals but also working towards filling key sustainable development gaps left by the goals (see Box 2.1). The report proposed the following six essential elements for delivering sustainable development goals:

- *dignity*: to end poverty and fight inequalities;
- *people*: to ensure healthy lives, knowledge and the inclusion of women and children;
- *prosperity*: to grow a strong, inclusive and transformative economy;
- *planet*: to protect our ecosystems for all societies and our children;
- *justice*: to promote safe and peaceful societies and strong institutions;
- *partnership*: to catalyse global solidarity for sustainable development (UN 2014b).

Sustainable development: from principle to practice

Controversially perhaps, Dresner (2002: 67) observes that 'sustainable development is not such a vague idea as it is sometimes accused of being'. Nevertheless, it undoubtedly remains a contested concept, largely as a result of problems in its operationalization and measurement. Moreover, as noted in the previous section, it is also susceptible to varying ideological interpretation. In other words, both putting the principles of sustainable development into practice and assessing their effectiveness has proved to be a difficult task. Indeed, from a development studies context, there has been a shift from the idealism of sustainable development to a more people-focused approach of human development (see Chapter 1 and Telfer 2015a). Nevertheless, from a political perspective, it remains the dominant development paradigm, and hence an understanding of the debates surrounding the concept is essential in the context of this book. In other words, it is important to highlight the key questions or criticisms of sustainable development as these may be equally levelled at the concept of sustainable tourism development in particular.

A full critique of sustainable development is, of course, beyond the scope of this chapter (see Elliott 2013). However, the principal difficulty facing the operationalization of sustainable development is the long-recognized ambiguity and inherent contradictory nature of the concept (Redclift 1987). That is, the twin objectives of

Box 2.1

Sustainable Development Goals: beyond the 2015 UN Millennium Development Goals

The UN Millennium Development Goals are presented in Box 1.1. In the process of developing the post-2015 development agenda, a wide range of conferences, consultations and committee work has taken place. The Report of the UN Open Working Group of the General Assembly on Sustainable Development Goals from August 2014 presented the following list of Sustainable Development Goals for consideration by the UN. Each of these goals is further elaborated on in the report, with targets that have indicators focused on measurable outcomes (UN 2014c). These 17 Sustainable Development Goals are now part of the 2030 Agenda for Sustainable Development building on the Millennium Development Goals.

> The agenda and goals should also be received at the country level in a way that will ensure the transition of the Millennium Development Goals to the broader more transformative sustainable development agenda, effectively becoming an integral part of national and regional visions and plans.
>
> (UN 2014b)

The new goals were adopted by the UN on 25 September 2015.

Goal 1. End poverty in all its forms everywhere.

Goal 2. End hunger, achieve food security and improved nutrition, and promote sustainable agriculture.

Goal 3. Ensure healthy lives and promote well-being for all at all ages.

Goal 4. Ensure inclusive and equitable quality education and promote lifelong learning opportunities for all.

Goal 5. Achieve gender equality and empower all women and girls.

Goal 6. Ensure availability and sustainable management of water and sanitation for all.

Goal 7. Ensure access to affordable, reliable, sustainable and modern energy for all.

Goal 8. Promote sustained, inclusive and sustainable economic growth, full and productive employment and decent work for all.

Goal 9. Build resilient infrastructure, promote inclusive and sustainable industrialization and foster innovation.

Goal 10. Reduce inequality within and among countries.

Goal 11. Make cities and human settlements inclusive, safe, resilient and sustainable.

Goal 12. Ensure sustainable consumption and production patterns.

Goal 13. Take urgent action to combat climate change and its impacts.*

Goal 14. Conserve and sustainably use the oceans, seas and marine resources for sustainable development.

Goal 15. Protect, restore and promote sustainable use of terrestrial ecosystems, sustainably manage forests, combat desertification, and halt and reverse land degradation and halt biodiversity loss.

Goal 16. Promote peaceful and inclusive societies for sustainable development, provide access to justice for all and build effective, accountable and inclusive institutions at all levels.

Goal 17. Strengthen the means of implementation and revitalize the global partnership for sustainable development (UN 2014c).

* Acknowledging that the United Nations Framework Convention on Climate Change is the primary international, intergovernmental forum for negotiating the global response to climate change (UN 2014c).

Source: UN (2014b, 2014c)

'development' and 'sustainability' are considered by many to be an oxymoron: a contradiction in terms. How can development (necessitating resource exploitation) be achieved at the same time as sustainability (minimizing resource depletion)? Moreover, different interpretations of the phrase itself, and its two constituent elements, further complicate matters. Is sustainable development literally development that can be sustained, giving precedence to development (however defined), or is it development that is restricted by environmental sustainability (again, however defined)? Indeed, even if objective assessments of sustainability could be made, development inevitably remains a subjective concept. Thus, for example, while many consider tourism in Bhutan to be environmentally sustainable, the extent to which it is contributing to the country's sustainable development remains debatable (see Box 2.2).

In addition to its inherent ambiguity, the concept of sustainable development is seen as encouraging hypocrisy and delusion (Robinson 2004), while the objective of achieving a balance between environmental sustainability, economic growth and social development and well-being has more recently been increasingly challenged. It argued that according equal value to these three pillars of sustainable development not only risks 'tackling issues of sustainable development in a compartmentalized manner' (Giddings *et al.* 2002: 189), but also obscures the inescapable fact that social development is, to an extent, dependent on economic growth, but that both society and the economy depend on environmental sustainability. Moreover, the very ideas of economic and social sustainability have come to be criticized as politically attractive but difficult to define: 'many people just believe "social sustainability" to be important . . . without being able to actually articulate its meaning' (Mundt 2011: 86). In other words, the notion of

Box 2.2

Tourism and sustainable development in Bhutan

The Himalayan Kingdom of Bhutan remained relatively isolated from the rest of the world up until the 1960s. Indeed, it was not until the mid-1970s that small numbers of tourists began to visit the country, although at that time both gaining the necessary visas and actually travelling to Bhutan (there was no airport, so access was by road from India) was a complex and lengthy process. Consequently, relatively few tourists chose to visit Bhutan and it was not until the opening of the international airport in 1983 and the subsequent extension of the runway in 1990 that greater numbers (in Bhutanese terms) began to arrive.

From the outset, tourism in Bhutan was introduced with the principal objective of generating foreign exchange earnings. Given its mountainous terrain, the country has few natural resources upon which to build its economy, although the generation and export of hydroelectric power to neighbouring India accounts for around 50 per cent of all exports and more than 40 per cent of GDP. Agriculture and forestry are major activities employing 62 per cent of the workforce, yet just 7 per cent of the land area is suitable for cultivation. Moreover, although the country is often portrayed as a Shangri-La, remote mountain communities follow a harsh existence; there is a lack of access to services, and although poverty levels have fallen significantly in recent years, from 23 per cent in 2007 to 12 per cent in 2012, it remains included on the UN's list of least-developed countries. However, the country possesses a rich and diverse cultural and natural heritage, providing the principal attraction for tourists. It is home to a number of rare Himalayan species, it boasts spectacular scenery and its vibrant culture is manifested in its distinctive architecture and its spectacular festivals.

Despite its potential attraction to visitors and for high earnings from tourism, tourism development in Bhutan has, since its inception, continued to focus on high-yield/low-volume tourism, although more recent arrivals figures increasingly appear to contradict this official policy. Nevertheless, in order to protect the country's cultural and natural heritage for local communities and visitors alike, the scale and nature of tourism has been tightly controlled through charging a minimum day rate for international visitors. From 1991, this was set at US$200 in the high season and US$165 in the low season, but was recently increased to US$250 and US$200, respectively (with per person surcharges up to US$40 per day for groups of less than three). All aspects of the local tourism industry, from the number of local tour operators to the supply of accommodation, are also closely regulated. Initially, just 200 international tourists a year were permitted entry, but this figure has gradually risen over the years.

As can be seen from Table 2.2, there has been a remarkable growth in arrivals over the last decade; indeed, the 2013 figure was 116,209 (TCB 2014). However, it should be noted that this includes regional arrivals, primarily from India, who

Table 2.2   Bhutan: international tourist arrivals, 1990–2012

Year	Arrivals	Year	Arrivals
1990	1,538	2002	5,599
1992	2,763	2004	9,249
1994	3,971	2006	17,365
1996	5,138	2008	27,636
1998	6,203	2010	40,873
2000	7,559	2012	105,407

collectively accounted for 63,426 arrivals in 2013. Regional tourists pay neither visa fees nor the minimum tariff, to an extent undermining the low-volume/high-value policy. Nevertheless, direct international tourism receipts amounted to almost US$64 million, with total income from all tourism estimated to be US$220 million. Moreover, the sector provides employment for more than 26,000 people in Bhutan.

The activities of tourists within the country are also strictly controlled – although the industry has been 'privatized', every aspect of tourism is controlled or regulated by the Tourism Council of Bhutan – the purpose being to minimize the negative impacts of tourism. However, while this has resulted in Bhutanese tourism being widely seen as an example of successful sustainable tourism practice, questions must be raised over the extent to which tourism is genuinely contributing to sustainable development. In particular:

- Regulation restricts the extent to which local people can access the industry and, hence, benefit financially.
- Tourism to Bhutan is highly seasonal, impacting on permanent employment and income levels.
- Tourism activity is limited to a few western and central valleys; much of the country does not receive tourists and therefore does not share directly the benefits of tourism.
- Local community involvement in tourism is extremely limited.
- Despite relatively low numbers of tourists, the natural environment has suffered significant impacts, particularly deforestation, the erosion of delicate vegetation and the creation of 'garbage trails'.
- An increase in the number of licensed tour operators to facilitate greater numbers of tourists has led to price competition among operators, thereby challenging the high-yield/low-volume policy.

Thus, as tourism plays an increasingly important role within the Bhutanese economy, appropriate policies will be needed to ensure that its contribution to sustainable development is improved.

Source: Brunet et al. (2001); Dorji (2001); TCB (2014)

sustainability has been extended to the domains of economy and society, thereby rendering it increasingly meaningless.

Further criticism is also directed at the concept of sustainable development in that it is considered by some to be, in the context of global development, yet another manifestation of Western hegemony. Or, perhaps, if the process or goal of development is no longer possible in an era of globalization, does this suggest that sustainable development is unachievable? On the one hand, the so-called neo-liberal school argues that economic globalization, achieved through free trade and capital mobility, will eventually lead to wealth and prosperity for all; conversely, the radical school argues that globalization is increasing the rift between the wealthy 'core' and the less-developed 'periphery'. What is certain is that the concept of sustainable development is problematic. Not only is it, in a sense, a luxury that many countries, particularly those suffering from the 'pollution of poverty', simply cannot afford, but also it gives rise to a number of questions, including:

- *What should be developed sustainably?* Personal wealth, national wealth, human society, ecological diversity?
- *For how long should it be sustained?* For a generation, a century, 'forever'?
- *Against what baseline can sustainable development be assessed?* No further environmental degradation, limits of acceptable change?
- *Who is responsible for sustainable development?* Individuals, national governments, the international community?
- *Under what political-economic conditions is sustainable development possible?*

Answers to these questions, and others, remain elusive. Nevertheless, although the achievement of sustainable development or, more generally, global environmental sustainability, undoubtedly faces enormous challenges, it is recognized that the alternative (that is, unsustainability) is not an option.

Sustainable tourism development

As observed in the introduction to this chapter, the roots of sustainable tourism development lie in the strategies for the development of alternative forms of tourism that emerged during the late 1980s. As concern grew over the negative consequences of the rapid and uncontrolled growth of mass international tourism, epitomized perhaps by the development of the Spanish 'Costas' from the 1960s onwards, attention was increasingly focused on alternative approaches to tourism development. Variously labelled as, for example, green, responsible, appropriate, ethical, low-impact, soft or ecotourism, these are styles of tourism that collectively represent, literally, an alternative to mass tourism development. Designed to minimize tourism's negative impacts while optimizing the benefits to the destination, they share a number of characteristics that, collectively, lie in opposition to those of conventional mass tourism (see Table 2.3).

Table 2.3 *Characteristics of mass versus alternative tourism*

Conventional mass tourism	Alternative forms of tourism
General features	
Rapid development	Slow development
Maximizes	Optimizes
Socially/environmentally inconsiderate	Socially/environmentally considerate
Uncontrolled	Controlled
Short term	Long term
Sectoral	Holistic
Remote control	Local control
Development strategies	
Development without planning	First plan, then develop
Project-led schemes	Concept-led schemes
Tourism development everywhere	Development in suitable places
Concentration on 'honeypots'	Pressures and benefits diffused
New building	Reuse of existing building
Development by outsiders	Local developers
Employees imported	Local employment utilized
Urban architecture	Vernacular architecture
Tourist behaviour	
Large groups	Singles, families, friends
Fixed programme	Spontaneous decisions
Little time	Much time
'Sights'	'Experiences'
Imported lifestyle	Local lifestyle
Comfortable/passive	Demanding/active
Loud	Quiet
Shopping	Bring presents

Source: Adapted from Butler (1990); Lane (1990)

Alternative tourism is considered by some to be synonymous with sustainable tourism, and there are, of course, many contemporary examples of such tourism development in practice. Typically, they tend to be small-scale and appropriate to the area, with the emphasis on protecting and enhancing the quality of the tourism resource. Ownership and control of tourism development is largely in the hands of local communities and is directed towards optimizing the long-term benefits to visitors, the destinational environment and local people (see Box 2.3).

Box 2.3

Chambal Safari Lodge, Uttar Pradesh, India

The Chambal Safari Lodge in Uttar Pradesh, a northern state of India bordering Nepal, is located in some 36 acres of farmland and forest that has belonged to the same family for more than six centuries. For many years, the site had hosted a biannual cattle fair but was also coming under pressure as a resource for sand mining, and so, in 1999, the Chambal Conservation Foundation was established in order to develop sustainable projects that would not only maintain the natural beauty and wildlife resources of the valley, but would also lead to an improvement in the lives of the local communities. Fundamental to this was the development of ecotourism based around the conversion of the Mela Kothi, or main building, into an ecolodge boasting 12 luxurious rooms. From there, visitors can explore the area, going on village or nature walks to experience and learn about local flora, fauna and culture.

In addition to placing Chambal on the international birding and wildlife map, the Foundation contributes to the protection and regeneration of natural habitats, particularly the reforestation of the area, as well as developing a rainwater-harvesting project. It also works with local communities not only to raise awareness of environmental issues, but to provide employment. The majority of people working for the Foundation are from the local community, while specific activities, such as running camel safaris for tourists, provide much-needed income. Moreover, the lodge is furnished from locally crafted furniture and all meals prepared for tourists are made with ingredients supplied by local farmers or grown in the lodge's own farm. The Foundation is also active in promoting conservation in the region and supports efforts to reduce poaching.

In 2014, Chambal Safari Lodge won the Responsible Tourism Awards silver award for wildlife conservation.

Source: www.chambalsafari.com/about-chambal-safari-lodge.html; www.responsibletravel.com/awards/categories/wildlife.htm

However, alternative tourism has also been widely criticized for being just that: an alternative, rather than a solution, to the 'problem' of mass tourism. At the same time, the focus on alternative forms of tourism development, particularly on ecotourism (see Plates 2.1 and 2.2) – itself a controversial concept (see Fennell 2008) – has served to amplify the distinction between mass, implicitly 'bad' tourism (and tourists, perhaps), and alternative, 'good' forms of tourism. Consequently, although the concept of sustainable tourism development, as originally conceived, applied the principles of sustainable development to tourism more generally (Hunter 1995), attempts to define and implement sustainable tourism development have largely been undermined by the continuing mass–alternative dichotomy. As one commentator once proposed, for tourism development to be sustainable, it should be based upon 'options or strategies considered preferable to mass tourism' (Pigram 1990). Despite the World Tourism Organization's assertion that 'Sustainable tourism development guidelines and management practices are applicable to all forms of tourism in all types of destinations, including mass tourism' (UNWTO 2015a), this is still very much the case.

Plate 2.1 *South Africa, Dikhololo Resort near Pretoria: tourists preparing to go on a game-watching outing*

Source: Photo by D. Telfer

Plate 2.2 *The Gambia: floating accommodation rooms at an ecolodge*

Source: Photo by R. Sharpley

What is sustainable tourism development?

It is probably true to say that over the last 25 years, the concept of sustainable tourism development has been the dominant issue in both the study and practice of tourism (see Telfer 2013 for an examination of the Brundtland Report on tourism). Certainly, the subject has spawned innumerable academic books and articles, as well as two dedicated journals (*Journal of Sustainable Tourism* and *Journal of Ecotourism*). Even by 1999, an annotated bibliography of sustainable tourism published by the WTO listed 96 books and 280 articles on the subject (WTO 1999), and, since then, the relevant literature has grown considerably. Moreover, although support for the concept has, perhaps, begun to wane, it still attracts significant academic interest; a major *Encyclopaedia of Sustainable Tourism* (Cater and Garrod 2015) has, for example, recently been published. Nevertheless, despite the degree of attention paid to it, it has continued to prove difficult, if not impossible, to achieve consensus on a definition of sustainable tourism development.

In the editorial to the first issue of *Journal of Sustainable Tourism*, Bramwell and Lane (1993) described sustainable tourism as a 'positive approach intended to reduce tensions and friction created by the complex interactions between the tourism industry, visitors, the environment and the communities which are host to holidaymakers', though arguably this reveals neither the developmental objectives of sustainable tourism, nor the processes by which such processes might be achieved and measured. Rather, it suggests that sustainable tourism is a more benign or soft approach to tourism development. However, logic would suggest that if sustainable tourism development represents a sector-specific application of the 'parental paradigm' of sustainable development, then it should also share its principles and objectives. In other words, the essential role of tourism development is its contribution to wider economic and social development within the destination; therefore, sustainable tourism development is, or should be, seen simply as a means of achieving sustainable development through tourism. Thus, not only should tourism itself be environmentally sustainable, but also it should contribute indefinitely to broader sustainable development policies and objectives.

Such an approach was, in fact, adopted by the Globe 90 Conference in Canada, where, in recognition of tourism's developmental role, three basic principles to guide tourism planning and management were proposed (Cronin 1990):

- Tourism must be a recognized sustainable economic development option, considered equally with other economic activities.
- There must be a relevant tourism information base to permit recognition, analysis and monitoring of the tourism industry in relation to other sectors of the economy.
- Tourism development must be carried out in a way that is compatible with the principles of sustainable development.

This approach has a number of important implications in terms of conceptualizing sustainable tourism development. First, it demands that tourism's developmental potential is assessed against other economic sectors (if the potential to develop other economic activities exists) – that is, tourism is considered within a broader socio-economic and environmental context. This, second, implies that tourism should not compete for scarce resources; rather, emphasis should be placed on the most efficient and sustainable shared use of resources. Third, the mass–sustainable tourism dichotomy referred to earlier becomes irrelevant. In other words, the challenge is to ensure that all forms of tourism (including mass tourism) are planned and managed in such a way as to contribute to sustainable development. Finally, and pointing to a major and as yet unresolved problem, a holistic approach that embraces the entire tourism system is required. That is, while it is important to consider the contribution of tourism to development in the destination, it is no less important to consider the tourism-generating region and, in particular, the environmental consequences of travel to the destination (Høyer 2000).

Subsequently, many tourism planning and policy documents embraced the principles of sustainable development. The typical principles and guidelines for sustainable tourism initially set out in such documents are summarized in Table 2.4.

In particular, the sustainable use of natural resources and the development of tourism within physical and sociocultural capacities are of fundamental importance, while consideration is given to equitable access to the benefits of tourism. The concept of futurity is also implicit within these guidelines. Moreover, the principle of community involvement appears to satisfy the specific requirements of self-reliance and endogenous development that are critical elements of the sustainable development paradigm.

In a more recent document, *Sustainable Tourism for Development*, the World Tourism Organization (UNWTO 2013b) defines sustainable tourism as 'tourism that takes full account of its current and future economic, social and environmental impacts, addressing the needs of visitors, the industry, the environment and host communities', and identifies three key processes:

● make optimal use of environmental resources that constitute a key element in tourism development, maintaining essential ecological processes and helping to conserve natural heritage and biodiversity;
● respect the sociocultural authenticity of host communities, conserve their built and living cultural heritage and traditional values, and contribute to intercultural understanding and tolerance; and
● ensure viable, long-term economic operations, providing socio-economic benefits to all stakeholders that are fairly distributed,

Table 2.4 *Sustainable tourism development: a summary of principles*

....

- The conservation and sustainable use of natural, social and cultural resources is crucial. Therefore, tourism should be planned and managed within environmental limits and with due regard for the long-term appropriate use of natural and human resources.

- Tourism planning, development and operation should be integrated into national and local sustainable development strategies. In particular, consideration should be given to different types of tourism development and the ways in which they link with existing land and resource uses and sociocultural factors.

- Tourism should support a wide range of local economic activities, taking environmental costs and benefits into account, but it should not be permitted to become an activity that dominates the economic base of an area.

- Local communities should be encouraged and expected to participate in the planning, development and control of tourism with the support of government and the industry. Particular attention should be paid to involving indigenous people, women and minority groups to ensure the equitable distribution of the benefits of tourism.

- All organizations and individuals should respect the culture, the economy, the way of life, the environment and political structures in the destination area.

- All stakeholders within tourism should be educated about the need to develop more sustainable forms of tourism. This includes staff training and raising awareness, through education and marketing tourism responsibly, of sustainability issues among host communities and tourists themselves.

- Research should be undertaken throughout all stages of tourism development and operation to monitor impacts, to solve problems and to allow local people and others to respond to changes and to take advantage of opportunities.

- All agencies, organizations, businesses and individuals should cooperate and work together to avoid potential conflict and to optimize the benefits to all involved in the development and management of tourism.

Source: Adapted from ETB (1991); Eber (1992); EC (1993); WTO (1993); WTO/WTTC (1996)

including stable employment and income-earning opportunities and social services to host communities, and contributing to poverty alleviation (UNWTO 2013b: 19–20).

Both the early and more recent sets of principles for sustainable tourism clearly adhere to those of sustainable development more generally. The question that must be asked, however, is: Does sustainable tourism development remain, in Butler's (2013: 223) words, a 'fantasy'? Certainly, in practice, sustainable tourism development has more typically reflected what has been described as a 'tourism centric and parochial' perspective (Hunter 1995). That is, rather than considering tourism within a broader developmental context, the principal objective has been the development of sustainable tourism or, more precisely, sustaining tourism itself.

Consequently, rather than optimizing its contribution to the broader sustainable development of the destination as a whole, attention has been focused primarily on the preservation of the natural, built and sociocultural resource base upon which tourism depends in particular settings, thereby permitting the longer-term 'survival' of tourism as an economic sector. In short, sustainable tourism development is most commonly taken to represent 'tourism which is in a form which can maintain its viability in an area for an indefinite period of time' (Butler 1993: 29).

To an extent, of course, this is sound business practice. All industries strive to maintain their resource base for longer-term survival while sound environmental policies may significantly enhance profitability. It does not, however, equate with sustainable development. Thus, while there is evidence of many successful sustainable tourism projects, these tend to be localized and small-scale; conversely, there is little or no evidence of sustainable tourism development on a wider scale. This, in turn, suggests that 'true' sustainable tourism development (that is, tourism development that is consistent with the tenets of sustainable development) is unachievable. Indeed, it is now generally accepted that sustainable tourism development, though highly desirable, is an idealistic concept that is difficult, if not impossible, to put into practice.

This is not to say efforts have not been made (and continue to be made) to implement sustainable tourism practices. Indeed, as we shall see shortly, many organizations and businesses throughout the tourism system have attempted to adopt or promote the principles of sustainable tourism development. However, it is important to recognize that, as an approach to tourism development, sustainable tourism faces a number of significant challenges.

Sustainable tourism development: weaknesses and challenges

Despite the lack of consensus over the definition and viability of sustainable tourism development, it is probably true to say that, in principle, few would argue against its objectives. Equally, however, few would also argue against the fact that, as a specific economic and social activity, tourism displays a number of characteristics that appear to be in opposition to the concept of sustainable development. These characteristics, referred to in a widely cited paper as tourism's 'fundamental truths' (McKercher 1993), may be summarized as follows:

- As a major, global activity, tourism consumes resources, creates waste and requires significant infrastructural development.
- The development of tourism may, potentially, result in the over-exploitation of resources.
- In order to survive and grow, the tourism industry has to compete for scarce resources.
- The tourism industry is predominantly made up of smaller, private businesses striving for short-term profit maximization.
- As a global, multi-sectoral industry, tourism is impossible to control.
- 'Tourists are consumers, not anthropologists'.
- Most tourists seek relaxation, fun, escape and entertainment; they do not wish to 'work' at being tourists.
- Although an export, tourism experiences are produced and consumed 'on site'.

Collectively, these 'truths' point to three major issues. First, tourism is not a 'smokeless' industry. As has long been recognized, the development of tourism may result in significant environmental and social impacts for destinations, the effective management of which is fundamental to the sustainability of tourism. Second, the consumption of tourism is of direct relevance to its (sustainable) developmental contribution – the scale, scope and nature of the demand for tourism represent significant challenges to sustainable development. However, recent research by Miller et al. (2010) has demonstrated both the lack of awareness among tourists of the impacts of tourism and their unwillingness to adopt more responsible behaviour as tourists. Third, the structure, scale and inherent power relations of the tourism industry raise important questions about the likelihood of a collective, uniform commitment to the principles of sustainable development on the part of the industry.

These issues are returned to shortly and are also considered in more detail in subsequent chapters. However, a further general weakness of the concept of sustainable tourism development deserving comment is that its principles and objectives have tended to be manifested in numerous sets of guidelines that, together, represent a relatively inflexible and, arguably, Western-centric 'blueprint' for the development of tourism (Southgate and Sharpley 2015). In other words, they propose a relatively uniform approach to tourism development, usually based upon managing the limits (according to Western criteria) of acceptable environmental and social change, which is unable to account for the almost infinite diversity of tourism developmental contexts. That is, all destinations differ in terms of a variety of factors, including local developmental needs, local environmental attitudes and knowledge, local governance and planning systems, the maturity and diversity of the local economy, and so on. For example, some least-developed countries have, arguably, yet to reach the 'take-off' stage referred to in Chapter 1. As a result, not only are they unable to take advantage of some of the opportunities offered by the development of tourism, but also they face a set of development priorities, such as widespread poverty and malnutrition, different to those addressed by sustainable tourism development. Thus, in some countries, such as Zanzibar (see Box 2.4), an alternative approach to tourism and development may be required.

Interestingly, more recent approaches to sustainable tourism do in fact focus on the specific issue of poverty, and particularly the need to develop policies that spread the benefits of tourism development to the poorest members of destination societies who are unable to establish a formal position in the local tourism system (UN 2003; Mitchell and Ashley 2010; Scheyvens 2011). Referred to as pro-poor tourism, this is considered in more detail below.

Tourism as sustainable development?

In addition to the general concerns about the validity of the concept of sustainable tourism development discussed above, there are a number of specific areas in which tourism, as a particular vehicle of development, is unable to meet the principles and required processes of sustainable development. These reflect some of the 'truths' of tourism and are considered in detail elsewhere (Sharpley 2000). Nevertheless, a brief review of these points of divergence between sustainable tourism development and the broader sustainable development paradigm is useful, for two reasons:

- it goes some way to explaining the inevitability of the tourism-centric, localized and small-scale focus of most sustainable tourism projects in practice; and
- it points to ways in which tourism may, conversely, make a contribution to the sustainable development of destination areas.

Box 2.4

Tourism and development in Zanzibar

Lying some 32 km off the coast of East Africa, Zanzibar is an archipelago comprising two main islands, Unguja and Pemba, as well as around 50 smaller islets, and is a semi-autonomous part of the United Republic of Tanzania. With an economy traditionally dependent on agriculture in general, and the production of cloves in particular, it remains a poor country; in 2009, per capita GDP in Zanzibar amounted to $548, representing a per capita average daily income of just $1.50. Moreover, 13 per cent of the population of 1.1 million live below the food poverty line and 49 per cent below the basic needs poverty line, while other indicators, such as life expectancy, secondary school enrolment and so on, point to continuing severe underdevelopment. However, the country is blessed with extensive and beautiful beaches, and hence, since the mid-1980s, tourism has come to play an increasingly important role in the economy of Zanzibar. It is estimated, for example, that direct tourism expenditure accounts for some 25 per cent of GDP and around 80 per cent of export earnings, while, remarkably, some 80 per cent of government revenues accrue from the tourism sector.

However, the economic significance of tourism in Zanzibar has not been translated into wider socio-economic development, though this arguably reflects problems of policy and governance rather than a lack of success in developing tourism itself. Indeed, at first sight, Zanzibar appears to have enjoyed sustained growth in tourism over the last three decades.

Table 2.5   Tourist arrivals in Zanzibar, 1986–2010

Year	Arrivals	Year	Arrivals
1986	22,846	2000	97,165
1988	32,119	2002	87,511
1990	42,141	2004	92,161
1992	59,747	2006	137,111
1994	41,433	2008	128,440
1996	69,159	2010	132,836
1998	86,455		

As can be seen from the table, Zanzibar has enjoyed relatively consistent growth in tourist arrivals, the great majority of visitors staying on the main island Unguja (also known as Zanzibar Island), typically splitting their visit between the capital, Stone Town, and a beach resort. Nevertheless, the nature of tourism in Zanzibar itself limits

its potential contribution to development. Specifically, roughly 50 per cent of tourists arrive on package tours and spend the majority of time (and money) in a resort; the other 50 per cent are independent travellers, typically younger backpackers, who are relatively low spenders. In both cases, average length of stay in Zanzibar is between just five and seven days, further limiting the extent to which the country benefits from tourist expenditures. At the same time, however, the country suffers a high level of leakages (80 per cent of all food and beverages consumed by tourists in Zanzibar is imported), establishing a tourism business is both complex and expensive, and employment in the sector among Zanzibaris is relatively limited. Indeed, it is estimated that approximately half of all tourism jobs are held by mainlanders (primarily Tanzanians and Kenyans). These issues reflect what some consider to be a lack of appropriate policies for and governance of the tourism sector in Zanzibar, in particular:

● Lack of investment: despite significant revenues accruing to the government, little is reinvested in infrastructure, facilities, training programmes or business development/support schemes.
● Planning and policy failures: despite the existence of master plans, no effective planning or zoning to control the development of tourism has been implemented.
● Institutional confusion: the governance of tourism is 'shared' between disparate and sometimes competing government departments.
● Political interference: even if formal tourism planning policies and mechanisms were in existence, it is likely that these could be overruled through political interest and interference.

Hence, the lack of development in Zanzibar reflects, to a greater or lesser extent, the failure of the state to fulfil its responsibilities to translate the opportunities presented by tourism into social and economic development.

Source: Sharpley and Ussi (2014)

With reference to Table 2.1, the development of tourism is unable to meet sustainable development's fundamental principles, nor its development and sustainability objectives:

Holistic approach. All tourism development should be considered within a global socio-economic, political and ecological context. More simply stated, all elements of the tourism experience should be sustainable. However, given the breadth of the tourism system and the fragmented, multi-sectoral character of the tourism industry, such an approach is difficult, if not impossible.

Within this context, a particular issue is the unsustainability of most modes of transport (Høyer 2000; Hall *et al.* 2013). By definition, tourism involves travel/transport by land, sea or air. The environmental impacts of air travel (both infrastructural developments and aircraft emissions) have emerged as a particular area of contemporary concern and, as a result, some organizations have introduced innovative schemes. For example, in 2005, British Airways announced a scheme whereby its passengers can offset the environmental cost of their journey by making voluntary donations, equivalent to that environmental cost, to invest in projects that seek to reduce global carbon dioxide levels. More recently, in 2011, Virgin Atlantic introduced the first onboard/online scheme, supporting projects in India and Indonesia. However, such carbon offset initiatives remain voluntary and have little effect on worldwide emissions related to transport and travel given relatively limited take-up by the travelling public (Mair 2011).

Futurity. The tourism industry is not only diverse and fragmented; it is also largely comprised of small, private sector, profit-motivated businesses. Therefore, although there is undoubtedly evidence of tour operators and other organizations that have adopted a 'responsible', longer-term perspective (Mann 2000; www.responsibletravel.com), it is likely that most businesses are more concerned with short-term profit considerations rather than the long-term sustainable development of the destination.

Equity. Although tourism development by no means leads inevitably to destinational dependency, the structure, ownership and control of the tourism industry on a global scale, as well as the regionalized and polarized nature of international tourist flows, suggests that inter- and intra-generational equity is unlikely to be achieved through tourism. In other words, although there are many examples of community-based tourism projects, the overall political economy of the tourism system is such that tourist flows and the tourism industry itself are generally dominated by Western- and, more recently, Asian-owned global networks (Bianchi 2015). Moreover, the tourism system within destinations or countries also tends to be dominated by the local elite, restricting equitable access to the benefits of tourism.

Development objectives. Tourism has, of course, long been considered an effective means of achieving development. However, 'development' in the tourism context usually means traditional economic development rather than the more contemporary interpretations of its objectives. That is, tourism undoubtedly represents an important source of income, foreign exchange and employment. However, it remains unclear to what extent broader (sustainable) developmental goals, such as the satisfaction of basic needs, self-reliance and endogenous development, or indeed the contemporary objective of human development (see Chapter 1) can be achieved through tourism. Moreover, not only are the benefits of tourism usually restricted to particular geographic areas and/or sectors of local destinational communities – as noted below, so-called 'pro-poor tourism' attempts to meet this challenge – but also the very nature of tourism as a discretionary and fragile form of consumption is such that the ability of destinations to manage tourism development may be compromised by factors beyond their control. Thus, it is increasingly recognized that while tourism contributes to economic growth, it does not necessarily lead to 'development'.

Sustainability objectives. A fundamental principle of all sustainable tourism development policies is that the natural, social and cultural resources upon which tourism depends should be protected and enhanced. Furthermore, most, if not all, sectors of the tourism industry have a vested interest in following such a policy. This may result from either a genuine commitment to sound environmental practice or for more pragmatic business reasons. Either way, however, the extent to which sustainability objectives are achievable remains questionable. Resource sustainability is dependent upon all sectors involved directly and indirectly in the tourism industry working towards common goals, and although different organizations and industry sectors have, to a lesser or greater extent, adopted environmental management policies (such as carbon offset schemes offered by numerous airlines), sustainability in tourism will only be achieved when the industry as a whole accepts the need for such policies.

Most significantly, the achievement of sustainable tourism development is dependent upon a number of prerequisites, in particular the adoption of a new 'social paradigm' regarding the consumption of tourism (or, more simply stated, the need for all tourists to become 'good' or 'responsible' tourists), and the emergence of global political and economic systems dedicated to more equitable resource use and development. Both of these issues are considered in more detail later in this book, although, as this chapter goes on to conclude, perhaps the greatest challenge facing progress towards the achievement of sustainable tourism development (and, indeed, sustainable development more generally) are global political-economic structures and the trend towards globalization.

However, it is important to note that despite the undoubted difficulties in implementing 'true' sustainable tourism development, various approaches have been adopted by businesses and organizations within the tourism system, from accreditation schemes and 'eco-labelling' to more direct intervention in tourism development processes, in promoting sustainable tourism development. These are by no means universal; for example, Forsyth (1995) found little evidence of widespread adherence to sustainable business and development principles within the UK travel and tourism industry, although a number of new initiatives have evolved over the last decade. Nevertheless, they serve to demonstrate a commitment to the objectives of sustainable tourism development within different levels and sectors of the tourism system.

Sustainable tourism development in practice

A complete review of sustainable tourism development in practice is well beyond the scope of this chapter. Innumerable examples exist of codes of conduct and/or ethics, tourism development guidelines, specific sustainable/ecotourism projects, industry sector initiatives, NGO activities and so on, while some projects or initiatives enjoy wider recognition than others. For example, Zimbabwe's CAMPFIRE programme (Communal Areas Management Programme for Indigenous Resources), established in the late 1980s, grew to involve some 30 communities throughout the country by 2000, and is well known as a pioneering and generally successful project that has promoted wildlife conservation and community development through tourism. Despite political upheaval in the country, not only has it continued to generate income for the local community and support wildlife preservation (Balint and Mashinya 2008), but similar projects have been established elsewhere in southern Africa (Ngwira and Mbaiwa 2013). Conversely, there are many smaller-scale local initiatives that have not attracted such wide attention. For example, the Tumani Tenda ecotourism camp in the Gambia, first established in 1999, supports a village community of 300 but is virtually unknown outside that country (Jones 2005; www.tumanitenda.co.uk).

Moreover, the promotion or implementation of sustainable tourism development occurs at different levels within the tourism system (international, national, local), in different sectors (private, public, voluntary) and in a number of different ways. At the same time, various activities and schemes may also be interrelated. For example, the International Federation of Tour Operators (IFTO) established the *Tourism for Tomorrow* awards in 1989 in order to encourage the tourism industry as a whole to adopt environmentally sound practices. Sponsorship was taken over by British Airways in 1992 to extend global awareness of the awards and, since 2004, they have been run under the auspices of the WTTC. The aim of these awards is to recognize:

> best practice in sustainable tourism within the industry globally, based upon the principles of environmentally friendly operations; support for the protection of cultural and natural heritage; and direct benefits to the social and economic well-being of local people in travel destinations around the world.
>
> (WTTC 2015)

Similarly, the Responsible Tourism Awards, founded in 2004:

> rest on a simple principle – that all types of tourism, from niche to mainstream, can and should be organised in a way that preserves, respects and benefits destinations and local people. We want to celebrate the shining stars of responsible tourism – the individuals, organisations and destinations working innovatively with local cultures, communities and biodiversity.
>
> (Responsible Travel 2015)

Thus, it is difficult to establish a full and clear picture of how sustainable tourism development is manifested in practice.

Nevertheless, for the purposes of this chapter, it is useful to explore the application of the principles of sustainable tourism development from the perspective of different policies/initiatives and, where relevant, different sectors of the tourism system. Many specific examples are, of course, provided in the tourism literature, while a recent document published by the United Nations Department of Social and Economic Affairs provides a number of contemporary case studies of best practice in sustainable tourism (Wei 2013; see also UNEP/WTO 2005). The principal areas of activity are as follows:

Sustainable tourism development guidelines

Perhaps the most common way in which the principles of sustainable tourism development are promoted is through the publication of guidelines for development or through actual policy documents. Guidelines or principles for sustainable tourism development have long been published at the national, regional and international levels and by both the public and voluntary sectors. One of the earliest sets of guidelines, for example, was published by the World Wide Fund for Nature (WWF) in collaboration with the pressure group Tourism Concern (Eber 1992), while the UNWTO, WTTC and other global organizations have produced guidelines, statements of good practice and other documents extolling the virtues of sustainable tourism and, more specifically, ecotourism. Inevitably, of course, such documents are simply guidelines; there is no requirement, legal or otherwise, for businesses or destinations to adopt them. Conversely, a number of countries, such as Costa Rica, have produced tourism development policies that explicitly embrace the principles of sustainable development and, in particular, ecotourism. One of the first, and better known, was Australia's National Ecotourism strategy, published in 1994, subsequently followed in 2003 by a Tourism White Paper that embraced sustainable tourism more broadly. Importantly, the National Ecotourism strategy was supported by grant funding totalling AUS$10 million, ensuring that its policies would be actioned. Guidelines on sustainable tourism have also been produced for particular types of destination, such as national parks (Stevens 2002).

Perhaps one of the best examples of a guidelines document is *Making Tourism More Sustainable: A Guide for Policy Makers* (UNEP/WTO 2005). This is significant in two respects. First, in recognition of the fact that it is governments that develop and implement policy, it is aimed specifically at regional or national tourism policymakers. Second, it signifies a departure from the rigid, 'blueprint' approach to sustainable tourism development by stating that 'it must be clear that the term "sustainable tourism" – meaning "tourism that is based on the principles of sustainable development" – refers to a fundamental objective: to make all tourism more sustainable' (UNEP/WTO 2005: 11). It goes on to identify 12 aims that represent an agenda for sustainable tourism (Table 2.6).

Table 2.6 *Agenda for sustainable tourism*

1	Economic viability	7	Community well-being
2	Local prosperity	8	Cultural richness
3	Employment quality	9	Physical integrity
4	Social equity	10	Biological diversity
5	Visitor fulfilment	11	Resource efficiency
6	Local control	12	Environmental purity

Source: Adapted from UNEP/WTO (2005)

However, the extent to which such an agenda, closely reflecting the objectives of sustainable development, is achievable remains questionable.

Accreditation schemes

A popular means of promoting sustainable practice within the tourism industry is through voluntary accreditation schemes and, more specifically, eco-labelling or other certification schemes that identify products or services that meet specific environmental or quality benchmarks. Many such schemes exist, particularly at the national level; one international accreditation scheme for tourism is Green Globe. Originally launched by the WTTC in 1994 as a membership and commitment-based programme (www.greenglobe. com), the scheme has evolved into a benchmarking and certification system for sustainable tourism. Environmental improvements are monitored through annual benchmarking and the organization also provides technical support in a number of countries. At the national level, funding was provided in Australia under the National Ecotourism Strategy mentioned above to support the development of the National Ecotourism Accreditation Programme (NEAP), now called the EcoCertification Programme. This certifies the environmental credentials of products (tours, hotels, etc.) rather than the businesses providing them. Although there is evidence to suggest that such accreditation schemes encourage higher environmental standards in the tourism industry, the overall benefit certification of eco-labelling schemes remains the subject of debate.

Codes of conduct

Codes of conduct have long been used as means of trying to influence the behaviour or practices of individuals and organizations without recourse to laws and regulations (see Plate 2.3). In tourism, such codes are numerous and are aimed at tourism businesses, visitors and policymakers, and are formulated by international agencies (e.g. UNWTO Global Code of Ethics for Tourism), governments, NGO/voluntary organizations or even bodies representing particular sectors of the industry as a means of self-regulation (Mason and Mowforth 1995). Codes of practice are utilized where it is inappropriate or difficult to impose regulations; however, they rely on voluntary action, and there is, of course, no

Plate 2.3 *Russia, St Petersburg: codes of conduct for tourists visiting the Peter and Paul Fortress*

Source: Photo by D. Telfer

mechanism for monitoring or ensuring compliance. Nevertheless, they are widely used as a means of trying to encourage tourists to behave more responsibly, although some see this as evidence of the creeping 'moralization' of tourism (Butcher 2002).

Industry initiatives

Sustainable tourism development initiatives are, arguably, most effective when implemented by the tourism industry itself. In other words, corporate social responsibility (CSR) is considered to be a key element in the achievement of sustainable development in general; for the tourism industry, this means adopting environmental and responsible practices in the day-to-day running of the business and, in the case of tour operations and other international activities, addressing sustainability issues at the destination. An enormous variety of initiatives are in evidence in the tourism industry, from individual organizations or tour operators offering holidays that are designed to involve tourists in local community or environmental projects to local, community-run projects (Mann 2000). Nevertheless, it should be noted that de Grosbois (2012) examined the online CSR reports of 150 top hotel companies in the world and found that while there are a large number of hotels reporting commitment to CSR goals, a smaller number reported specific details on their initiatives and even fewer reported on actual performance achieved. There are also a number of schemes that involve groups of businesses in working towards sustainable tourism development, an early example being the ECOMOST project, established by the International Federation of Tour Operators, which sought to increase the sustainability of tourism in the Balearics (IFTO 1994). Contemporary initiatives include:

The International Tourism Partnership (ITP)

This has developed from the successful International Hotels Environment initiative originally launched in 1992 and, after 20 years as a programme of the International Business Leaders Forum, is now hosted by Business in the Community. ITP comprises 18 hospitality corporate members representing 23,000 properties and 1.5 million employees in over 100 countries worldwide, and seeks to share best practice globally within the hospitality sector (see www.tourismpartnership.org).

Tour Operators' Initiative (TOI)

Founded in 2000, the TOI was a voluntary, non-profit initiative open to all tour operators that, in December 2014, merged with the Global Sustainable Tourism Council (GSTC), an international body that seeks to establish and maintain standards for sustainable tourism. With a diverse membership including tour operators, tourism boards, UN agencies and communities, it sets 'minimum requirements that any tourism business or destination should aspire to reach in order to protect and sustain the world's natural and cultural resources, while ensuring tourism meets its potential as a tool for conservation and poverty alleviation' (GSTC 2015).

Other initiatives include the Global Partnership for Sustainable Tourism, a UN-backed initiative launched in 2011 that seeks to promote sustainable tourism practices around the world, and the Travel Foundation (www.thetravelfoundation.org), a UK-based charity that works with outbound tour operators to manage tourism more sustainably in the destination. It does so by developing projects that boost the benefits of tourism to destination communities, promote the conservation of the local environment and culture, and enhance tourists' experiences. There also exist a number of organizations that market 'responsible' holidays; for example, Responsible Travel is an online travel agent that provides holidays 'for people who've had enough of mass tourism' (www.responsibletravel.com).

Voluntary sector/NGOs

The concept of sustainable tourism development, and indeed efforts to promote it, have long been underpinned by a number of voluntary organizations and pressure groups. Since 1989, the London-based Tourism Concern has campaigned to raise awareness of tourism's negative consequences and worked with the industry and destinations to enhance tourism's developmental contribution. Similar groups operate in other countries, such as Studienkreis für Tourismus und Entwicklung in Germany, while research has revealed the extent to which grass-roots movements in some countries oppose inappropriate tourism developments (Kousis 2000). Volunteer tourism and NGOs are explored in more detail in Chapter 5.

Pro-poor tourism

One of the more recent approaches to tourism development is so-called pro-poor tourism (Scheyvens 2011, 2015). Given its specific focus on poverty reduction, pro-poor tourism does not necessarily reflect the broader developmental aims of sustainable development, although the UNWTO links the two concepts under their Sustainable Tourism-Eliminating Poverty (ST-EP) initiative. Holden (2013) notes that by putting priority on the poor, pro-poor tourism may differ from other alternative forms of tourism where the strategy may be on nature conservation with economic benefits going to the community rather than the poor specifically. Pro-poor tourism seeks to achieve greater equity by providing the poorest members of destination societies with the opportunity to benefit from access to tourism markets. Pro-poor tourism is not a specific type of tourism (all types of tourism can be made to be more pro-poor); rather it is an approach to tourism that seeks to incorporate those less fortunate in the benefits of tourism. In other words, the poor are frequently excluded from the local tourism sector and, thus, are unable to sell locally produced products or provide other services. The objective of pro-poor tourism, therefore, is to open up access for the poor to the tourism sector, thereby providing them with a vital source of income. Although there are many examples of successful pro-poor tourism in practice, the need to specifically target the poor in tourism development points to failure of previous development policies to spread the benefits of tourism equitably. Moreover, as an interventionist scheme that is frequently dependent on project management and funding, its viability as a means of developing capacity among the poor is frequently questioned (Harrison 2008; see also Chapter 5).

There is no doubt, then, that attempts to apply the principles of sustainable development are occurring in different ways and in different sectors across the tourism system. There is also no doubt that at the level of individual initiatives or projects, significant progress has been made in achieving greater sustainability in tourism development. However, tourism is a global activity; only when the majority of travel and tourism businesses and organizations (and, indeed, tourists themselves) around the world adopt the principles of sustainability will sustainable tourism development become a realistic objective. Therefore, the final question to be addressed is: What are the implications of transformations in the global political economy, and globalization in particular, for sustainable tourism development?

Sustainable tourism and globalization

In recent years, increasing attention has been paid by academics to the relationship between tourism and globalization, both generally (Knowles *et al.* 2001; Wahab and Cooper 2001) and in the specific context of tourism and development (Bianchi 2015). With reductions in trade barriers, the tourism market is becoming more globalized. There is pressure for countries to open their borders if they want to take part in the global economy. Multinationals have the ability to operate across borders with fewer restrictions. As such, the multinational corporations hold a great deal of power in where they choose to locate, how they operate and where they get the resources they need to operate. This raises some very profound issues in terms of the three corners of sustainability (economic, environmental and social). Reid (2003) cites the work of Van den Bor, Bryden and Fuller (1997: 3), who state that in the new global economy, capital can locate 'wherever the costs of production are the lowest and where social and environmental restrictions are fewest'. In some regard, there has been a shift in power from the state to the corporation. Governments in developing countries are forced to compete against each other offering various incentives in trying to attract multinationals, as these companies bring with them name recognition and access to tourist markets that developing countries need. Bianchi (2015) addresses the central concern of the political economy of tourism in not merely whether incomes are rising from tourism or whether the large multinationals provide a decent wage, but rather the extent to which different modalities of global tourism are leading to an increase or reduction in inequality of access to power and resources. He also raises the question as to whether new areas of discussion, such as fair trade and related development programmes, as explored in Chapter 5, represent isolated examples of endogenous development or they begin to challenge the very asymmetrical structures of tourism production and exchange. The following chapter will explore the complex relationship between tourism and globalization in greater detail.

Discussion questions

1 Why is there so much controversy surrounding sustainable tourism development?
2 Does sustainable development represent a form of Western imperialism?
3 Are new evolving alternative forms of tourism development more sustainable?
4 How can a specific tourism development or a tourism industry/organization be made to be more sustainable?
5 What roles do governments, NGOs, industries and individuals have in the implementation of sustainable tourism development?

Further reading

Blewitt, J. (2008) *Understanding Sustainable Development*, London: Earthscan.

A comprehensive, accessible text that explores the concept of sustainable development from a variety of perspectives.

UNEP/WTO (2005) *Making Tourism More Sustainable: A Guide for Policy Makers*, Paris/Madrid: United Nations Environment Programme/World Tourism Organization.

This document (which can be downloaded from the UNEP website: www.unep.fr/shared/publications/pdf/DTIx0592xPA-TourismPolicyEN.pdf) provides a comprehensive and contemporary review of appropriate policies and processes for the achievement of sustainability through tourism development. A number of excellent, in-depth case studies of sustainable tourism 'in action' supplement the main text.

Websites

www.gstcouncil.org
The Global Sustainable Tourism Council is a non-profit organization with a global membership of travel and tourism businesses, UN agencies, hospitality companies, national tourism organizations and others that seeks to promote the widespread adoption of global sustainable tourism standards.

www.internationaltourismpartnership.org
The Tourism Partnership, hosted by Business in the Community, is a partnership of leading international hospitality organizations. It provides and shares with the industry the knowledge and ability to achieve more responsible practices in travel and tourism.

www.responsibletravel.com/awards
This site provides details of the annual Responsible Tourism Awards.

www.thetravelfoundation.org
The Travel Foundation is a UK-based charity that works with tour operators to establish and support projects in the destination that, through links with the tourism sector, contribute to sustainable development.

www.tourismconcern.org.uk
For almost 25 years, Tourism Concern has campaigned for more ethical and just forms of tourism.

www.wttc.org/tourism-for-tomorrow-awards
This site provides details of the annual Tourism for Tomorrow Awards.

3 Globalization and tourism

Learning objectives

When you finish reading this chapter, you should be able to:

- **understand the influences of globalization on tourism development**
- **identify the power of multinational tourism corporations in tourism development**
- **be familiar with the technological, economic, political and cultural aspects of tourism and globalization**
- **recognize the political economy of tourism and the links to globalization**

Tourism is very much an agent of globalization, with multinational tourism corporations providing services to people as they move across the globe bringing with them money, values and patterns of consumption. Globalization represents greater integration at a global level in a wide variety of areas such as trade, finance, communication, information and culture. Facilitated by improvements in transportation and information technology, it has led to time–space compression where people, goods and information move greater distances, crossing political borders in shorter periods of time. Globalization is manifested in 'trans-state' processes that operate not merely across borders, but as if the borders are not there (Taylor *et al.* 2002). Corporations conduct business at international scales presenting challenges for nation states (Reid 2003), and these global interactions have impacts in terms of economies, cultures, politics and environments. Mansbach and Rafferty (2008: 744–6) identify the following features of globalization:

- spread of global communication technologies that shrink the role of geographic distance;
- spread of knowledge and skills and an explosion in political participation;
- emergence of a global market that transcends state boundaries and limits states' control of their own economies;
- worldwide diffusion of a secular and consumerist culture;
- emergence of English as a global language;
- widening demand for democratic institutions and norms; and
- networking of groups to form a nascent civil society.

Globalization is a much-contested term with both supporters and critics, resulting in what de la Dehesa (2006) refers to as winners and losers. Some argue that trade liberalization opens borders and promotes economic growth, while others argue the gap between rich and poor is increasing and globalization results in a loss of national identity, culture and control. The tourism development dilemma means that countries interested in pursuing tourism as an agent of development must enter a very competitive global market where the processes of globalization are unevenly distributed, complex and volatile (Momsen 2004).

Globalization and tourism interact in a variety of dimensions, as revealed in the following example of a tourist on a mass tourism holiday package in a developing country. A key element to keep in mind is the power and control issues that accompany the globalization process associated with tourism. A tourist departing on a low-cost airliner or charter flight for a two-week holiday at an all-inclusive beach resort in a developing country will, in all likelihood, have made the booking through a travel agent selling holiday packages for a multinational tour operator based in a developed country or self-booked through the Internet. Both the travel agent and the individual tourist will have used the Internet to connect to evolving and dynamic global distribution channels in which consolidation in online travel companies has been occurring in recent years. The tour company selected by the tourist will typically be vertically integrated, owning or having strategic alliances with airlines, travel agents and hotels, and little of what the individual has paid in his or her home country for the holiday will go to the destination. The accommodation will probably be reserved in a multinational hotel company with a recognizable brand name that, in the tourist's mind, offers a standardized product with the services and food he or she is used to while at home. At the hotel, the tourist will be able to browse the Internet and watch global news coverage on channels such as BBC World Service or CNN International. Shopping and eating away from the hotel may very well be at internationally recognized brand name stores and restaurants. While in the destination, the tourist will pay for most things with one of the credit cards issued by the few multinationals that control global personal finance (Sklair 1995). With global banking, the hotel will be able to transfer profits out of the developing country, and with international courier services, the executive chef at the hotel, who is likely a citizen of a developed country, can import products from any part of the globe.

In a very unequal power relationship, developing countries are often at the mercy of multinational corporations to market their destination, construct and/or manage hotels and fly in the tourists, hence the significance of the political economy of tourism. While the globalization of the tourism industry may have many advantages that facilitate the arrival of money-carrying guests, there are also many concerns as the local is quickly brought into contact with the global. Conflicts result from clashes in cultures, religion and family values as shifts occur in the structure of the labour force (Wall and Mathieson 2006). It has also been argued that gender inequalities are reinforced where women are often portrayed as either domestic help or as the desirable exotic 'other' as part of the sex tourism industry (Momsen 2004). These impacts will be discussed in greater detail in Chapter 7. The power of tour companies lies in their ability to simply change destinations. If a given hotel in a destination does not keep up with the demands of the tourists and or the tour operator, the tour operator may in fact choose quickly to take their business elsewhere, leaving the hotel scrambling to fill vacant rooms. Tourism is a very competitive industry and tour operators need to be able to make a profit. While tourism is clearly not responsible for all of the negative effects of globalization, it is certainly part of the process. Part of the development dilemma within the framework of sustainability is how the industry can operate so that the destination's developmental needs are taken into account and how the power that multinationals hold can be used to that end.

The purpose of this chapter is to outline the relationship between tourism and globalization. The chapter will first explore the nature of globalization and identify some of the processes facilitating greater global interconnectedness. As Hall (2005) suggests, globalization is a far-reaching idea that embraces nearly all aspects of contemporary culture, identity, governance and economy. Globalization is highly complex and controversial, with some arguing that benefits mainly accrue to multinationals in developed countries. However, in many cases, there is limited choice for developing countries other than to participate in the global economy, and many governments welcome the multinationals. The chapter will begin by examining the processes of globalization before addressing the relationship between tourism and globalization in terms of technology, economics, politics and society.

Processes of globalization

Globalization is about increasing mobility across frontiers in terms of the mobility of goods and commodities, information and communication, products and services, and the mobility of people (Robins 1997). The process of globalization has been facilitated by a number of trends, including increased market liberalization and trade, foreign investment, privatization, financial deregulation, rapid technological change, automation and changes in transportation and communication, standardization and immigration (Momsen 2004; Weinstein 2005; Bianchi 2015). Scholte (2005) highlights rationalism, capitalism, technological innovation and regulation that facilitate globalization as the four primary forces that have generated the emergence and expansion of supra-territorial spaces. Harvey (1989) introduced the concept of space–time compression whereby advances in technology, transportation and communication have sped up the pace of life, overcoming spatial barriers where the world seems to collapse inwards upon us. In terms of transportation, the Airbus A380, the double-decker airliner with a capacity from 544 seats up to as many as 853 seats, depending on the configuration, made its first commercial flight in 2007 (Airbus 2015). The rise of the low-cost or no-frills airlines, such as EasyJet and Ryanair in the UK, and GoAir in India, have been very successful opening up secondary airports operating more on short-haul flights. EasyJet, for example, has been opening up routes to Eastern Europe as well as making flights to Morocco and Tunisia in northern Africa. In Canada, WestJet flies to Nassau, The Bahamas. Key to the success of low-cost airlines have been low fares, the use of Internet booking (90 per cent of their business), short turnaround times, high plane usage and paperless ticket distribution (Leinbach and Bowen 2004). There has been a rapid growth in Internet sites selling discount airline tickets, hotel rooms and rental cars, which is reinventing and expanding the concept of distributed data that allowed computer reservations systems to revolutionize air travel (Leinbach and Bowen 2004). The 24-hour global news media process and package news events from around the world and deliver them over IP (Internet Protocol) networks, cable, fibre and through satellite and cell phone transmissions to viewers (Rain and Brooker-Gross 2004). Noting the complexity of technology, Wilbanks (2004: 7) suggests that technology 'underlines many structures for exercising power and control from the expanding reach of globalisation to new opportunities for local empowerment'. In this age of the media, we are also constantly surrounded by commercial and informational messages in our homes, workplaces and our landscapes, from product placement in movies to outdoor advertising to network news packaged for long-distance jet passengers (Rain and Brooker-Gross 2004). The information we receive through advances in technology not only makes us aware of new destinations and travel products, but also of natural disasters, war, health concerns and terrorist attacks, all of which turn tourists away from destinations. For example, in October 2005, when Hurricane Wilma hit the Mexican resort area of Cancún, images were rapidly sent out over the Internet and through news services alerting potential tourists of the situation in the destination. These brief examples in tourism reveal the nature of the increasing connectivity in our world.

Within the processes of globalization, there has been a transformation in the means of production and consumption. It has been suggested that there has been a shift in industrial production systems from Fordism to post-Fordism, also known as flexible specialization, starting in the 1970s (Harvey 1989). Fordism is focused on economies of scale characterized by mass production of standardized goods. In the context of tourism, Fordism is evident in mass standardized package tours where the tourism industry determines the quality and type of product typically sold to mass tourists (Ioannides and Debbage 1998; Mowforth and Munt 2003). While evidence of Fordism is still strongly recognizable (and in many destinations the dominant form of development), there is evidence that many parts of the tourism industry are adopting more flexible production systems characteristic of the trend in post-Fordism. The focus in post-Fordism is on economies of scope or network-based economies and high levels of product differentiation through focusing on small batch production of specialized commodities targeting niche markets (Ioannides and Debbage 1998). In tourism, there has been a trend towards specialized operators with tailor-made holidays undertaking niche marketing to more independent or experienced tourists. There is reliance to a greater degree on information technology, integration, networks and strategic alliances. As production has changed, so too has consumption. Urry (1990; see also 2012) outlines in some detail the characteristics of post-Fordist consumption where consumption is dominant over production:

- there are new forms of credit for consumers, allowing higher levels of purchasing and debt;
- most aspects of life are being commodified;
- there is greater differentiation of purchasing patterns by different market segments;
- there is greater volatility of consumer preferences; and
- there is a reaction of consumers against being part of a 'mass', and so there is a need for producers to be consumer-driven (Urry 1990).

These shifts in consumption have been characterized by Poon (1989) as changes from 'old tourism', which is typified by the standardized mass package holiday, to 'new tourism', which is based on flexible and customized holidays targeted at specific segments who are often more experienced travellers. These changes in production and consumption have taken tourists to the far corners of the globe as some tourists are seeking to stay away from the mass tourism destinations. Several authors (Harvey 1989; Urry 1990; Uriely 1997; Mowforth and Munt 2003, 2009) have linked the changes outlined above to the shift from modernism to postmodernism, which involves dissolving the 'boundaries, not only between high and low cultures, but also between different cultural forms such as tourism, art, education, photography, television, music, sport, shopping and architecture' (Urry 1990: 83). It is about breaking down barriers and accepting other ways or different perspectives and so the mass tourism package is shifting to customization. Uriely (1997) suggests that trends such as an increase in the number of small and specialized travel agencies, the growing attraction of nostalgia and heritage tourism along with the growth in nature-oriented tourism and simulated tourism-related environments are labelled aspects of postmodern tourism. As the means of production and consumption (see Chapter 6) are changing to be more flexible and customized, tour companies are outsourcing components of the production process to different companies sometimes located in different countries. The shift in the production process is not necessarily uniform in destinations, as Torres (2002) discovered in Cancún where she found evidence of both Fordist and post-Fordist elements. These elements were manifest in different 'shades' of mass tourism, 'neo-Fordism' and 'mass customization'. While these elements may represent examples of the complex functional aspects of globalization, there is also the realm of policy, politics and trade, which are also key elements in facilitating globalization.

Globalization is not new; throughout history, there have been waves of global economic integration that have risen and fallen (Dollar 2005). With the rise of economic neo-liberalism, a wave of economic integration developed with a dramatic shift in the nature of international trade from exports of primary products to exports of manufactured products and, along with this, service exports including tourism and software have increased enormously (Dollar 2005). One of the key elements of globalization is market liberalization that has opened national economies to the global market (McMichael 2004). The shift towards market liberalization is evident in the comment by Dollar (2005) that increased levels of integration have been facilitated partly by deliberate policy changes, where the majority of the developing world (measured by population) has moved from an inward-focused strategy to a more outward-oriented one, reflected in huge increases in trade integration. In fact, in some cases, there are corporations in developing countries that are expanding their operations beyond their borders; these have been referred to as globalization's offspring (*The Economist* 2007). For example, Chinese investment in Australian tourism has been on the increase, with a Chinese developer set to invest up to AUS$4 billion in Australian tourism, infrastructure and agriculture as Chinese outbound tourism numbers are expected to surge (Xinhau 2014). Some low-cost airlines are also expanding into neighbouring states as well as adding in more longer-haul flights. Pegasus Airlines, Turkey's low-cost carrier, is, for example, adding flights to London Gatwick and Istanbul with connections to other destinations in Turkey (eTN Global Travel Industry News 2015). McMichael (2004) suggests that liberalization involves downgrading social goals of national development along with upgrading participation in the world markets through policies such as tariff reduction, export promotion, financial deregulation and relaxation of foreign investment.

It is important to note that a reduction in global trade followed the arrival of the 2008 Global Financial Crisis. World trade dropped by almost 11 per cent in 2009 and exports of advanced economies fell almost 12 per cent, while exports of emerging and developing economies dropped approximately 8 per cent (Al Sadik and Servén 2012). While global trade has since started to recover, albeit at a subdued rate (Constantinescu et al. 2015), the realization of the instability of global markets after the financial crisis has led to increased government intervention and raised significant doubts about overreliance on the free market. However, as pointed out in Chapter 1, tourism as a component of global trade continues to grow and often works best in terms of economics in a free-market economy.

Influence of the Bretton Woods meetings

In addition to the changes in the economies of countries seeking to participate willingly or otherwise in the global market, there has been a rise in global financial institutions governing world trade and providing loans to countries, which are also driving market liberalization and globalization (McMichael 2004). A number of key institutions and regulatory frameworks were established as a result of the Bretton Woods meetings after the Second World War. Sometimes referred to as the Bretton Woods institutions (named after the ski resort in New Hampshire, USA, where representatives of 45 countries met 1–22 July 1944), these institutions include the International Monetary Fund (IMF) and the International Bank for Reconstruction and Development (IBRD), which would be one of the institutions of the World Bank Group. Subsequent meetings led to the signing of the General Agreement on Tariffs and Trade (GATT) in Geneva in 1947. GATT was a multilateral treaty negotiated in 'rounds' to promote trade and limit protectionism (Roberts 2002). Out of the GATT Uruguay Round, the World Trade Organization was formed in January 1995. The World Trade Organization is often seen as the successor to GATT and, in 2014, it had 160 member countries, up from 149 in 2005. The World Trade Organization is different from a treaty as it has independent jurisdiction and the ability to enforce its rules on member states. It can set rules on the movement of goods, money and productive facilities across borders, thereby restricting countries from passing legislation or policies that discriminate against such movement (McMichael 2004). It also releases Ministerial Declarations from its meetings, and the declaration from Doha in 2001 stated that:

> International trade can play a major role in the promotion of economic development and the alleviation of poverty. We recognize the need for all our peoples to benefit from the increased opportunities and welfare gains that the multilateral trading system generates.
>
> (World Trade Organization 2001)

The World Trade Organization also administers the General Agreement on Trade in Services (GATS), which includes tourism under the category of 'Tourism and Travel Related Services'. The aim of GATS is to liberalize trade in services so that member countries have to allow foreign-owned companies free access to their markets and display no favouritism towards domestic companies (Scheyvens 2002). This, however, gives foreign-owned companies with greater economic power an advantage over small domestic companies.

Globalization, structural adjustment and new international development banks

The shift towards greater market liberalization has also been part of the conditions applied to many of the loans to developing countries in the structural adjustment programmes (SAPs) issued by international lending agencies, such as the World Bank and the International Monetary Fund, in the 1980s and 1990s. Conditions of these loans involved receiving governments adjusting their economic structures and political polices to be more open to global trade, as well as reducing the size and level of involvement of the government in the economy. A more recent trend in these loans has been the inclusion of the condition of state effectiveness and good governance (McMichael 2004) along with a focus on poverty reduction. In 1999, the IMF and the World Bank introduced Poverty Reduction Strategy Papers (PRSPs) as a follow-up to the SAPs (Mowforth and Munt 2003). The PRSPs are a participatory process involving local governments and donor agencies to map out the 'macroeconomic, structural and social policies and programs that a country will pursue over several years to promote broad-based growth and reduce poverty, as well as external financing needs and the associated sources of financing' (IMF 2006). The aim of the program was to bring development more in line with the United Nations Millennium Development Goals (see Chapter 1). Tourism has been adopted in a number of developing country PRSPs, including those not readily associated with tourism, such as Bangladesh, the Central African Republic, Sierra Leone and Guinea (PPT 2004) (see Chapter 4). The PRSP written by the IMF (2008) for Haiti included tourism as one of the four growth vectors along with agriculture and rural development, infrastructure modernization and science, technology and modernization. Mowforth and Munt (2003, 2009) note, however, that critics claim that PRSPs are really another form of SAPs with language about poverty reduction. These programmes have been criticized for the reduction in the power of the state, including the reduction of state social welfare programmes. A recent trend in the world of international finance has seen China come to the forefront in terms of loans to developing countries while it has also been the lead in developing two international development banks, including one representing BRICS (Brazil, Russia, India, China and South Africa) nations and the other representing 21 Asian nations titled the Asian Infrastructure Investment bank (S.R. 2014). The movement of China towards being a major international lender for development projects has provided another source of finance for nations not wanting to take on some of the required policy changes of the SAPs and PRSPs often required by the IMF or World Bank.

Economic trading blocs

Economic integration at a more regional level can also be seen in the form of free trade zones and economic unions or partnerships. Examples include the North American Free Trade Agreement (NAFTA) and the European Union, both of which include the integration of developing or transitional economies with developed economies. Other examples include the Association of Southeast Asian Nations (ASEAN), Mercosur (South American trading bloc), Southern African Development Community (SADC), CARICOM (Caribbean nations) and the Singapore, Indonesia and Malaysia Growth Triangle. Within this Growth Triangle, the resort of Bintan, Indonesia (a 55-minute ferry ride from Singapore) offers seven hotels, four golf courses and five spas (Bintan Resorts 2006). This greater shift towards integrated economies suggests that the economic and social well-being of countries, regions and cites everywhere depends increasingly on complex interactions framed at the global scale where systems of production, trade and consumption have become global in scale and scope (Knox *et al.* 2003). Much of these enhanced interactions are facilitated by advancements in information and communications technology (ICT), which will be addressed in the next section in terms of the technological aspects of tourism and globalization.

Technological aspects of tourism and globalization

Technological innovation has permeated all aspects of the tourism industry, from advances in materials for lightweight design of commercial airlines (for example, Boeing's Dreamliner) to green technologies such as solar power in hotels and to advances in information and communication technologies. Tourists are able to use mobile devices to plan their vacations, find and reserve a restaurant, pay for their meal and accommodation, record it all (photos and video) and post their experiences to the Internet. In short, all sectors of the tourism industry have been impacted by technology. However, in the context of this book, it is important to consider how these technologies have impacted or been adopted in developing countries. The focus of this section is primarily on the spread of information and communication technologies (ICT) and the influence of e-commerce on tourism, the first item listed in the features of globalization by Mansbach and Rafferty (2008) outlined earlier.

The Internet has revolutionized information provision and purchasing patterns in the tourism industry globally, creating not only opportunities, but also threats to tourism intermediaries (Buhalis and Ujma 2006; Buhalis and Jun 2011). Access to the Internet continues to grow at a rapid rate. In 1996, there were approximately 16 million Internet users (0.4 per cent of the global population); by 2014, there were over three billion users (approximately 42 per cent of the global population) (Internet World Stats 2015). Table 3.1 illustrates Internet users by region and selected countries within the regions. What these figures illustrate is that while Internet access continues to grow, there are countries that continue to have very limited access to the Internet, contributing to what is referred to as the digital divide. Minghetti and Buhalis (2011: 6) describe the digital divide in terms of (i) the global digital divide: 'divergence of IT access and usage between industrialized and developing societies, caused by: differentials in GDP and income; human capital and digital skills; telecommunication infrastructure and connectivity; policy and regulatory mechanisms;

socio-demographic characteristics' and (ii) the domestic or social digital divide: 'gap between information "included" and "excluded" in each country, dependent on socio-psychographic and economic characteristics'. If developing countries are to take advantage of new revenue streams generated by online tourism marketing and financial transactions, efforts need to be made to extend Internet service and access, breaking down the digital divide so it is not only the local elite that have access. If there is low ICT access in a developing country, then tourism businesses in those destinations will lose out on potential tourists browsing the Internet and also become dependent on external traditional intermediaries (Minghetti and Buhalis 2011). An overview of some of the developments and trends in e-tourism follows.

As a result of speed and connectivity, boundaries between various distribution channel players are becoming blurred, with some travel intermediaries disappearing while other new ones are emerging. Distribution channels are being revolutionized, from linear proprietary electronic distribution systems with a focus on CRSs (computer reservation systems) and GDSs (global distribution systems) to those that are more dynamic and becoming dominated by websites of suppliers, travel agents, Web intermediaries, CRSs, GDSs and DMSs (destination management systems) (O'Connor *et al.* 2001). The Internet has allowed consumers to package their own trips by collecting inventory from various suppliers supporting B2C (business-to-consumer) e-commerce (Buhalis and Ujma 2006). With the shift towards customers assembling their own holidays online, also referred to as dynamic packaging, there have been damaging consequences for European charter airlines (O'Connell and Aurélise 2015). Online travel agencies such as Expedia emerged offering multiple products from various companies/organizations, and nowadays even online travel agencies are being searched by other online intermediary companies offering meta-searches looking for the best deal. Suppliers around the world are setting up their own websites avoiding traditional GDSs. At the beginning of this chapter, an example was provided of a single tourist on a mass package tour to a developing country to illustrate the linkage between tourism and globalization and issues of power and control. There is no doubt that mass package tours still dominate the market and that the companies that operate them continue to hold a great deal of power. However, if the example is shifted to focus on a traveller on an independent holiday, the diffusion of information communications technologies (ICT) has provided an opportunity for small tourism providers, including indigenous tourism operators in developing countries, to launch dynamic Internet sites attracting the more independent traveller. This development has started to challenge the hitherto unique position of tour operators regarding packaged products (Buhalis and Ujma 2006) and opened a window for more control from small operators and destinations in developing countries. A further evolution of the business model is the C2C (consumer-to-consumer) model and the C2B (consumer-to-business) model. In the C2C model, consumers can interact with each other through platforms such as eBay and sell items including booked hotel rooms. In the C2B model, consumers provide a product or service to a company. The producer of a blog or vlog (video blog) can generate

Table 3.1 Selected Internet access figures, 2014*

Region	% of Internet users in the world	Example countries	Internet penetration as % of population	% of users in the region
Asia	45.7	South Korea	92.4	3.3
		China	47.4	46.3
		Cambodia	6.0	0.1
Europe	19.2	Finland	97.1	0.9
		Russia	61.4	15.0
		Romania	51.4	1.9
All Americas	20.7	Bermuda	95.3	0.0
		USA	86.9	44.0
		Brazil	54.2	17.4
		Cuba	28.0	0.5
Africa	9.8	Madagascar	74.7	5.8
		South Africa	51.5	8.4
		Niger	1.7	0.1
Middle East	3.7	UAE	95.7	7.9
		Iran	55.7	40.2
		Iraq	9.2	2.7
Oceania/Australasia	0.9	New Zealand	94.6	15.5
		Tonga	37.7	0.1
		Vanuatu	11.3	0.1

* Data to 30 June 2014

Source: www.internetworldstats.com

income from a business through making recommendations or referrals. A ruling by the Advertising Standards Authority in the UK requires bloggers to show on their websites if they are being paid (Bugsgang 2015). Some online companies also give credit or payments for online referrals that result in a purchase.

A recent development that has spread into the tourism industry facilitated by the Internet is collaborative consumption or the 'sharing economy'. These concepts share two commonalities, including the reliance on the Internet, especially Web 2.0, and the use of temporary access, non-ownership models of using consumer goods and services (Belk 2014). While Web 1.0 was a one-directional provider of information to consumers who did not interact, Web 2.0 has been referred to as the people-centric Web, the participative Web and the read/write Web, allowing users to utilize the Web in a more collaborative and interactive manner, emphasizing social interactions and collective intelligence with users posting content (Murugesan 2007). Through Web 2.0, social media has been widely

> adopted by travellers to search, organise, share and annotate their travel stories and experiences through blogs and microblogs (e.g. Blogger and Twitter), online communities (e.g. Facebook, RenRen and TripAdvisor), media sharing sites (e.g. Flickr and YouTube), social bookmarking (e.g. Delicious), social knowledge sharing sites (e.g. Wikitravel) and other tools in a collaborative way.
> (Leung *et al.* 2013: 4)

Critical to the tourism industry is the fact that tourists are able to access the Internet through multiple mobile communications devices (for examples, smartphones) with camera, audio and video recording capabilities throughout their trip. Mobile applications (apps) for these devices continue to be developed, and online travel agencies such as Expedia and Priceline are examining ways of using apps not only to enhance customer experience, but to also leverage other mediums to gain market share (Johnson 2014). Web 2.0 has also been used for crowdsourcing (generating content by users), and crowdfunding (raising funds) for tourism development projects. The use of social media has become of strategic importance for tourism competitiveness (Leung *et al.* 2013), with new online firms helping tourism companies use social media to better target their markets, as well as monitor and improve their online reputations.

How, then, can collaborative consumption or the 'sharing economy' via Web 2.0 help in the development process and bridge the digital divide? Airbnb is such an example of an online site where people can rent out a room in a person's home. The site has over one million listings in 190 countries and is valued at over $13 billion (Basen 2015). In April 2015, Airbnb announced it would be one of the first US companies to offer accommodation for licensed US travellers in Cuba since President Obama eased travel restrictions in January 2015. Cuba is known for its network of *casas particulares*, or traditional private home-stays operated by micro-entrepreneurs, and there are now over 1,000 listings in Cuba on Airbnb. For those operators with limited Internet access in Cuba, they are working with hosting partners in arranging booking requests (Airbnb 2015). Another example is Uber, an online taxi service that has operators in 56 countries, including developing countries. In March 2015, Uber and UN Women launched a partnership, with Uber committing to create one million jobs for women on the Uber platform by 2020 (Uber 2015). A criticism of the sharing economy, however, is that operators are able to avoid regulations such as paying for licensing and taxes, and may be taking income from those in the more formal sector.

Firms are utilizing a variety of strategies to make profits in a highly competitive, dynamic and global environment. As Papatheodorou (2006) suggests, there is a dualism evolving in the market with a small number of dominant players. Even on the Internet, mergers have occurred in online travel agents so that the bulk of online travel sales are concentrated within a few companies (Buhalis and Ujma 2006). A question that needs to be examined is to what degree tourism companies in the developing world are able to participate in the Internet, and if they do, are their websites picked up through the main search engines? One of the main criticisms with respect to this dualism is that, for developing countries, the control of multinational tour companies is often located in the developed world and profits are often repatriated, leaving little for the destination.

Economic aspects of tourism and globalization

Hoogvelt (1997) suggests that there are three key features to economic globalization and these are evident in tourism. First, there is a global market discipline whereby, with the awareness of global competition, individuals, groups and national governments have to conform to international standards of price and quality. The second feature is flexible accumulation through global webs. Companies are organizing through networks or global webs where production activities and services are brought in (often managed electronically) for the short term and, rather than owning suppliers, they are treated as independent agents. However, control is still maintained by the parent company. As an example, services and/or production is contracted out across the globe where production costs (including labour) are cheapest. The system of contracting out externalizes risk as far as possible for the parent company (Bianchi 2015). The third key feature is financial deepening, which relates to the fact that money quickly moves across borders and money is increasingly being made out of the circulation of money. As indicated by Scholte (2005), the movement of money has been facilitated by regulation. Scholte (2005: 105) states that regulation has promoted globalization in the following four ways:

- technical and procedural standardization;
- liberalization of cross-border movements of money, investments, goods and services (but not labour);
- guarantees of property rights for global capital; and
- legalization of global organizations and activities.

While some of these regulations have been put in place freely, in some situations countries are under pressure from a variety of sources to implement them (Scholte 2005). Prasad et al. (2003) examined the relationship between financial globalization and growth in developing countries. In principle, financial globalization could influence the determinants of economic growth through the augmentation of domestic savings, reduction in the cost of capital, transfer of technology from developed countries and the development of domestic financial sectors. Indirectly, the factors could be felt through increased production specialization owing to better risk management, along with improvements in macroeconomic policies and institutions induced by the competitive pressures of globalization (Prasad et al. 2003). While the average per capita income for more financially open developing economies grows at a more favourable rate than less financially open economies, whether it is a causal relationship remains a question (Prasad et al. 2003). Prasad et al. (2003) stress that it is important to understand other factors, such as governance and rule of law, but to also acknowledge that some countries with capital account liberalization have experienced output collapses connected to costly banking or currency crises (for example, the 1997 Asian Financial Crisis and the 2008 Global Financial Crisis).

Multinational tourism corporations seek to make profits across political boundaries, and it is often the price of production of the holiday, including wages, which is the driving force. Services such as hotel bookings are contracted out, and money or profit is moved from country to country and invested in money markets. Hjalager (2007) developed a model explaining the stages in the economic globalization of tourism. The first stage (missionaries in the markets) includes national tourism boards reaching out to new markets resulting in marketing efforts and market expansion of larger tourism corporations. The second stage (integrating across borders) outlines the integrations and incorporation of tourism business across borders, including developing brand profiles in foreign markets, mergers, franchising and licensing. The third stage (fragmentation of the value chain) focuses on the fragmentation and flexible relocation in space of the production processes though outsourcing, value chain splitting and taking advantage of international labour markets. The final stage (transcending into new value chains) sees the industry being challenged as new markets and business types appear. The tourism industry incorporates these shifts by incorporating knowledge industries, marketing business and brand extensions and the global media. Commenting on future trends in the tourism industry, Costa and Buhalis (2006) suggest that rapid liberalization and deregulation of markets will bring fierce competition, which is based not only on price, but also on quality and characteristics of the products supplied. A tourist wanting a vacation in a sunny and warm destination in the Caribbean will often base his or her decision on price. The idea of increasing competition and globalization in tourism was also commented on by Mowforth and Munt (2003: 12), whereby 'globalisation is about capitalising on the revolutions in telecommunications, finance and transport, all of which have been instrumental in the "globalisation" of tourism'. Those who are able to control or participate in these revolutions benefit while others are left behind. See Box 3.1 for a discussion of globalization and Cancún.

As firms move to position themselves in a globalized marketplace, Papatheodorou (2006) notes that a market dualism has emerged in tourism. There is a multitude of small producers (competitive fringe) that coexists with a small number of powerful transnational corporations. The size of these powerful corporations is important in terms of the scope and scale of economies and the imposition of barriers to market entry and exit through asset specificity and irreversibility (Papatheodorou 2006). The expansion and concentration evident in the tourism industry is cause for concern as it is consistent with anticompetitive practices (Papatheodorou 2006). Several sectors of the tourism industry will be explored briefly here to illustrate trends in the industry.

Box 3.1

Globalization and Cancún, Mexico

Cancún, built on a 14-mile island on the northern tip of the Yucatán Peninsula, is a large-scale resort complex that attracted 4.8 million visitors in 2014. Tourism is expected to continue to grow with upgrades in infrastructure, refurbishment of existing hotels, opening of new hotels, the development of new beach areas, as well as new flights and connections to the destination (Figueroa 2015). Originally financed by the Mexican government and the Inter-American Development Bank, it was built as a regional development project in an area of everglades, swamp and jungle to create new jobs outside traditional urban centres such as Mexico City. The Mexican government made major regulatory and tax concessions to investors in the early development phase and now Cancún has rapidly become an urbanized resort. Many of the international hotel chains are present, with foreign investment from countries such as the United States, Canada, Italy and Spain. Other chains, such as Pizza Hut, McDonald's, Subway, KFC, TGI Friday's, Outback and Walmart are present. Cooper (2003) argued that there has been intense competition and that notably Spanish hotel chains had been driving down prices of hotel rooms and cutting costs to squeeze out the competition. The all-inclusive policy has drastically reduced the income of the workers who depend on tips. 'Low wages, unstable markets, racism, a high cost of living and poor housing are some of the conditions would-be migrants have found in Cancún' (Hiernaux-Nicolas 1999: 139). The initial infrastructure, built for the surrounding community to house workers, has been well surpassed as the population has grown to approximately 500,000. Migrants who have moved into the area to seek work in the hotels or in construction have set up shanty towns. There are also concerns over environmental damage and groundwater contamination, as well as political corruption (Cooper 2003). Azcárate *et al.* (2014) argue that after Hurricanes Gilbert in 1988 and Wilma in 2005, hotel and real estate developers, aided by the government, privatized and commoditized Cancún's public lands and resources for the exclusive use of the global tourism market. This has transitioned the destination from low-density luxury to a mass, all-inclusive resort after Gilbert and the emergence of the contemporary timeshare high-rise after Wilma. What was set up as a planned resort has evolved into an urbanized centre exposed to the forces of global competition.

Source: Hiernaux-Nicolas (1999); Cooper (2003); Azcárate *et al.* (2014); Figueroa (2015)

Airlines and cruise ships

Airlines have undergone significant deregulation with market economies and subsequent competition coming to the forefront of aviation policymaking (Papatheodorou 2006). An increasing number of governments have privatized national airlines to reduce public expenditure and to encourage operating efficiency (Graham 2006). The trend towards deregulation started in 1978 in the US domestic market and similar trends occurred worldwide through bilateral and multilateral agreements (Graham 2006; Papatheodorou 2006). Open skies agreements between countries have liberalized airline operation and one of the major advancements in deregulation has been the rise in the low-cost carriers, which has given new momentum and triggered price wars (Papatheodorou 2006). Well-known UK-based low-cost carriers include Ryanair and EasyJet, while the number of low-cost carriers in Asia has grown tremendously. Spring Airlines, for example, is based in Shanghai, China, flying 90 routes across China and to international destinations such as Bangkok and Phuket (Spring Airlines 2015). In Latin America, GOL Linhas Intélignetes SA is the largest low-cost carrier, with close to 900 daily flights to 62 destinations (domestic and international) in 13 countries (GOL 2015). As with many of the transport sectors, there has been a trend towards concentration or horizontal integration with key perceived cost savings in economies of scale, marketing advantages and technological development opportunities (Graham 2006). An example of this is larger airlines being able to capitalize on their networks through strategic alliances with other airlines. Table 3.2 contains the members of three of the largest airline alliances, including Star Alliance, Oneworld and SkyTeam. The largest, Star Alliance, has 29 member airlines with 4,561 planes serving 1,321 airports in 193 countries. They have over 653 million annual passengers and employ 410,274 (Star Alliance 2015). The frequent flyer programmes that exist within these alliances, inducing passengers to fly with airlines that have global networks rather than using smaller or cheaper airlines, suggests that these partnerships can also create fortress airport hubs where new carrier entry is impossible (Papatheodorou 2006). In a similar vein, globalization is at work in the cruise ship industry (see Box 3.2) (Plates 3.1 and 3.2).

Table 3.2 *Members of Star Alliance, Oneworld and SkyTeam*

Star Alliance	Oneworld	SkyTeam
Adria Airways	airberlin	Aeroflot
Aegean Airways	American Airlines	Aerolineas
Air Canada	British Airways	AeroMexico
Air China	Cathay Pacific	Air Europa
Air India	Finnair	Air France
Air New Zealand	Iberia	Alitalia
ANA	Japan Airlines	China Airlines
Asiana Airlines	LAN	China Eastern
Austrian	TAM Airlines	China Southern
Avianca	Malaysia Airlines	Czech Airlines
Brussels Airlines	Qantas	Delta Airlines
Copa Airlines	Qatar Airways	Garuda Indonesia
Croatia Airlines	Royal Jordanian	Kenya Airways
EGYPTAIR	S7 Airlines	KLM
Ethiopian Airlines	SriLankan Airlines	Korean Air
EVA Air		Middle East Airlines
Lot Polish Airlines		Saudia
Lufthansa		TAROM
Scandinavian Airlines		Vietnam Airlines
Shenzhen Airlines		Northwest Airlines
Singapore Airlines		
South African Airlines		
Swiss		
TAP Portugal		
THAI		
Turkish Airlines		
United		
US Airways		
Varig		

Source: Oneworld (2015); SkyTeam (2015); Star Alliance (2015)

Box 3.2

Globalization and the cruise ship industry

In his analysis of the cruise ship industry, Wood (2004) argues that the cruise ship industry represents the coming of age of neo-liberal globalization. Cruise ships are getting larger ('Allure of the Seas' from Royal Caribbean can hold a maximum of 6,318 passengers) and overall passenger numbers continue to increase. Projected for 2015 the total worldwide cruise industry is estimated at $39.6 billion with 22.2 million passengers. The three cruise line companies, Carnival Cruise Lines, Royal Caribbean Cruise Lines and Norwegian Cruise Lines, and their respective brands account for 76.7 per cent of revenue and 81.6 per cent of passengers. Many cruise ships are registered in countries such as Panama, The Bahamas or Bermuda, which is referred to as flying flags of convenience. Under international maritime law, enforcement of labour, environmental, and health and safety laws are done under the flag the ship is flying. Countries that make it known that enforcement and fees will be minimal for things such as pollution violations are know as 'flags of convenience states' (Wood 2004). Flying flags of convenience and operating at times in international waters offers the industry a greater deal of flexibility than land-based operations. One aspect of globalization is minimizing costs, and the cruise lines have been criticized for labour policies. They are able to recruit their labour force from all over the world and pay wages that are a small fraction of prevailing wages in the Caribbean (Wood 2004). As the ships have no permanent home, they are free to add or delete stopovers from their itineraries. Caribbean islands are left competing, trying to attract the ships by offering low port passenger fees. There are debates over the economic impact cruise ships have on destinations and whether cruise ship passengers can be converted to return to the island they have visited as a stay-over guest. As the ships continue to grow in size, ports need to be adapted to accommodate the ships. Concerns have been raised over the environmental impacts of cruise lines through damage caused by anchors on coral reefs, cases of illegal dumping at sea and the impacts of thousands of cruise tourists disembarking for a few hours in a port of call. Wood (2004) notes, however, that the assistance that Caribbean intergovernmental associations have received from the World Bank and the European Union in dealing with cruise ships, along with efforts of several NGOs on environmental and labour concerns, are promising developments.

Source: Wood (2004); Bratcher (2014); Cruise Market Watch (2015)

Plate 3.1 *The Bahamas, Nassau: multiple cruise ships in port*

Source: Photo by D. Telfer

Plate 3.2 *The Bahamas, Castaway Cay: private island owned by Disney where their cruise ships dock*

Source: Photo by D. Telfer

Accommodation, tour operators and restaurants

An example of a corporation that has both horizontal and vertical integration across national boundaries is the Accor Group. Based in Europe, it is one of the largest corporations in the tourism industry. It operates over 3,700 hotels worldwide, ranging from economy to upper-scale service hotels. The hotels in the company's portfolio include Sofitel (luxury), Pullman (upscale), Novotel, and Mercure, (mid-scale), and Ibis (economy) (Accor 2015). Strategies used by hotel chains to extend their reach across the globe include acquisitions, mergers and joint ventures, along with one or more of franchises and management contracts and consortia (Go and Pine 1995) (see Plate 3.3). Vertical integration is becoming increasingly common as the same company will control tour operators, retail agencies, airlines and hotels, thereby securing inventory and the distribution system (Buhalis and Ujma 2006). Tour operators are another sector of the industry that has seen consolidation. Following a series of mergers and acquisitions over recent years, the European

Plate 3.3 *Indonesia, Lombok: sign indicating where a future Holiday Inn will be built*

Source: Photo by D. Telfer

market is dominated by a small number of large firms. The top
four European tour operators by revenue in 2012 were TUI
(€18.3 billion), Thomas Cook (€11.3 billion), Kuoni (€4.8 billion)
and DER Touristik (€4.65 billion) (fvw 2015a). At that time, TUI
and Thomas Cook had a combined market share of over 50 per cent

Table 3.3 *Locations of Hard Rock Cafés and Hotels in developing and transitional countries*

Mexico/Central America	Asia	Pacific Rim
Cancún*	Almaty	Fiji
Cozumel	Bali	Guam
Guatemala City	Bali*	Saipan
Rivera Maya*	Bangkok	
San Jose	Bengaluru	**Middle East/Africa**
Vallarta*	Chennai	Bahrain
	Goa*	Dubai
Caribbean	Guam	Hurghada
Aruba	Gurgaon	Johannesburg
Cayman Islands	Hyderabad	Nabq
Nassau	Ho Chi Minh City	Sharm El Sheik
Punta Cana	Hong Kong LKF	
Punta Cana**	Jakarta	**Europe**
San Juan	Kota Kinabalu	Bucharest
Santo Domingo	Kuala Lumpur	Budapest
St Maarten	Macu	Gdansk
	Macu**	Istanbul
South America	Makati	Krakow
Asunction	Melaka	Moscow
Bogotá	Mumbai Andheri	Prague
Buenos Aires	Mumbai Worli	Warsaw
Buenos Aires Aeroparque	New Delhi	
Caracas	Pattaya	
Cartagena	Pattaya*	
Curitiba	Penang	
Margarita	Penang*	
Medellin	Phuket	
Lima	Pune	
Santa Cruz	Siem Reap	
Santiago		

* Hard Rock Hotel, ** Hard Rock Hotel & Casino
Source: Hard Rock Café (2015)

in major source markets, including the UK, Scandinavia, the Netherlands and Belgium, and over 30 per cent of the German market (fvw 2015a). In December 2014, TUI AG and TUI PLC merged to become TUI Group, the world's largest integrated leisure tourism business. Further illustrating the power of vertical integration, the merged TUI Group has 35 million customers, operates 300 hotels, 141 aircraft, 12 cruise ships and 1,800 travel agents (fvw 2015b). Transat A.T. Inc is an integrated international tour operator specializing in organizing, marketing and distributing vacation travel and packages with more than 60 destination countries. The company operates in Canada, France, the Caribbean and Mexico, as well as the Mediterranean. Air Transat is a business of Transat A.T., flying three million passengers a year to 60 destinations in 30 countries (Air Transat 2015). There has also been consolidation in the online travel agencies, where the largest companies include Expedia, Kayak, Orbitz, Priceline and, Travelocity (Johnson 2014); recently, Priceline purchased Kayak. The largest online agency in China is ctrip, having over 54 per cent market share (Tani 2015).

Plate 3.4 *Cambodia, Siem Reap: Hard Rock Café is not far from Angkor Wat, a UNESCO World Heritage Site*

Source: Photo by R. Sharpley

Much attention in globalization has been paid to airlines, hotels, tour operators and fast food restaurants, such as McDonald's, in terms of expansion and investments abroad. Another example is the Hard Rock Café chain, and Table 3.3 displays the locations of Hard Rock Cafés and Hotels in developing and transitional countries. The investment decisions taken by Hard Rock Café clearly illustrate a connection to capital cities and well-known tourism destinations, and there has been an expansion in Asia (see Plate 3.4).

Political aspects of tourism and globalization

The political dimensions of globalization and tourism are very much interconnected to the economic dimensions. As global corporations wield global economic power, they also wield global political power. This section of the chapter will focus on interrelated concepts, including the changing nature of the state and the perceived loss of national control in the face of globalizing markets, along with a brief discussion of the political economy of tourism. One of the traditional criticisms of globalization is that as nations open their doors to global markets, state interest is often pushed to the side in favour of the priorities of multinational firms. Countries that have embraced willingly or otherwise the liberalization agenda embedded within globalization will have to open their doors to foreign competition. Those countries that have signed up to the World Trade Organization or have borrowed money from international financial organizations face, to differing degrees, measures that seek to reduce the role of the state. In his book *Confessions of an Economic Hit Man*, John Perkins (2004) takes the argument to an even more controversial level, suggesting that developing countries are granted loans placing them in debt to global banking corporations based in the developed world that then require the developing countries to support Western policies on a variety of issues. The rise in Chinese international lending mentioned earlier has, however, started to shift this dynamic.

The globalization process has impacted governance. Sovereign states are key to international political regulation, yet governments face the situation where policy options are influenced by other governments or international organizations, especially in the area of economic policies (Lane 2005). Scholte (2005) suggests that globalization has facilitated the emergence of post-sovereign governance whereby the state is still important but there has been a change in the main attributes. With the rise of supra-terrritoriality, there has been a move towards mulitlayered governance involving substate, state and supra-state agencies (Scholte 2005). This is evident through organizations such as regional trading blocs and the Bretton Woods institutions mentioned previously in this chapter, and trans-world organizations such as various agencies of the United Nations. Some of these organizations have had an influence on tourism development through funding, such as the United Nations Development Programme (UNDP), the UN Steering Committee on Tourism for Development and the European Union. Daher (2005) notes the contributions of the World Bank and the Japanese International Cooperation Agency as international donors for various urban regeneration/heritage tourism projects in Jordan, commenting that the funds were channelled through complicated donor agencies' tendering procedures yet modest outcomes were achieved while foreign debt continued to accumulate. One other change of note with respect to the alterations of the role of the state is privatized governance, whereby there is regulatory growth occurring outside the public sector (Scholte 2005). This involves, for example, a range of business associations, NGOs, think tanks, foundations and even criminal syndicates (Scholte 2005). International to local NGOs focusing on a variety of issues have been exerting greater political influence (for example, environment – Greenpeace; human rights – Amnesty International; children – World Vision).

The changing nature of governance and the rise of governing structures that stretch across political boundaries raise issues of power and control. The section now turns to examining the political economy of tourism. Early discussions of the political economy approach to tourism focused on the idea that tourism has evolved in a manner that closely resembles the historical patterns of colonialism and economic dependency (Lea 1988). 'Metropolitan companies, institutions and governments in the post colonial period have maintained special trading relationships with certain elite counterparts in Third World countries' (Lea 1988: 12). These groups from the developed world, through their close association with local political and commercial classes, are able to encourage decisions over economic policy, labour legislation and commercial practices that are consistent with their interests (Britton 1982). The overall tone of political economy tends to be negative about the impacts of tourism as it is a way for wealthy metropolitan nations to develop at the expense of less fortunate nations (Lea 1988). Power and control issues raised by the political economy approach to tourism can also be observed within dependency theory, as outlined in Chapter 1. National economic autonomy in the developing world becomes subordinate to the interests of foreign pressure groups (including multinational tourism companies) and privileged local classes rather than relecting a broader political consensus in the developing country (Britton 1982, 1991). Britton (1982) developed an enclave model of Third World tourism, which is a three-tiered hierarchy. At the top are the metropolitan market countries where the headquarters of the major tourism companies such as transport, tour companies and hotels are located, and they dominate the lower levels in the hierarchy. In the centre of the model are the branch offices and associated commercial interests of the metropolitan companies located in the destination that operate in conjunction with local tourism counterparts. At the base of the pyramid model are the small-scale tourism enterprises that are marginal but dependent upon all of the organizations located higher up in the model. Foreign companies then come to dominate the system and capital accumulation travels up the hierarchy so it is the multinationals that benefit the most. While there are certainly dimensions of this model still at work in the industry, it is interesting to note that the investment in tourism is not only coming from developed countries, but also from firms based in developing or transitional countries. For example, an analysis of hotel investment in 2015 found that with changing rules for outbound investment in China, there will be an estimated US$5 billion worth of Chinese capital spent on outbound hotel investment (JLL Hotels and Hospitality Group 2015). There has been an increase in Chinese investment in various sectors in Africa and tourism has been one of those sectors in countries such as Ghana and Kenya (Xiao 2013).

In reviewing the literature, Cleverdon and Kalisch (2000) suggest that far from bringing economic prosperity to developing countries, tourism has great potential to reinforce economic dependency and social inequality. While some fortunate sections of society such as ruling elites, landowners, government officials or private businesses may benefit, the poor, landless and rural societies are getting poorer not only materially, but also in terms of their culture and resources. Examples noted by Cleverdon and Kalisch (2000) include eviction and displacement for resort construction, rising land, fuel and food prices, and commoditization of culture. Lea (1988) outlines four main reasons for the successful invasion of developing countries.

They seldom invest large amounts of their own capital in developing countries, but rather seek funding from private and government sources in the destination. The infrastructure required to support new resorts (roads, power lines, etc.) is also funded locally or through foreign loans. A viable visitor flow is maintained through global marketing campaigns. Finally, transnational corporations participate in profits through charging management fees, limited direct investment, and various franchise, licensing and service agreements. Their power comes from their ability to withdraw from these agreements, leaving the developing country in a very vulnerable position. Drawn by the attractiveness of the tourism industry and its perceived economic windfall, governments actively pursue multinationals to come and set up operation in their countries. What this has led to is that countries are forced to enter a bidding war offering multinationals a range of things such as forgivable loans, reduced taxes or complete tax holidays, along with the failure to enforce environmental laws, in hopes of attracting the multinational (Reid 2003). In an interview with a hotel manager of a beach resort in Tunisia in 1996 by one of the authors of this book, the manager indicated that the demands of the tour operator they had been working with for the past several years for extra services had been increasing, yet the payment made to the hotel had not increased.

While the traditional model of the political economy of tourism tends to focus on the controlling actions of multinational companies in developed countries, Bianchi (2015), however, argues that through economic globalization and market liberalization, there is an increasing complex and differentiated geography of tourism production. This dynamic change in geographic production challenges the straightforward north–south power and control balance presented in the neocolonial/dependency model of international tourism. The challenge to the state-centric approach is the increasing dominance of transnational tourism corporations, and the growing structural power of market forces at a global and regional level (Bianchi 2015). The international political economy (IPE) approach argues that orthodox international relations work is deficient as it focuses exclusively on the relations of the government (Preston 1996). The IPE approach draws on development economics, historical sociology and economic history, and one of the key ideas is that the basic needs of any polity are related to wealth, security, freedom and justice (Preston 1996). In summary, Preston (1996: 291) suggests that:

most broadly, the IPE approach offers the model of a world system comprising a variety of power structures within which agent groups, primarily states, move and where the specific exchanges of agent groups and global structures generate the familiar pattern of extant polities.

Regionally based economies, such as the Singapore, Indonesia and Malaysia Growth Triangle (Telfer 2015b), for example, have become more autonomous actors competing for mobile tourism capital (Bianchi 2015).

While there is a great deal written about the dominant forces of globalization, it has also been argued that it is too simplistic to accept globalization as an 'all-encompassing unilateral and hegemonic force' (Teo 2002: 459). Hazbun (2004) argues that many of the studies of globalization focus on the loss of the power of the state resulting in deterritorialization. However, Hazbun (2004) argues, using examples from the Middle East, that it can also lead to reterritorialization, where the state can exercise power and control over territorial assets. Using the example of Tunisia, the country has diversified from beach tourism to promoting its urban and Islamic heritage, as well as creating a new market by focusing on its less-developed southern desert region. The desert product was an attempt to develop a territorial-specific place image that had previously been identified with a low-cost Mediterranean beach destination. The efforts included marketing a more exotic image of Tunisia, with its urban markets and desert landscapes, where movies such as *Star Wars* and *The English Patient* were filmed. While acknowledging the difficulties, Hazbun (2004: 331) suggests that by 'asserting control over tourism spaces and processes that convert places, cultures and experiences into territorial defined tourism commodities', actors such as states, local communities or transnational corporations are able to drive the reterritorialization and thereby control the conditions that draw tourists and capital investment.

Global and local forces do interact in the destination. There are a number of authors who suggest that it is important to recognize the global–local nexus (e.g. Robins 1997; Teo 2002). As Robins (1997) argues, the global economy cannot simply override existing social and historical realities. Global entrepreneurs need to negotiate and come to terms with local contexts, conditions and constraints. As the globalization of the tourism industry continues, more parties are becoming involved. Each group, whether it be tourists, hotels, travel agents, private enterprises, local governments, NGOs or individuals, has interests to protect (Teo 2002). Globalization has established new and complex relations between global spaces and local spaces (Robins 1997). The key question is: What is the strength of the local institutions with respect to the forces of globalization? This will be explored in the next section.

Cultural aspects of tourism and globalization

The stereotype of cultural globalization has Western lifestyles and forms of consumption spreading across the globe, resulting in a convergence of culture that is defined by capitalism (Allen 1995). This process has also been referred to as cultural imperialism and can lead to assimilation. The so-called 'global products', such as Coca-Cola, Levi jeans, McDonald's hamburgers or Benetton clothes, are reflections of the idea that as people gain access to global information, they become globally aware (Robins 1997). Moreover, tourism is one such source of global information. The processes of Disneyization (Bryman 2004) and McDonaldization (Ritzer 2010) demonstrate not only changes in production, but also represent potential cultural changes. Global media sends idealized images of destinations to potential tourists in the developed world while those in developing countries receive images of Western consumption patterns. Advances in technology and communication have taken people to more remote areas on adventure or ecotourism holidays. Travel writer Bert D'Amico (2005) refers to the 'Coke Line', which is the line that one crosses to find a community not yet touched by outside forces. Specifically, he is referring to indigenous communities that have not yet been exposed to Coca-Cola. This line is rapidly disappearing. The further people travel, the fewer and fewer places there are left that have not been exposed to tourists. For example, the Chinese government spent US$4.2 billion to construct a train route into Lhasa, Tibet. The Sky Train travels at such a high altitude (3,700 metres) that the rail cars are pressurized like an airliner. The train began operation in 2006. Proponents of the plan argued that it would help the economy of Tibet, including doubling Tibet's annual tourism income to US$725 million by 2010. Critics, however, raise concerns over environmental damage and the impact this will have on the culture of Tibet (Olesen 2006). How communities and individuals react to outside forces may be an indication as to the strength of the host culture. Is there a resilient local culture or will it be overrun, not only by the culture of the tourists, but also by the rapidly evolving global information networks? Part of the tourism development dilemma associated with culture is that if a culture is frozen in time to be put on display for tourists, can development still occur? Cultural brokers will seek to make a profit while others try to avoid the tourists. The remainder of this section will focus on different aspects of culture, tourism and globalization.

Cross-cultural contact

Tourism brings cultures from around the world together, and a variety of studies have been conducted on the nature of guest–host relationships (Smith 1977 is an important early work) and the social and cultural impacts of tourism (Wall and Mathieson 2006; see also Chapter 7 in this book). Reisinger and Turner (2003) have explored cross-cultural behaviour in detail. The launch of the *Journal of Tourism and Cultural Change* is also a reflection of the interest in the area. Culture is a highly contested concept with many differing definitions. Peterson (1979) suggests the culture consists of four kinds of symbols that include values (choice statements ranking behaviour or goals), norms (specifications of values relating to behaviour in interaction), beliefs (existential statements about how the world operates, which often serve to justify values and norms) and expressive symbols (any and all aspects of material culture). To what extent do tourism and the processes of globalization change any of these four 'symbols' of culture? A number of concepts have been explored in the tourism literature on cultures coming together, including assimilation (where one culture loses out and is brought into another culture), acculturation (where cultures come together and share attributes) and cultural drift (where the host temporarily transforms their behaviour while in contact with tourists) (Wall and Mathieson 2006). Tourism is inevitably linked to globalization, and so does the host culture take on attributes of the tourists' culture or is there a broader emerging global culture that tourism is helping to spread? Caution, however, must be used when considering a convergence towards a global culture. Robins (1997) suggests there are three main aspects to consider in terms of culture and identity with respect to globalization. The first is homogenization, whereby cultural globalization brings about the convergence of cultures. The second is that there are cosmopolitan developments with cultural encounters across frontiers, resulting in cultural fusion and hybridity. The third is a reinforcement of cultural diversity, often against the perceived threats to homogenization or cosmopolitanization. There are those who react to the global changes by returning to their roots and reclaiming their traditions, thereby reinforcing the diversity of cultures. The reclamation of tradition may stimulate cultural tourism as will be discussed below. The impact of globalization on cultures is complex, pulling cultures in different, contradictory and conflicting ways (Robins 1997).

Conflicts can result from clashes in cultures, religion and family values as shifts occur in the structure of the labour force. In the context of tourism, much has been written on the demonstration effect associated with international tourism. Tourists model certain types of behaviour through action and dress, which are copied by those living in the destination. A fisherman turned supplier for the Sheraton Senggigi Beach Resort in Lombok, Indonesia, changed his manner of dress from the traditional Indonesia sarong to Western-style clothing. At the same hotel, a local young female hotel employee was prevented from returning to work by her father after only a short time of being employed at the hotel as her father could not get used to his daughter mixing with foreign tourists and making more money than he was. Workers in the tourist industry also typically gravitate towards the language used by tourists, and so while English is often the main language in tourism, as the Chinese outbound market continues to grow, there is growing need to accommodate the Chinese language.

Tourism results in immigrants moving into the destination to take up jobs in the tourism industry, bringing along their own culture. There is concern in the Galapagos Islands that migrants from Ecuador's mainland, hoping to get a job in tourism, are putting a strain on the island and its residents, already dealing with an influx of tourists (Schemo 1995; Ragazzi *et al.* 2014). These migrants may also bring their own customs and traditions, which interact with the local culture (Bookman 2006). The conditions of employment in developing countries have raised concern. Tourism Concern, the British NGO, ran a campaign against employment conditions at hotels in developing countries where many employees do not have contracts and are trapped in poverty through low wages. In some developing countries, sex tourism is part of the tourism product, and this is emphasized through the global media, where residents of the destination are presented as the desirable exotic 'other' (Momsen 2004). Mitchell (2000: 88) explores the political economy of culture and suggests that 'geography – spatiality, the production of spaces and places, and the global doings of the economy – is front and centre in the processes of cultural production, and hence in systems of social reproduction'. A new area of research in tourism is that of slum tourism in and around large urban centres. Frenzel (2014) comments on the symbolic factors in slum tourism, noting a concern is how the potential positives including empowering residents, or attitude changes among tourists towards poverty are balanced with the negative aspects, such as spectacular display of poverty for the voyeuristic tourists as well as the perpetuation of the myths about slums. These issues all raise concerns as to how cultures are portrayed around the world.

As globalization has facilitated travel and interactions between hosts and guests, there is also concern over the rapid spread of infectious diseases. Tourists who participate in sex tourism are at risk of contracting a variety of sexually transmitted diseases to be brought back home. In 2002, SARS (Severe Acute Respiratory Syndrome) originated in Guangzhou, China, and quickly spread to a number of countries. As Kimball (2006: 45) suggested, 'the unknown virus was going global, travelling by airline'. Similarly, the 2014 Ebola outbreak in West Africa caused a reduction in tourism well beyond the Ebola zone (Stone 2014), further illustrating the impact diseases can have on visitor numbers. Kimball (2006) argues that with increased travel and trade, there is clearly a risk with globalization that is creating a new ecology of infectious diseases.

Interactions between people of different cultures can also have a positive outcome whereby knowledge and understanding are generated between the groups. New ideas on topics such as empowerment, human rights and peace may be shared, thereby contributing to broader notions of development. Macleod (2004) raises the issue of identity in the context of tourism and cultural change. He suggests that it is valuable to consider people having a complex set of identities that form the overall picture. Where tourism occurs, people may experience change in identity on a personal and public level as individuals and groups. Fishermen become shopkeepers or tour guides and young women seek financial independence or pressure groups revitalize identities. Macleod (2004) goes on to further suggest that in the context of globalization, people may encounter a broader range of role models, networks of associates and opportunities, as well as undergoing a new self-perception. People or groups may also experiment with traditional roles, reinterpret events, reinvent themselves or gain confidence in their lifestyles and values (Macleod 2004: 216).

Cultural tourism

Destinations present themselves through many different cultural factors such as 'entertainment, food, drink, work, dress, architecture, handicrafts, media, history, language, religion, education, tradition, humour, art, music, dance, hospitality and all the other characteristics of a nation's way of life' (Reisinger and Turner 2003). People travel to either see or immerse themselves in 'authentic' local culture, and heritage tourism is receiving greater attention in the literature (Daher 2005) (see Chapter 4). Some argue that culture and tradition can be saved through dance, art, festivals and souvenirs, while others claim that this commoditization of culture for the tourists destroys the original culture. Culture sells, and it is packaged or commodified for the tourists, raising debates about what is authentic. Crafts are modified so that they are attractive and portable for tourists. This shift, however, may very well generate a great deal of income and hence lead to development. Azarya (2004) examined the challenges and contradictory consequences of indigenous groups being part of the tourism industry, with reference to the Massai pastoral groups in Kenya and Tanzania. The positive outcomes were associated with new sources of income through working in the industry, as well as exhibiting themselves as tourism exhibits, being photographed, selling souvenirs, and opening their villages and putting on dances. These benefits come at the cost of 'freezing' themselves at the margins of society, as it is their marginality which is the essence of the tourism product.

Tourist cultural enclaves

While some tourists seek out the local culture when they travel, others want to avoid it and choose to remain inside the confines of the hotel. A tourist cultural enclave may emerge for those that choose to remain within the resort area, and therefore there is very little contact with the host community. A lifestyle is transplanted from the developed world to the developing world, and this provides few opportunities for locals to make money from the tourists if they remain behind the wall of the resort. From a solely economic perspective, it may be in the tourism resort's best interests to keep their tourists inside the resort so that they will spend money in the restaurants and on souvenirs rather than facing the competition from outside the doors of the hotel. From a planning perspective, the development of enclaves may be a conscious choice in order to limit cultural contamination through tourist–local encounters.

Business culture

One other area to be explored, in terms of culture, is the influence of globalization on business culture. Previously in this chapter, examples of various multinationals have been given that illustrate the global reach of these firms. With these firms also come their operating procedures, which, in most cases, have elements of standardization across the multinational chain, not only for efficiency reasons, but also for the tourists' peace of mind so that they can expect the same level of service from hotel to hotel operating under the same brand. The question is whether the multinational is able to adapt their standardized operating policies and procedures to fit within the local destination. For example, the restaurant chain McDonald's adapts its menus and hotels, and often will incorporate local architectural designs into their construction. While this may be relatively easy, what may be more of a challenge is to incorporate Western-style management systems into non-Western settings.

Globalization has not only opened the cultures of the world to tourists; it has also opened up cultures of the world to residents of the destinations. Globalization has the potential to magnify cultural impacts as so many groups are moving around the world and people are rapidly exposed to mass media. Culture is, however, dynamic, and as Wall and Mathieson (2006) suggest, cultural change is influenced by both internal and external factors and cultures would change in the absence of tourism. Culture is not only an attribute of a destination; it is also a product, and therefore culture in its various forms has the potential to have significant impact on tourism planning and development, management and marketing (Reisinger and Turner 2003). These interactions between tourism and culture are further explored in Chapters 5 and 7.

Conclusion

Globalization is a highly complex and multifaceted process. However, stereotypes associated with globalization are often oversimplifications of a highly dynamic process. Globalization is not an equal process, and there are winners and losers. Reflecting on globalization, McMichael (2004: 152) states that 'it is tempting to think of globalisation as inevitable, or as destiny, until we notice that it only includes one-fifth of the world's population as beneficiaries'. The dilemma for developing countries is whether or not they decide to open their doors to let the multinational tourism companies into their country. However, do they really have a choice? The international companies bring a recognizable brand and image associated with them along with the necessary infrastructure to bring international tourists with them. It does come at a cost. The proponents of the global capitalist system have argued the growing world economy, of which tourism is a major player (although not always readily acknowledged), is vital to global development. However, just as critics have signalled the end of development as a concept, so too are critics raising the spectre of the collapse of globalism (Saul 2005). While acknowledging the growth in global trade as a success story, the promises of globalization are not materializing as there is ongoing Third World debt and poverty (Saul 2005). The 2008 global economic crisis also demonstrated the problems with the market economy. Tourism is 'quintessentially linked to globalisation and the phenomenon of time–space compression. In the sociocultural realm, tourism is emblematic of globalisation, hyperreality, fantasy and post modernity' (Potter *et al.* 1999: 95). The spread of ICT has facilitated the globalization process and has been embraced by the tourism industry, and ways must be found to reduce the digital divide. If tourism is to be used as an agent of development, then there are calls for the implications of globalization for destinations to be studied less as negative versus positive but more as a process that is dynamic, contingent and contested (Teo 2002). The following chapter considers the tourism development process, which is highly influenced by the globalization process.

Discussion questions

1 What power do developing nations have in the face of multinational tourism corporations?
2 How has information technology changed the way in which tourists and the tourism industry behave?
3 What are positive and negative consequences of consolidation in the tourism industry?

4   Can you identify ways in which multinational corporations can improve their linkages with destination communities?

5   What is the impact of international tourism on local culture?

Further reading

Hall, C. M. (2005) *Tourism: Rethinking the Social Science of Mobility*, London: Pearson Prentice Hall.

This book provides a very good examination of key concepts related to mobility. Tourism is examined through key concepts, including globalization, localization, identity, security and global environmental change.

Stubbs, R. and Underhill, G. (eds) (2006) *Political Economy and the Changing Global Order*, 3rd edn, Don Mills, Ontario: Oxford University Press.

The book is structured into the following four sections: (a) changing global order; (b) global issues; (c) regional dynamics; and (d) responses to globalization. There is a good analysis of the power and control issues in globalization.

Websites

www.ihgplc.com

This is the website of the InterContinental Hotel Group and illustrates the nature of a global tourism corporation. There is a wide variety of information on the various brands within the company and investor information.

www.staralliance.com/en

The Star Alliance is an alliance of airline companies with global reach. Background information on the formation of the alliance and information on the partners are available on this site.

www.unep.org/tools/default.asp?ct=tour

This is the website to the United Nations Environment Programme (UNEP) activities in tourism. The site outlines the organization's sustainable development resources and illustrates how a global-level organization can influence tourism development.

4 The tourism planning and development process

Learning objectives

When you finish reading this chapter, you should be able to:

- understand the tourism planning and development process
- recognize government policy on tourism developments
- identify the changes occurring in tourism planning
- debate the pros and cons of various forms of tourism development

Tourism development is a complex process involving the coming together of domestic and international development agents and key stakeholder groups with state policy, planning and regulations. The resulting tourism form not only has impacts in the host destination, but there are potential broader developmental outcomes benefiting the destination. Cruise ship docks, beach resorts, urban heritage centres, national parks, ecotourism resorts, casinos and village tourism represent a few of the very different and diverse tourism products created to attract tourists to developing countries. Decisions need to be made as to what form or forms of tourism are best suited to a destination for the long term in order to meet national developmental goals. For example, does concentrated large-scale mass tourism development, with high levels of multinational involvement such as a beach resort, do more in achieving developmental goals, or does small-scale dispersed community-based tourism generate more benefits? In a highly competitive global market, most developing countries are choosing policies to diversify their products. Cuba, for example, in addition to its beach resorts, is rapidly expanding its ecotourism sector, and with tensions easing with the US in 2015, the tourism sector may be on the verge of expansion. Previous debates on form and function tended to be polarized around the idea that one form of tourism is more sustainable than another, and, more recently, it has been recognized that all forms of tourism need to be made more sustainable. Each type of development will bring different tourists, with varying levels of disposable income and expectations, along with different opportunities for locals to participate in the tourism economy. State policy and planning on tourism often dictates the nature of development in a destination, and, given the complexities of the development process, what works in one destination may not work as well in another. In other destinations, it may be the industry and the market determining development. In order to address the questions of form and function of tourism, one needs to first examine a series of politically oriented questions that are central to the tourism and development debate:

- What is the desired outcome of the development?
- What are the tourism policy and planning regulations in the destination?
- What are the institutional arrangements and the political realities in the destination?
- What are the values of the key actors and institutions involved in the development process?
- Who is in control of the decision-making process?
- What project is selected, how is it financed and who operates it?
- Who benefits from the development?
- Can tourism development contribute to national development goals?

In developing countries, tourism planning and development often occurs through a top-down planning approach. Liu and Wall (2006) argue that decision-making with respect to tourism developments is predominately based on the interventions of government agencies and large tourism firms, resulting in the dominance of external, often foreign capital and the marginalization of local people. Within sustainability, there are calls for increased local involvement in the planning process; however, it is important to consider to what degree this is possible and how it can be facilitated. The focus of this chapter will be on the overall tourism planning and development process, along with an examination of selected strategies of tourism development as potential tools to help meet the developmental goals of a destination.

Tourism and development process

In the tourism and development debate, one of the first questions to be considered is: What is the desired outcome of any tourism expansion or refurbishment? Gunn and Var (2002) identify the goals for better tourism development as enhanced visitor attraction, improved economy and business success, sustainable resource use, and community and area integration, while more recently, destination planning has also focused on place branding, strategic partnerships and creative destinations. When the question is asked, 'Why get into tourism?', economic arguments are often presented first, with host governments trying to establish a strategy that has the correct mix of tourism developments to generate the greatest amount of foreign exchange, the argument being that once economic benefits start to accrue, then other spin-off benefits will arise for the host society in terms of other developmental goals. It is difficult to say to what extent tourism, pursued from an economic perspective, will lead to broader development goals such as greater self-reliance, endogenous growth, fulfilment of basic needs, environmental sustainability or other goals related to the 2015 UN Millennium Goals (see Boxes 1.1 and 2.1). The selection of any particular tourism development strategy or combination of strategies is a very complex process. It needs to be placed within the context of the politics of place and an understanding of the institutional arrangements in the destination, as well as an understanding of the values and power of those with the investment funds or those that can influence development. For example, the World Bank's Africa Region Tourism Strategy is titled *Transformation Through Tourism: Harnessing Tourism for Growth and Improved Livelihoods*. The four pillars of the strategy are policy reform, capacity building, private sector linkages and product competitiveness (Christie *et al.* 2013). In some cases, immediate profitability from a new cruise ship dock or resort may take precedence over environmental protection or other broader development goals, and governments are often forced to make a trade-off between competing concerns. Other destinations may adopt a tourism policy guided by sustainable development and, as part of that policy, reduced numbers of visitors may well be a target. These potential trade-offs illustrate the difficulty in implementing sustainable development, and echo comments by Mitchell (2002) that change, complexity, uncertainly and conflict are central in resource and environmental management and, it is argued here, they are also central in planning tourism development.

Figure 4.1 illustrates the complex nature of the tourism and development process as it relates to form and function. A brief explanation is provided here in advance of more detailed analysis in the remainder of the chapter. The first set of elements to be considered are the values, ideologies, goals, priorities, strategies and resources of tourism development agents. Figure 4.1 contains a list of some of the main tourism development agents, which includes various government, private industry and not-for-profit organizations. Some of those with influence are in the destination, while other organizations operate from abroad, such as multinational corporations, international funding agencies or regional trading blocks, as illustrated in Chapter 3. The interaction of these various groups occurs through a policy, planning and politics filter. This is referred to as a filter here, as any development project must work its way though the layers of political and bureaucratic structures in the destination. The political scope of a tourism development can potentially extend beyond the local political situation, incorporating multiple dimensions and interest groups across local, regional, national and even international scales that may be in support of, or against future development. Policies are established, plans are written, and regulations are put in place by governments to either promote or restrict development. It is in this environment that governments may try to offer various incentives to attract tourism development. Corruption may also subvert regulations. Proposals for tourism development are put forward through the planning and regulatory processes in the destination and, if the project is externally driven, the local meets the global. Key factors here include the type of political structure that exists in the developing country, institutional arrangements, and to what extent local participation is permitted in the planning process.

What is ultimately physically constructed will be greatly influenced by the tourism policy of the state, the values of those in charge and the resources they have to work with. The design of a building may relate to whether a destination has been branded or strongly associated with an established type of tourism (for example, ecotourism in Costa Rica, safari tourism in Kenya or beach tourism in Fiji). As illustrated in the centre of Figure 4.1, there are debates in the literature over the scale that tourism development should take and the merits of enclave tourism versus integrated tourism, which typically has more opportunity for links with local communities. Later in this chapter, the pros and cons of a number of different tourism developments will be explored in the context of contributing to overriding development goals. One of the important considerations in the overall development process is the strength of linkages to the local economy, both in terms of the formal and informal sectors. Those linkages may be at local, regional or national level; however, as the linkages extend beyond national borders, leakages start to increase. The final section of Figure 4.1 illustrates the overall development outcomes in terms of the economy, environment and society.

Values and power

The first box in Figure 4.1 contains the values, ideologies, goals, priorities, strategies and resources of the various tourism development agents. It is important to consider the interrelationship of these concepts for the various individuals and groups involved in tourism development, as well as their level of power and control in the overall development outcomes in the destination. Goldsworthy (1988) argues that all development theories, policies, plans and strategies either consciously or unconsciously express a preferred notion of what development is, and it is these preferences that reflect values. Values have been regarded as an 'enduring belief that a specific mode of conduct or end state of existence is personally or socially preferable to an opposite or converse mode of conduct or end state of existence' (Rokeach 1973 in Gold 1980: 24). Peterson (1979) suggests that values, then, are choice statements ranking behaviour or goals. Differing development agendas will reflect different goals, and these goals reflect social, economic, political, cultural, ethical, moral and even religious influences (Potter 2002). As the nature of development has widened its scope beyond economics, the literature on development ideologies has changed to also emphasize political, social, ethnic, cultural, ecological and other dimensions of the broader processes of development and change (Potter 2002). A great deal of development thinking also remains politically uninformed and more attention needs to be given to the

Figure 4.1 *Tourism development process*

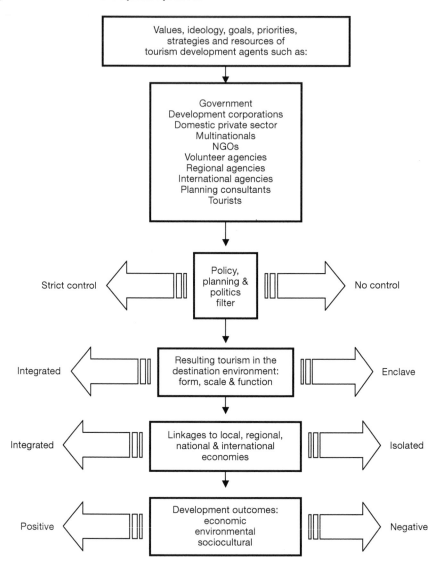

ideological underpinnings of development theory (Goldsworthy 1988). A few examples of broad categories of ideological underpinnings of development theory include positions such as conservative, liberal and radical (Goldsworthy 1988). Peet and Hartwick (1999: 3) suggest that 'developmentalism is a battle ground where contention rages between bureaucratic economists, Marxist revolutionaries, environmental activists, feminist critics, postmodern sceptics, radical democrats and others'. Development has also been criticized for its role in Western imperialism (see Said 1978).

In examining tourism policies in developing countries, Jenkins (1980) notes the dichotomy in aims between private investment and government investment. Private investment criteria centre on profitability with other considerations taking secondary positions. Government investment, although concerned with profitability, must also consider other non-economic implications such as social and cultural impact and land-use policies. Depending on their priorities, governments may take a 'tourism first' approach, focusing efforts on the industry itself, or a 'development first' approach, where planning is framed by broader national development need (Burns 1999a). As mentioned in Chapter 2, with the shift to sustainability, there has been increased concern in issues such as corporate social responsibility. Robbins (2001) refers to firms with a traditional corporate culture focused on generating wealth and firms with a social-environmental culture where the focus is not only on wealth, but also on being socially and environmentally active.

Tourism development agents have values and ideologies that need to be considered within the context of power. There is a wide diversity of approaches to the study, meaning and definition of power (see Coles and Church 2007), and several authors have stressed the importance of understanding the power relations in tourism (Hall 1994, 2007; Mowforth and Munt 1998, 2003; Coles and Church 2007; Yang *et al.* 2014; Bianchi 2015). Questions such as how much power tourism development agents hold, where they get their power base from, and to what extent they are able to use their power to generate their preferred outcome are important considerations in the study of tourism planning and development. Todaro (1997) argues that most developing countries are ruled directly or indirectly by small and powerful elites to a greater extent than developed nations (see also Todaro and Smith 2011).

Power comes from a variety of sources and is exercised in different fashions in the tourism development process. In examining social power, Mann (1986) suggests that a general account of societies, their structure and history can be best given by the interrelations of four sources of social power: ideological, economic, military and political. In China, for example, there has been rapid tourism growth, and tourism policies and practices are changing, reflecting efforts to create a distinctly Chinese approach to tourism policy and planning that seeks to blend traditional approaches, command economy approaches and Western concepts, including sustainable development (Yang *et al.* 2014). Specifically, in areas with ethnic minorities, China's tourism policies have attempted to blend top-down decision-making with some grass-roots involvement.

In the context of tourism, governments may have power based on laws and regulations, while industry may have power through wealth, information or technology, and citizen groups may have power though local participation. Similarly, international organizations may have power through finances, such as the lending of money, or through policy through platforms on items such as climate change, poverty reduction or sustainable development. The institutional arrangements will influence, in part, how power can be used; however, corruption may get around laws and regulations. Duffy (2000), for example, found that in Belize there is an expansion of organized crime, and the enforcement of environmental regulations, on which ecotourism is based, has become problematic. The government regulations on the environment are 'rendered ineffective when a junior arm of the state is opposed by more powerful interest groups that lie within and outside the state apparatus' (Duffy 2000: 562). Conversely, Jayoti and DiRienzo (2010) found that a reduction in corruption levels has a positive impact on tourism competeveness across nations and that the marginal gains in tourism competitiveness from reduced corruption levels are greater in developing countries than in developed countries.

The political economy of tourism was examined in the context of the globalization of the tourism industry in Chapter 3. In examining the changing global order, Underhill (2006) suggests that the complex relationship between power and wealth (or between the political and economic domains) in international politics can be seen in the links between political authority, on one side, and the systems of production and distribution of wealth referred to as the market, on the other. As for tourism, to what extent does a government set policies to open their borders to multinational corporations? Bianchi (2015) suggests that conceiving market behaviour in isolation from ideologies and values of the various actors and groups involved, as reflected in the free-market notion of *comparative advantage*, underplays the political nature of markets whereby the state has historically conditioned activities of the economic classes and, furthermore, ignores the uneven consequences of unlimited market competition. Bianchi (2015: 291) summarizes the political economy of tourism as 'the examination of the systematic sources of power which both reflect and constitute the competition for resources and the manipulation of scarcity, in the context of converting people, places, and histories into objects of tourism'.

In laying out the background of international political economy, Underhill (2006) reviews the contributions from three main approaches in international relations, including the realist tradition (state as unified autonomous actor), liberal approaches (interaction of individuals in economic sphere) and radical approaches (for example, Marxist approaches), as well as noting the literature on feminism, environmentalism and postmodernism. The exercise of power in international political economy occurs in a setting characterized by complex interdependence among states, their societies and economic structures at a domestic level and at an international level (Underhill 2006). This takes place through an integrated system of governance 'operating simultaneously through the mechanisms of the market and multiple sovereignties of a system of competitive states' (Underhill 2006: 19). Milner (1991) in Underhill (2006) suggests that examining the international system as a web of interdependencies necessitates a focus on linkages among actors. As discussed in Chapter 1, the emergence of a global development theory places the emphasis on new and evolving international forms of governance (for example, on climate change), which implies new power arrangements beyond the state level. The chapter now turns to examining various actors in the tourism development process.

Actors in the development process

A wide variety of actors representing the public, private and not-for-profit sectors come together in the tourism development process, as illustrated in Figure 4.1. In this section, a variety of examples will be used to illustrate the different roles. As outlined above, each of these actors has a set of values, ideologies, goals, priorities, strategies and resources that they bring to the tourism development process. At times, these various groups will have similar end goals, while at other times they may be in conflict with one another. A private developer may seek maximum profitability through the construction of a high-rise resort, while the state may chose to regulate building heights. All actors contribute in some fashion to the supply of tourism products, which Page and Connell (2006) summarize as national and regional tourism organizations, destination organizations, local authorities, attractions, transport, accommodation, retail sector (shops), travel agents, tour operators and the hospitality industry (for example, restaurants and cafés).

State involvement

One of the most important actors in the tourism development process is the state. Through various government ministries or state-sponsored tourism development corporations, governments establish the framework, policies, plans and regulations to both attract and control tourism development. As pointed out in Chapter 3, with the rise of globalization, there has been a change in the power of the state. In some regards, the state has lost a degree of power in the face of multinational corporations; however, in another sense, there is a more complex multilayered governance involving substate, state and supra-state agencies (Scholte 2005). In many developing countries, the state has an important role in facilitating tourism and there is a top-down approach to planning (Liu and Wall 2006). The government shapes the economic framework for the tourism industry, provides infrastructure and education requirements for tourism, establishes the regulatory environment that businesses operate in and takes a role in promotion and marketing (Hall 2005). Hall (1994) suggests that there are seven roles of government in tourism. These are coordination, planning, legislation and regulation, entrepreneurship, providing stimulation, social tourism, and interest protection. In Mexico, the Secretariat of State for Tourism formulates and carries out tourism policy and supervises the operations of Fonatur, the national fund for the development of tourism and the Mexican Tourism Board. Fonatur arranges financing for tourism enterprises along with real estate transactions, while the tourism board looks after promotion (see Box 4.1). A key element in the success of tourism development is the policies and regulations in place to attract both domestic and foreign tourism investment to a country or region. In India, for example, the government allows 100 per cent foreign direct investment in the construction of hotels and related projects, including airport expansion. They also allow a five-year tax holiday for organizations to set up hotels, resorts and convention facilities at specific destinations (Dezan Shira & Associates 2013). In Mexico, a target of US$9 billion for private investment (domestic and foreign) in the tourism sector for the period 2001–2006 was passed in June 2005. Forty-eight per cent of private investment went into Mexican beach destinations, and the United States is the largest foreign investor in Mexico's tourism infrastructure (Visit Mexico Press 2006a). In the period from 2007 to 2012, private investment for the development and maintenance of tourism projects in Mexico exceeded the target of US$20 billion (Mexico Premiere 2012). States also actively compete in attracting mega-events such as the World Cup or the Olympics. South Africa was the first country in Africa to host the World Cup in 2010 and Brazil will host the Olympics in 2016. While globally much of the foreign investment comes from developed countries, there has been a rise in south–south investment, with a number of transnational corporations from developing economies, including Singapore, Hong Kong, the UAE, Cuba, Malyasia, Poland, South Africa and Mauritius, investing in other developing destinations (UNCTAD 2007).

At the other end of the spectrum, the state can provide access to micro-credit schemes to help facilitate pro-poor tourism (see Chapter 5). At a regional or state level, the West Bengal government in India appointed a consultant to attract Indian and international investors for the state's tourism sector and to formulate future tourism policy. The areas for expansion identified include tea tourism, river tourism, Raj-nostalgia tourism and ecotourism (India eNews 2006). The state also has a role to play in destination marketing and creating an image attractive to tourists. The Internet has become vital as a platform for place-based marketing for national and regional tourism boards. Howie (2003), however, notes the tensions between the traditional planning approaches, guided through concepts such as carrying capacity, conforming land uses, acceptable functional uses and sustainable development, with the largely economic-driven market perspective of local chambers of commerce, major investors and influential bodies representing the private sector business. The state has to become increasingly entrepreneurial in a very competitive market. There has been a growth in the state entering partnerships with industry or other states in marketing or providing a facilitator role for the formation of partnerships for smaller companies.

Private sector involvement

Private sector development ranges from small-scale entrepreneurs in the informal sector, such as local tour guides, to small and medium-sized enterprises (SMEs), up to domestic hotel chains and multinational corporations, such as tour operators or hotel companies. It is important to note that while the larger corporations often draw

Box 4.1

Fonatur and tourism development in Mexico

Fonatur is Mexico's National Tourism Development Trust created in 1974 to develop planned integrated resorts and not only promote regional development, but also create projects that will have national impact. The organization creates master plans for development areas, arranges for financing of tourism enterprises, conducts real estate transactions and looks after infrastructure projects related to tourism development. Fonatur has integrated resort developments in Cancún, Los Cabos, Ixtapa, Loreto, Huatulco Bays and Nayarit. Together, these resorts are home to 40 per cent of the country's five-star hotels, receive 54 per cent of Mexico's foreign revenue from tourism and 46 per cent of all foreign visitors. Fonatur has invested $US2.6 billion in these resorts and actively seeks out investment for tourism development and provides a favourable environment for foreign investors. The regulatory framework for investment in Mexico supports foreign ownership in most economic fields and activities including real estate, allowing 100 per cent participation in shared capital. In addition to offering legal guarantees and security to foreign investors, the system in place allows unrestricted repatriation of profits, bonuses, dividends and interest payments. Fonatur holds an annual conference to attract developers and investors in tourism. In addition to redevelopment and expansion plans for the existing major resorts, Fonatur has a number of other current projects underway to diversify and develop new tourism products, including two mega-projects. The first is the construction of a premium integrated resort in the western coastal state of Nayarit designed to relieve pressure from Acapulco. It is the first time in over 20 years that Fonatur has been involved in the development of a new city, which is expected to have 14,500 rooms. The second large-scale project is the three-pronged Sea of Cortez project, which is focused on enhancing marina development along the Baja Peninsula. The mission statement for Fonatur in part states that it will be the institution responsible for the planning and development of sustainable tourism projects of national impact. Part of the vision statement includes a statement that development should occur with a social awareness that favours regional development and create permanent jobs that are appropriately remunerated. The challenge that Fonatur faces is not only in the actual development of these large-scale projects, but in ensuring that the benefits circulate to as many as possible. As Torres and Momsen (2004) discovered, in the case of Cancún, there are numerous challenges to overcome in strengthening backward economic linkages between these large resorts and other sectors of the local economy such as agriculture.

Source: Torres and Momsen (2004); Fonatur (2006); Visit Mexico Press (2006a, 2006b); US-Mexico Chamber of Commerce (2011)

the most attention, the small tourism business and the informal sector are crucial in terms of employment and generating opportunities for locals to participate in the tourism industry. These organizations will be explored in more detail in the next chapter. Criticism has been aimed at multinational tourism corporations in developing countries, including economic leakages, inappropriate forms and scale of tourism development, sunk costs and investment risks, overdependence on multinationals and foreign domination. However, these corporations also bring investment funds, know-how, expertise, managerial competence, market penetration and control, and demonstration effect for local entrepreneurs (Lickorish 1991; Kusluvan and Karamustafa 2001). Based in Brazil, Atlantica Hotels is a large multi-brand hotel administrator in South America, including 80 hotels in 43 Brazilian cities (Atlantica Hotels 2015). Tsogo Sun is a leading hotel group based in South Africa with over 90 hotels covering 15 hotel brands and 14 casinos in countries across Africa (Tsogo Sun 2015). At a global scale, the InterContinental Hotel Group owns, manages, leases and franchises through various subsidiaries over 4,800 hotels and 710,295 guest rooms in almost 100 countries and territories. The company's portfolio includes hotels such as InterContinental, Crown Plaza, Holiday Inn, Staybridge Suites, Candlewood Suites, Hotel Indigo and EVEN Hotels, and it has one of the largest loyalty programmes, Priority Club Rewards (InterContinental 2015). Lickorish (1991) argues these corporations are bound by company law (national and international), by the interests and demands of their shareholders and always by the marketplace, which determines their profits.

As investment opportunities may be less risky in developed countries, government agencies in developing countries often must invite in these corporations and pay for joint scheme partnerships with private corporations. In an analysis of foreign direct investment (FDI) and poverty alleviation in the Gambian hotel sector, Davidson and Sahli (2015) found that transnational corporations investing in larger and upmarket hotels contributed to overall accommodation capacity, as well as providing flow-on benefits to those depending on the tourism industry as well as to the wider economy. Other FDIs, such as smaller-scale (semi)resident businesses, offered niche-style accommodations that tended to have more focus on community and environmental responsibility. Local ownership was concentrated in the mid-range and budget hotels that demonstrated close links to local communities that was in line with cultural expectations (Davidson and Sahli 2015).

Much of the attention in the tourism literature centres on multinationals. However, small domestic firms and the informal sector in developing countries contribute a great deal to the industry. Gartner (2004) suggests that estimates of the number of small firms in many developing countries is far larger than the number of medium or large firms and accounts for between 40 and 90 per cent of non-government employment. 'Small firms which are the dominant form of tourism business at the destination level are the backbone of a destination's tourism economy' (Gartner 2004). In a study in Ghana, Gartner (2004) found these firms faced many challenges, not only from global factors and issues of dependency, but also from their inability to raise capital, lack of managerial skills and cultural obligations, which require them to hire friends and family. In addition to small firms are the individuals who work in the informal sector, such as tour guides; these positions are very significant to the local economy in many destinations, though not counted in formal employment figures.

International agencies

International agencies also have a role to play in the tourism development process. Tourism consultants are often called upon to offer their expertise in the design and development of specific tourism projects. Developing countries have been assisted by agencies such as the World Bank or United Nations agencies (Lickorish 1991). As illustrated in Box 4.2, the World Tourism Organization offers a tourism consultant service. In addition to planning assistance, international agencies have provided financial assistance. As indicated in Chapter 3, tourism is increasingly being written into Poverty Reduction Strategy Papers as part of loan applications to the IMF and the World Bank. In a study of World Bank involvement in tourism projects in the 1970s, Davis and Simmons (1982) identified 24 projects covering the Mediterranean basin, Mexico and the Caribbean, Africa and Asia. Involvement covered a range of elements, including infrastructure, construction and rehabilitation, lines of credit and technical assistance. Hawkins and Mann (2007) update the role of the World Bank in tourism development to 2006 and note there has been a shift to a more micro-level and policy intervention targeted at outcomes such as raising the livelihoods of local people, reflecting the focus on livelihoods at the centre of the World Bank's 2013 report on tourism in Africa (Christie *et al.* 2013). This shift is also indicative of the focus on poverty reduction and the UN Millennium Development Goals. The African Report develops recommendations for countries in sub-Saharan Africa at various stages of tourism development based on a 'tourism destination pyramid', as indicated in Table 4.1. At the base of the pyramid are the pre-emergent and potential/initiating tourism countries. The second level is emerging/scaling-up tourism countries, and the top of the pyramid comprises the consolidating/deepening and sustaining success countries. Along with the recommendations in Table 4.1, the authors also stress all countries need to address political support and capacity building for tourism.

The Multilateral Investment Guarantee Agency (MIGA), which is part of the World Bank Group, is a global political risk insurer for private investors and companies that are investing outside their home country. MIGA guarantees investments against the risks of currency transfer, expropriation and war and civil disturbance. In Costa Rica, Conservation Ltd and the Bank of Nova Scotia, Canada, went through MIGA to provide investment insurance for their ecotourism investment in the Rain Forest Aerial Tram that takes tourists on a 90-minute ride through the treetops in a rainforest near San Jose (MIGA 2006). The challenge for developing countries in borrowing money from international agencies is considering to what degree they have to conform to the mandate of the funding agency. The funds may come with strings attached.

Another international organization, the Global Sustainable Tourism Council, is trying to establish and manage standards for sustainable tourism as well as developing sustainability criteria. Two sets of criteria have been developed, one for hotel and tour operators and the second for destinations. The Global Sustainable Tourism Criteria for Destinations are divided into the following sections: (a) demonstrate sustainable destination management; (b) maximize economic benefits to the host community and minimize negative impacts; (c) maximize benefits to communities, visitors, and culture, and minimize negative impacts; and (d) maximize benefits to the environment and minimize negative impacts (GSTC 2013). Each of the main criteria is broken down into sub-criteria along with suggested indicators. The shift towards global criteria is also evident in TOURpact.GC launched by the UN Global Compact and the UNWTO. The programme aligns the principles of the UNWTO's Global Code of Ethics for Tourism

Box 4.2

The UN World Tourism Organization and tourism planning consultants

As tourism continues to grow, the tourism consultant business also grows. If a government or business organization decides to pursue a tourism development project, they can typically use their own respective staff members to do the designs and planning for the project or they can hire external consultants. These organizations provide a wide range of consulting services, from creating national tourism master plans to specific site plans with architectural designs. National governments can also obtain the services of tourism consultants through the United Nations World Tourism Organization (UNWTO), Technical Cooperation Service. The purpose of the service is to support member countries in their efforts to 'develop and promote the tourism industry as an engine for socio-economic growth and poverty alleviation through the creation of employment' (UNWTO 2015b). At the request of a member government, the UNWTO Technical Cooperation Service can coordinate individual consultants or larger consultancy firms to go to the destination and work on the technical missions and projects. Funding for these projects comes from a variety of agents, including the United Nations Development Program (UNDP), the World Bank, the European Union, the Asian Development Bank, bilateral donors and others. The areas that the service covers include:

- identification and assessment of potential tourism development areas;
- establishment of coherent frameworks for long-term sustainable tourism development;
- preparation of national and regional Tourism Development Master Plans;
- development of community-based tourism;
- alleviation of poverty through tourism;
- development of rural and ecotourism;
- development of human resources for tourism;
- formulation and implementation of appropriate marketing and promotional strategies;
- strengthening of institutional capacities of National Tourism Administrations;
- adjustment and improvements in existing tourism regulations in accordance with international standards;
- stimulation and promotion of public-private partnerships;
- establishment of hotel classification systems; and
- deployment of information technology in tourism.

The Technical Cooperation Service is also responsible for the ST-EP (Sustainable Tourism for Eliminating Poverty). In 2011, the *UNWTO Technical Product Portfolio* was released, further illustrating the services offered, and the report is divided into the following categories: (a) policy planning and economic development; (b) statistics and quality standards; (c) sustainable development; and (d) product development, marketing and promotion. The guidelines for creating a tourism policy suggest a time frame of three to six months with a budget of €50,000 to €100,000. Individuals and tourism consulting firms can apply on the UNWTO Technical Cooperation Service website to be included in the database as potential experts who the service would call upon to carry out the work. Projects are carried out in specific countries, as well as at regional scales. At the regional level, for example, work was conducted through the Service on the Silk Road Regional Programme (SRRP), which involves China, Kazakhstan, Kyrgyzstan, Tajikistan and Uzbekistan. Specifically, work was done on updating a study on visa requirements for the Silk Road countries, as well as preparing a report on the inventory of the Silk Road tourism resources. Elsewhere, with the support of the European Union, the UNWTO Technical Cooperation Service is helping the South Pacific Tourism Organization (Fiji, New Guinea, Tonga, Tuvalu and Vanuatu) to develop a standardized system for recording, classifying and analysing tourism statistics. In Mali, Africa, the UNWTO Technical Cooperation Service sent a mission to prepare the guidelines to develop a future Tourism Master Plan. The mission evaluated the current state of the tourism sector and made many recommendations, including promoting cultural tourism (the Grand Mosque at Djenne is a UNESCO World Heritage Site), as well as ecotourism and adventure tourism. Recommendations were also made on the development of individual tourism sites, administration of the tourism sector, accessibility, accommodation, environmental and economic impacts, and marketing. As with all plans, an important element is the need for monitoring and ongoing evaluation and adjustment. One of the challenges with any report or plan that is generated is to what extent it will be implemented and later evaluated over time.

Source: UNWTO (2006, 2011, 2015b)

and the Global Compact's focus on social responsibility in business in support of the UN Millennium Development Goals (UN Global Compact 2008).

Regional-level organizations are also influential in tourism development. The Association of Caribbean States (ASC), which includes 25 Caribbean and Latin American countries, is taking an active role in tourism, including improving air and sea transport infrastructure and services between member states (Timothy 2004). A special committee of the ACS was responsible for the Convention on the Sustainable Tourism Zone of the Caribbean (STZC). The ACS sustainable tourism plan includes a strategy to encourage community members and other stakeholders to participate in the tourism planning process, stimulate entrepreneurial activity, and promote cooperation

Table 4.1 *Recommendations for tourism development in sub-Saharan Africa by the World Bank*

Stage of development	Recommendations
Pre-emergent	• No recommendations. • Fourteen countries are either war-torn, suffering civil strife or emerging from these situations. • Little can be done until responsible democratic governments are empowered to create stable political regimes and ensure safety and security of local population.
Initiating tourism	• Make land available for tourism and improve transport policies and infrastructure. • Focus on tourism asset with highest potential. • Focus scarce resources on locations and market segments with highest potential. • The above strategies reduce critical constraints (infrastructure, security, lack of skills) to attract world-class investors. • Targeting areas of high potential allows focused promotion of iconic attractions, piloting of policy reforms on land and air transport and creating appropriate institutions with coordination mechanisms. • Success of these initial projects developed with public sector and donor support attracts new investors to other areas.
Scaling up	• Political support, airline access and land availability must be assured. • Investment and destination promotion are critical. • Marketing campaigns, source-market awareness campaigns, positive image enhancement. • Investor promotion to build confidence and streamline process of investment. • Create central source for investor information. • Attend professional investment conferences where investors, lenders, insurance companies, real estate agencies meet. • Attract investors through incentives such as direct financial assistance (bonds, special purpose taxes), indirect assistance (zoning), fiscal measures (tax breaks).
Sustaining and deepening tourism successes	• As growth increases develop strategies to disperse tourists to different areas. • Distribute arrivals more evenly throughout the year. • Growth needs to be managed through things such as brand management, certifications, etc. • Increase visitor arrivals in off-season through pricing incentives, product diversification, special events/festivals.

Source: Christie *et al.* (2013)

between government agencies and the private sector (Timothy 2004).

Not-for-profit organizations

Not-for-profit organizations or non-governmental organizations are receiving greater attention for the role they play in the tourism development process and they range in scale from small local organizations to large-scale multinational NGOs. Their involvement may be in generating employment opportunities in community-based tourism initiatives or with volunteer tourism. Volunteer heritage conservation groups have assisted in protecting sites as well as offering interpretation to tourists. These organizations are also known for voicing concern over tourism development. Tourism Concern is based in the United Kingdom and has, for example, challenged Hilton for its claims of corporate social responsibilities (CSR). Hilton signed an agreement with a development company to manage Bimini Bay Resort and Casino in The Bahamas and the completed project was to include a casino, condominiums, a golf course and marina. Tourism Concern argued that the development threatened a nearby fragile ecosystem (Tourism Concern 2013) and, as a result of their joint campaign, the Bahamian government announced the establishment of the Bimini Marine Protected Area (although yet to be implemented) in order to protect the mangroves. Additionally, the resort project was reduced to half its planned size and the golf course dropped (Tourism Concern 2013). The role of NGOs in community-based tourism is examined in greater detail in the next chapter.

Tourists

The final group listed in Figure 4.1 as an agent of development are the tourists. Tourists themselves have a role to play in the development process. Whether that is a more passive role, namely lying on the beach and then paying their hotel bill at the end of their vacation, or a more active role, participating in a volunteer tourism project helping to construct a local school or assisting on a farm, both can potentially contribute to the development goals of the destination. In addition, the actions they take, the behaviour they exhibit, and their interactions with the locals and the environment can all potentially have an impact on tourism being a successful development tool (see Chapter 6).

Policy, planning and politics filter

Policy

Tourism development is about resource management, and Lickorish (1991: 160) argues that 'resource management starts with the preparation of an effective and realistic policy, and then a plan with related strategies for development and marketing which must be prepared together'. Politics and public policy are significant for tourism, whether at the local or global scales, as they regulate the tourism industry and tourist activity (Hall and Jenkins 2004). Hall (1994) argues that when a government adopts policies, they are selecting from different sets of values that can have a direct impact on the form of tourism that is developed. The political ideology of a government can determine whether the government favours large-scale resorts or backpacker hostels, ecotourism or casinos (Elliott 1997). It is increasingly evident that many tourism policies are being set within the framework of sustainable development.

How specific policies come about, as Hall (1994) suggests, can arise within the context of a policy arena where interest groups (industry associations, conservation groups and community groups), institutions (for example, government department and agencies responsible for tourism), significant individuals (for example, high-profile industry representatives), and institutional leadership (for example, ministers of tourism and government officials) interact and compete in determining policy choices. According to Figure 4.1, the agents of tourism development, such as a hotel developer, will encounter what has been identified as the policy, planning and politics filter. Identified as a filter to represent the various layers of political, bureaucratic and regulatory administration, the filter is the screen that individual tourism developments must go through before they are constructed. The nature of this filter will be different in each destination, depending on the local political and administrative conditions. In some cases, local community groups or environmental groups may have the opportunity to voice their opinions in this filtering process, potentially altering the nature of the tourism development proposed. In the case of Mexico (see Box 4.1), the destination is very open to foreign investment; however, in the case of Vietnam, Lloyd (2004) found that there has been a turbulent relationship between foreign investors and the Vietnamese Communist Party (VCP), which governs a socialist market economy.

In such an economy, there is tension between the state and its desire to retain a substantial role in the economy through control over privatization, monopolization of infrastructure and high-income generating investments, and the pressure from foreign companies and lending agencies such as the IMF or World Bank that advocate a reduced role for the state (Lloyd 2004).

A recent development in the context of policy and regulation is the study of tourism governance and, in particular, the importance of good governance for sustainable tourism development (Bramwell and Lane 2014). 'Governance involves the processes for the regulation and mobilization of social action and for producing social order' (Bramwell and Lane 2014: 2). A number of studies have examined the government-related challenges associated with developing and implementing sustainable development policy in tourism such as in China (Cao 2015) and Thailand (Muangasame and McKercher 2015). In China, Cao (2015) found that although there is enthusiasm for sustainable tourism, the relevant policy and regulatory frameworks are fraught with contradictory objectives, are often incoherent and have ambiguous legal provisions. Many of the relevant planning organizations have duplicate responsibilities, unclear responsibilities, interlocking activities and weak coordination owing to a complicated institutional infrastructure (Cao 2015). In Thailand, the Tourism Authority of Thailand adopted the '7 Greens' (green heart, green logistics, green service, green activity, green community, green attraction, green plus) sustainable tourism policy. Muangasame and McKercher (2015: 497) investigated its implementation and found that, again, while there is strong support for the sustainability policy, problems include: little effective buy-in; unwillingness to take leadership roles; and identification of weaknesses in the policy limiting its utility such as 'lack of clear objectives, failure to define terms, lack of collaboration between government departments, conflicts between local and national politics, crony capitalism and short-termism, together with the failure to identify measurable metrics and to implement clear evaluation procedures'.

One example of the impact of changes in ideology on policy and overall development was observed in Jamaica by Chambers and Airey (2001). During the 'socialist era' of 1972–1980, the Jamaican government pursued goals of self-reliance along with seeking to integrate tourism into the Jamaican way of life. In the subsequent period from 1980–1989, 'the period of capitalism', the emphasis shifted to reducing government intervention and pursuing foreign exchange. During the first era, there was some Jamaicanization of tourism, and the policies contributed to a decline in arrivals, occupancy and hotel provision and employment. During the second era, there was a recovery in tourism numbers but increasing tensions between locals and tourists.

Similarly, national tourism policy implementation in the Philippines from the 1970s to the latter part of the 2000s can be divided into three phases and be explained by the process of policy learning (Dela Santa 2015). Phase one, from the 1970s to 1986, is characterized by lesson-drawing, phase two, from 1986 to 1999, by technical and conceptual learning and phase three, from 1999 onwards, by the intensification of social learning. Dela Santa (2015) stresses the importance of understanding policy learning. In the Philippines, at the start of the twenty-first century, there has been greater integration of social learning mechanisms in sustainable tourism policy processes, resulting in even greater participation by civil society in tourism policy implementation by national and local governments (Dela Santa 2015). Government ideology then has the potential to impact tourism policy and development outcomes.

Types of planning

Once policies are established, then plans are written to oversee that developments reflect overriding policies. At a very broad level, Inskeep (1991: 25) defines planning as 'organising the future to achieve certain objectives' and it is carried out at different levels, from individuals planning everyday activities to corporate planning, to governments creating formal comprehensive national or regional plans. A more specific definition linked to sustainability is planning is a 'process which aims to anticipate, regulate and monitor change to contribute to the wider sustainability of the destination, and thereby enhance the tourist experience of the destination or place' (Page and Connell 2006: 477). Inskeep (1991: 25) identifies the major types of planning as economic development planning; physical land use planning; infrastructure planning for things such as transportation, water, electrical, waste disposal and telecommunications; social facility planning for educational, medical and recreation facilities and services; park and conservation planning; corporate planning; and urban and regional planning. Liu and Wall (2006) argue that more attention needs to be given to planning tourism employment in developing countries so that locals have the necessary training and skills to participate in the benefits of tourism. A recent development with respect to tourism planning is crisis management, and the UNWTO offers workshops in:

- emergency planning of tourism;
- tourism risk analysis, early warning and mapping;
- crisis coordination at the national and international level;
- crisis centre;
- crisis communication; and
- recovery techniques (UNWTO 2015c).

Most of the major types of planning discussed above are completed by governments. However, corporate planning deals with the strategies that a corporation puts in place to enhance their business profile and generate profit. An airline will select which country to fly to or an international hotel company will select a domestic partner to construct the hotel they will then manage.

Scales in planning

Tourism planning occurs across various scales and time frames. International organizations (for example, the United Nations World Tourism Organization), regional trading blocks (for example, Association of South East Asian Nations (ASEAN)) and international conservation and environmental laws (for example, World Heritage Convention) all have a potential influence on tourism policy and planning (Hall 2000). With increased competition, tourism planning is occurring across borders and the importance of partnerships continues to be recognized as vital for tourism development. The Growth Triangle of Singapore, Indonesia and Malaysia, for example, is promoted as a site for investment for multinational corporations, including those involved in tourism, and the Indonesian island of Bintan and Singapore are marketed together as tourism destinations (Timothy 2000). Individually, island states face a number of challenges, including economic, social, institutional and environmental constraints, and often exhibit dualism whereby there is a large-scale technologically progressive export sector along with a small-scale, fragmented and undercapitalized domestic sector (McElroy and de Albuquerque 2002). Baud-Bovy and Lawson (1998) outline the main emphasis planning takes at national, regional and local scales. At the national level, tourism master plans establish the broad framework for tourism, creating environmental, economic and social policies. At a regional level, development strategies and structural plans for the region are established, including a focus on regional infrastructure, protection areas and transport. At the local level, plans typically focus on local development plans, allocation of resources, conservation measures, zoning of land uses, densities, coordination and implementation of policies. Finally, at the project level, the focus of the plans is on market and financial appraisal, organization of investments, site acquisition, facility planning and construction, and the coordination of development and operational needs (Baud-Bovy and Lawson 1998). In national or regional plans, time frames typically extend over a 10- to 12-year period with specific development programmes set out within the broader framework with shorter time frames, such as three to six years (Baud-Bovy and Lawson 1998).

As noted above, there has been an increased focus on regional and destination planning stressing the importance of place branding and creative destinations through the development of networks, collaboration and clusters (Telfer 2015b). The presence of a cluster of tourism-related enterprises in a destination that are collaboratively marketing and developing tourism can increase the competitiveness of the destination. Clusters can generate agglomeration economies, potentially acting as growth poles, thereby further increasing tourism development (Telfer 2015b). Porter (1998) argues that clusters highlight externalities, linkages, spillovers and supporting institutions, thus affecting competition through increased productivity and capacity for innovation as well as cluster expansion. Thailand, for example, is benefiting from health tourism clusters (Komaladat 2009). Creative destinations are 'urban and rural places which enhance the well-being of their populations though tourism and embrace new ways of thinking and sustainable living; places which are attractive to live in and visit' (Morgan *et al.* 2011: 10). Indeed, for a time, the Ministry of Tourism in Indonesia was titled the Ministry of Tourism and Creative Economy. The successful building of a destination's reputation requires a productive coalition and partnerships between civil society, governments and businesses, which may have competing and conflicting goals. Hence, Morgan *et al.* (2011) argue that the Destination Management Organization (DMO) has a central role in coordination. A strong destination brand has six elements: the place's tone, traditions, tolerance, talent, transformability and testimonies (Morgan *et al.* 2011). The last item, testimonies, refers to the importance of what is told about the destination, and a key avenue for this is the Internet. India has been using the national branding campaign 'Incredible India' since 2002 and it has been credited for an increase in visitor numbers (Kerrigan *et al.* 2012).

In addition to scale, planning documents written for tourism have different purposes. Table 4.2 lists various types of plans for Indonesia as a whole and for the island of Lombok in particular. The national economic policy sets the framework for development in the country and tourism is identified as playing a key role. The national tourism plan establishes the development of new tourism destinations and gives responsibilities to the provinces for tourism development. More specific plans include the development of tourism on the island of Lombok, including village tourism, a feasibility study for an

Table 4.2 *Examples of tourism plans in Indonesia*

Type of plan	Title
National Economic Development Strategy	Master Plan Acceleration and Expansion of Indonesia Economic Development 2011–2025
	• tourism is one of eight main programmes along with agriculture, mining, energy, industrial, marine, telecommunication, development of strategic areas
	• need to invest US$14 billion to achieve tourism industry development (Global Business Guide Indonesia 2013)
National Tourism Policy	Master plan for National Tourism Development (Regulation No. 50 of 2011)
	• develop 50 national tourism destinations by 2050
	• local governments in each province to develop short-, medium- and long-term measures to develop tourism capacity (for example, transport links, qualified human resources) (UNDP 1992; Global Business Guide Indonesia 2013)
Island Tourism	Tourism Development Plan for Lombok (JCP 1987)
Village Tourism	Village Tourism Development Programme for Nusa Tenggara (WTO 1986)
Airport Feasibility Study	Feasibility Study for Airport Development in Lombok (Sofreavia *et al.* 1993)
	• the new airport was opened in October 2011
Resort Area	Lombok Island: The Manalika Resorts Area
	• integrated resort area by the Bali Tourism Development Corporation, which developed Nusa Dua in South Bali
	• 1,035 hectares, including 7.5 km of white-sand beaches, three protected coves, surrounding 3,000-hectare conservation zone (BTDC n.d.)

airport that opened in 2011, and a proposed large-scale resort. All are
linked to tourism yet have different focuses. In Malta, the Ministry of
Tourism, in cooperation with the Malta Environment and Planning
Authority, has just revised its height policy for hotels, allowing
hotels of three stars or more to build two additional floors over the
height limit permitted in certain areas in order to increase hotel
profitability (Malta Tourism Authority 2015). Strategic planning has
been one of the more popular management tools (Phillips and
Moutinho 2014) and a tourism strategic plan involves analysis (of
aspirations of stakeholders, surrounding environment and tourist
resources), formulation, implementation and performance evaluation
(Simão and Partidário 2012). Hall and Page (2006) highlight the
importance of strategic planning in the context of sustainability.
These planning documents outline a strategy, which is a means to
achieve a desired end. Strategic planning integrates planning and
management in a single process and guides future direction,
activities, programmes and actions on an ongoing basis. Strategic
planning is meant to be iterative, whereby planning systems adapt to
change and learn, and is becoming more important as a form of
tourism planning.

Changing approaches to planning

When examining changes in the approaches to tourism planning, it is
useful to refer to Table 1.2 on the evolution of development theory.
As various development paradigms and their respective ideologies
and implementation strategies have come to the forefront, they have
had an impact on tourism planning and policy. Traditionally, tourism
planning theory and practice has been associated with town planning
(Costa 2006). That is, the focus has been on land-use zoning or
development planning at local or regional levels relating to site
development, accommodation, building regulations, density of
development, and presentation of cultural, historic and natural
tourist features (Hall and Page 2006). However, in recent years,
governments and planners have had to adapt to include sustainability
(see Chapter 2), poverty reduction and climate change in tourism
planning. This section of the chapter will examine some of the
changing ideological influences, traditions and models in tourism
planning, which illustrate the shift towards collaboration. However,
as highlighted in Boxes 4.1 and 4.2, in developing countries, much
of tourism planning remains top-down in its approach with the
involvement of external agencies. As suggested in his book *Tourism*

and Politics: Policy, Power and Place, Colin Michael Hall (1994) illustrates the importance of understanding the power relationships within tourism planning and policy making. The section then concludes with an examination of a value chain approach as poverty reduction strategy in tourism planning.

Dredge and Jenkins (2011a) trace some of the ideological influences on tourism planning that have occurred in an era with demands for smaller government (Hall 2000; Dredge *et al.* 2011). In the 1980s, when Western governments were swayed by neo-liberalism, tourism planning departments began to focus more on market research rather than research focused on public interests. Marketing operations were often outsourced and responsibilities were offset through the creation of statutory corporations or commercial agreements with external providers. As a result, corporate interests with financial power and expertise began to have more influence on tourism policy (Dredge and Jenkins 2011a). By the late 1980s, however, sustainable development had firmly infiltrated tourism planning and development, although, as Dredge and Jenkins (2011a) argue, it has been used more as a rhetorical device than a set of clearly defined guiding principles and actions. From the late 1990s, increased criticism over neo-liberalism, including growing disparities in power and wealth, social and economic marginalization, poverty and terrorist activity, resulted in a reassertion of local issues, values and agendas (Dredge and Jenkins 2011a). These concerns were reflected in 'political modernization', whereby there is a shift in power away from central governments to new forms of coordination between politicians, political parties, bureaucrats and bureaucracy, corporate interests and civil society. This is linked to changes in public administration where there is a 'new public management', which places emphasis on facilitating and enabling tourism development by mobilizing the resources of others instead of government action and physical planning (Dredge and Jenkins 2011a). The result of these trends ('political modernization' and 'new public management') is that there has been an increased alignment of governments with the commercial sector through, for example, public–private partnerships, again giving tourism corporations more influence. While public interest, social justice, transparency and accountability are receiving increased attention in academia, they are outside the focus/interest of policy and practices in many situations (Dredge and Jenkins 2011a).

In addition to the ideological influences on tourism planning, it is important to consider the various traditions, models and emerging approaches. Hall (2002) examines the traditions of tourism planning by highlighting the work of Getz (1987). These traditions include boosterism; an economic, industry-oriented approach; a physical/spatial approach; and a community-oriented approach (see Chapter 5). To these, Hall (2000) adds the sustainable approach to tourism planning. With the shift towards sustainability, there has been an increased focus on how to make the concept more operational and, as a result, there has been increased attention on sustainable development management approaches and frameworks. Examples of this include the adoption of sustainability criteria and indicators, ecosystems management, impact assessments (economic, social, environmental), adaptive management, precautionary principle, carrying capacity and limits of acceptable change (see Telfer and Hashimoto 2006 for more detail). What these changing traditions mean is that those responsible for tourism planning would draft plans that have a specific focus as guided by the tradition under which they are operating. Both community and sustainable development approaches to planning tend to be more bottom-up planning, allowing for input from local residents, as opposed to top-down planning, where the planners are viewed as the experts. With shifts towards sustainability, there are increased calls for incorporating locals into the planning and development process; however, locals are frequently under-represented as investors and decision-makers as they have a lack of knowledge about tourism and associated skills, and priority is often given to economic growth by policymakers with little concern for equity (Cohen 1982 in Liu and Wall 2006). Liu and Wall (2006) argue that if the locals are to really benefit and participate, efforts must be taken to go beyond calls for more local involvement and move to incorporating human resource development into planning, thereby increasing the capabilities of the locals. If this is not done, benefits will continue to go to outsiders while the locals have to adjust to the changes that tourism brings (Liu and Wall 2006).

One of the challenges of tourism is that it is highly fragmented, which has caused problems for tourism planners and managers (Jamal and Getz 1995). A number of authors have proposed collaborative and integrative approaches. Jamal and Getz (1995) propose a collaborative community-based planning process as a way forward to getting groups to work together. It facilitates public and private sector interactions, as well as providing a mechanism for community involvement in tourism. Munanura and Backman (2012) also stress the collaborative approach in tourism planning; however, they note the importance of adopting this within the constraints and environment in the developing world. Timothy (1998) similarly calls for cooperative tourism planning, which requires cooperation between government agencies, between various administrative levels of government, between same-level autonomous polities, and between the private and the public sectors.

The integrative tourism planning approach brings together the separate components of tourism and integrates them with other agencies of development so that 'all development sectors and supporting facilities and services are interrelated' (UNEP 2009). This approach has strong links to sustainability (Inskeep 1991). Aspects of integration include:

● *geographical* – different territorial units;
● *systematic* – ensure all important interactions and issues are considered;
● *functional* – interventions by sectoral management bodies (for example, tourism) are harmonized with wider management strategies and objectives;
● *policy* – sectoral management (e.g. tourism) strategies and plans need to be incorporated into overall development policies, strategies and plans;
● *interdisciplinary* – disciplines to transcend sectoral boundaries;
● *vertical* – integration among institutions and administrative levels with the same sector;
● *horizontal* – integration among various sectors at the same administrative level;
● *planning* – among plans at various spatial levels and plans must not have conflicting objectives, strategies or planning proposals; and
● *temporal* – coordination among short-, medium- and long-term plans and programmes (UNEP 2009: 31).

It is not only important for integration to occur within the tourism sector, but also for tourism to be integrated into the overall development policies, plans and programmes of a country or region (UNEP 2009). Jamal and Jamrozy (2006: 168) present an integrated destination management framework that brings in the destination's ecological-human communities as equitable and integrated members of planning, marketing and goal setting. Within the framework, they advocate sustainable tourism marketing and they argue that based on an ecosystem network model, a sustainable planning-marketing orientation satisfies not only the needs and wants of individuals, but also strives to sustain ecosystems. Similarly, in the context of an integrated rural tourism approach (IRT), the emphasis is on the core elements of networking, scale, endogeneity, sustainability, embeddedness, complementarity and empowerment (Saarinen and Lenao 2014: 369). The challenge of any integrated approach, however, is the complexity involved and whether it is possible to achieve integration both within tourism and broader development plans that a government may have.

Similar to the ideological influences noted by Dredge and Jenkins (2011a) above, market-led approaches to tourism planning continue to receive attention. Costa (2006) predicts that tourism destinations will be planned and managed by models that place emphasis on the coordination and stimulation of private sector organizations that also bring public participation to the core of the decision-making process. Emerging approaches will give priority to improving coordination between private and public sector organizations and placing citizens (residents and visitors) at the core of decision-making. Costa (2006) also argues that the governance of tourism destinations, their capacity to be more sustainable, competitive and profitable, and the design of self-sustained development, will become the priority of tourism planning models. Organizations will be set up around tourism clusters ('product-space organizations') replacing old bureaucratic organizations ('space-product organizations') and tourism administrations will adapt to this new trend of following clusters. Within these clusters, groups will have to work collaboratively to also market their destination.

Importantly, poverty reduction has recently come to the forefront of development policies in general, and this trend has been reflected in tourism policy and plans in particular. One specific strategy within tourism planning utilized in poverty reduction and pro-poor tourism is the value chain approach (Beyer 2014). As outlined in Chapter 2, pro-poor tourism is a strategy that can be applied to all forms of tourism to help alleviate poverty. The value chain describes the complete product or service from the supplier to the producer or service provider, and then on to the end user (Beyer 2014). In the context of tourism, the value chain describes the complex network of horizontally and vertically linked organizations, including tourism service providers, accommodation companies and agricultural businesses (see Figure 4.2). The horizontal links connect firms directly involved in providing the tourism product from the tourists' arrival to departure and represent the direct economic benefits. The vertical links in the chain are companies that are indirectly involved, such as companies supplying food to hotels, which in turn create the indirect and induced economic multipliers (Beyer 2014). Economic multipliers are further discussed in Chapter 7.

> The primary objective of the value chain approach is to include poor and marginalised population groups in the dynamics of the tourism value chain and to rebalance the distribution of income within the value chain in favour of these groups.
>
> (Beyer 2014)

Steck and Wood (2010) used a value chain analysis in Zanzibar to map the financial flows within tourism and the values accruing to the poor in order to identify ways to increase the participation of

Figure 4.2 *Tourism value chain*

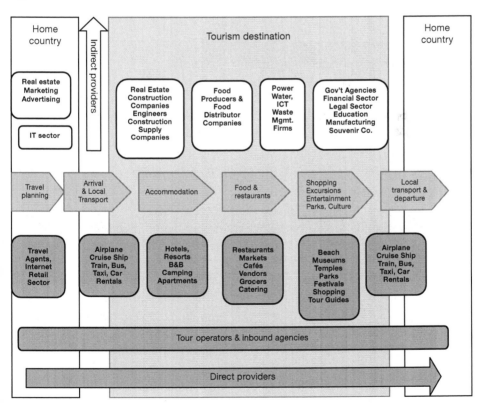

Source: After Beyer (2014)

those in poverty. Three main steps in pro-poor value chain assessment include: (1) diagnosis of current situation; (2) developing a list of potential projects; and (3) project planning so they can be assessed and implemented (Mitchell 2012). Once the value chain is mapped, plans can be developed to enhance benefits to those in poverty. There are, however, criticisms of pro-poor approaches. Saarinen and Lenao (2014: 369), for example, argue that it may be better to use existing government and other structures so local communities have greater control rather than relying on philanthropy, pro-poor tourism and corporate social responsibility programmes.

As tourism policy and planning continues to evolve, there is the recognition that tourism cannot be planned in isolation, but rather needs to be integrated into broader development plans and the public, private and not-for-profit sectors have to work together. As new planning paradigms evolve, Dredge and Jenkins (2011b) suggest greater understanding and debate is needed in the areas of the role of the state, structures, functions and operations of tourism agencies and institutions, distribution of power in planning and policymaking, governance structures and what is good policymaking. Recent planning developments have come in the area of creating policy and plans for climate changes, which are addressed in Chapters 8 and 9. To conclude this section, Saarinen and Lenao (2014) point out that many tourism planning and development models originate in the Global North, and careful consideration is needed when these models and approaches are applied to the Global South or developing countries.

Regulatory environment

In addition to various policies and plans, any proposed tourism development also encounters the regulatory environment in the destination. The regulatory environment covers a wide range of specific regulations or laws, such as building codes, labour laws, waste management regulations or environmental regulations that may be specific to the destination. For example, environmental impact assessments are typically required before development proceeds; however, different countries have different standards of regulations and degrees of enforcement. Some regulations will be strictly enforced and others will not, thereby possibly presenting some developers with the added advantage of choosing to locate their firm in one location over another. McElroy and de Albuquerque (2002) highlight Anguilla, Bermuda, St Lucia and the US Virgin Islands that have useful models of comprehensive development plans for small islands, and they note that most require a permitting process often involving public hearings, environmental impact statements (EIAs) and social impact statements (SIAs), along with cultural and historical inventories. The Balearic Islands (Majorca, Minorca, Ibiza and Formentera) introduced an ecotax on tourists in 2001; however, with a change of government, it was repealed in 2003. In Box 3.2, Wood (2004) indicates that cruise ships choose to fly flags of convenience in part to take advantage of reduced environmental and labour regulations. Different countries also have varying regulations with respect to contesting planning decisions approved by the state. Developers may want to consider if there are protests over a potential development, and if there is due process for opponents to have their claims heard. Lengthy delays with respect to public hearing regulations may also cause developers to select another location.

Tourism is very much influenced by politics. Potential tourism development projects encounter policies, plans and regulations covering a variety of different scales. To these elements, one must add the political situation in the destination, and which individuals or groups hold the balance of power, not only in terms of formal, but also informal, power. Do local elites control the industry and the government, thereby dictating what forms of tourism are preferred and where they should be located? The heading of this section, 'Policy, planning and politics filter', is meant to highlight the various complex layers that any development has to go through before any construction can begin.

Resulting tourism in the destination environment: form and function

Models of tourism development

Tourism development not only evolves from individual decisions on specific projects, but also within the context of multiple decisions, influencing over time the nature of development. A number of models have been presented in the literature that examine the changes in resort morphology in destinations over time. Miossec (1976), cited in Opperman and Chon (1997), developed a model that looks at the evolution of resorts, transport routes, tourists and hosts.

In the final stage of the model, there is a hierarchy of resorts that are highly connected and there are also excursion routes into the interior of the hypothetical island. As destinations open up, tourists have the potential to travel greater distances not previously accessible, thus increasing the level of contact with locals. Not only does improved infrastructure for tourism promote better access for tourists to the destination; it also induces migration of people looking for work in tourism. This migration often leads to sociocultural, economic and ecological change, as has been found in Zanzibar (Gössling and Schultz 2005) and Goa, India (Noronha *et al.* 2002). One of the most enduring models is the Tourism Area Cycle of Evolution, which is linked to the concept of a product life cycle. Developed by Butler (1980), the model suggests that tourism developments go through the stages of exploration, involvement, development, consolidation, stagnation, and decline or rejuvenation. In this model, as time passes, control of the industry tends to pass from local control to more external control as large-scale multinational tourism companies open resorts as the destination becomes more recognized. If a destination goes into decline, decisions need to be made as to what strategy should be used to rejuvenate the area. While some resorts have tended to follow the stages in the model, other destinations have jumped several stages, resulting in instantaneous development rather than development that occurs over a longer period of time. As in the case of Mexico, the government has purposely selected undeveloped areas and built large-scale resorts that are set up as regional growth poles (Telfer 2015b).

Wall (1993) developed a tourism typology consisting of attraction types (cultural, natural and recreational), location (water- or land-based), spatial characteristics (nodal, linear and extensive) and development strategies (highly developed, developed and developing). Accommodation type is regarded as a key element in the tourism system with implications on the characteristics of tourists, the built environment, economic impacts, degree of local involvement, and critical environmental and sustainability factors such as capital, land, water, energy and waste disposal systems. Wall (1993) suggests that a mix of tourist types (mass to explorer) and accommodation types (five-star to guest houses) can be integrated to promote sustainable development. In developing considerations for appropriate and sustainable tourism development, Telfer (2002b) examined (a) the scale and control of development and (b) local community and environmental linkages. Under each of these main categories, the following subcategories were examined:

(a) scale and control of development:

- focus of development, scale, rate of development, level of economic distribution, type of planning, local involvement, ownership, industry control, role of government, management origin, accommodation type, spatial distribution, tourist type, marketing target, employment type, infrastructure levels, capital inputs, and technology transfer;

(b) local community and environmental linkages:

- resource use, environmental protection, hinterland integration, intersectoral integration, cultural awareness, institutional development, and local compatibility.

An investigation of selected tourism development strategies

The selection of any specific tourism development strategy needs to be understood within the framework provided in Figure 4.1 and who is in control of the development process. For example, is the strategy for development based on a government tourism master plan or is the strategy being developed driven by a private entrepreneur as part of a broader investment strategy in tourism to satisfy shareholders? Private investment may or may not coincide with national, regional or local tourism plans, and various levels of government may have competing agendas. As Reid (2003: 225) suggests, 'local and regional development requires different processes and foci than does the goal of raising foreign currency in order to pay down foreign-held debt'. The options to development are as diverse as the number of different types of tourism; however, the selected strategy will also be governed by the politics of place and resources in the destination (natural, human, economic, infrastructure, technological). Some destinations will have an advantage in resort development, while others, for example, have, over time, been branded either formally or informally as ecotourism destinations. As suggested in the centre of Figure 4.1, the resulting tourism in the destination obviously has form, but it also has function linked to the development process. From large-scale to small-scale and from enclave resorts to integrated resorts, to what degree the tourism product establishes links to the destination will have a major bearing on tourism contributing towards the specified development goals. Those developments that are strongly integrated to local, regional or national economies will generate greater potential for more people residing in the destination to participate. However, those that are primarily linked to international economies rather than domestic economies may be more isolated economically from the local population placing the destination in a very dependent position. It is important to note, however, that even these types of development that are linked to multinationals can be strong generators of tourist numbers and foreign currency, and, in many cases, the best option for developing countries that do not initially have the capacity to effectively launch their own domestic tourism industry. Past debates tended to portray mass tourism as being exploitative and unsustainable while small-scale alternative types of tourism were preseted as being more sustainable. With the shift towards sustainability, an increasing focus on poverty reduction and the trend for destinations to diversify their tourism offerings, however, there is increasing pressure for greater integration of tourism within the local economy and for making all forms of tourism more sustainable. Weaver (2014) advocates enlightened mass tourism incorporating resolution-based dialectics whereby the assets of mass tourism (economies of scale, competitive innovation) and alternative tourism (ethical imperative, place sensitivity) are amalgamated so that the triple bottom line (social, environment, economics) is optimized (cited in Pegas *et al.* 2015). A fourth pillar of geopolitical sustainability was added by Weaver (2010), focusing on a state's mobilization of tourism to achieve geopolitical objectives, including internal stability and secure borders (cited in Pegas *et al.* 2015). While it is not possible to examine all types of tourism within the context of this chapter, this section will focus on a selected number of tourism developments to point out the strengths and weaknesses of each type in terms of their potential to contribute to development.

Resort development

Large-scale resorts targeting mass tourism often represent the main development strategy in many developing countries as they have the advantage of attracting a significant number of tourists, income and foreign investment (see Plates 4.1–4.3). Box 4.1, on Mexico, clearly illustrates the economic potential of large-scale beach resort development, and so this development option cannot be ignored. An example from Brazil highlights both domestic and international involvement in beach resort development. The Brazilian government, working to update the tourism sector, borrowed US$800 million from the Inter-American Development Bank to improve infrastructure in the northeast and created the Ministry of Tourism in 2003. Also from the Inter-American Development Bank, a US$75 million project is being proposed for 2016 in Brazil to create a tourism development programme for *Rio Grande do Norte State* that would include: 'tourism product strategy, tourism marketing strategy, institutional strengthening, tourism infrastructure and basic services and environmental management' (IDB 2015). In advance of the 2016 Olympics in Brazil, US$5.3 billion will be invested to build 422 new hotels (Fox News Latino 2013). The country is becoming a destination for European tour operators looking for something different from the Caribbean. Spanish and French multinational hotel companies have been opening resorts in the country, and there has also been domestic investment in resort development. In 2000, Previ, a pension fund for employees at Banco Brasil (the country's largest bank), partnered with Odebrecht, a large Brazilian construction firm, to spend US$200 million building five resorts, six smaller inns, conventions and sports centres, restaurants, stores, swimming pools, tennis courts and an 18-hole golf course on the Sauipe coast. Previ later bought out Odebrecht's stake in the project and then leased the five resort hotels to multinationals. Two were leased to the US chain Marriott, two were leased to France's Accor and one to Jamaica's SuperClubs (Kepp 2005). Oppermann and Chon (1997) point out that in many developing countries, prosperous hotel chains such as Atlantica Hotels from Brazil, which is a large multi-brand hotel administrator in South America, have emerged. Pegas *et al.* (2015) note the importance of domestic tourism in Brazil in developing their littoral pleasure periphery (LPP) dominated by almost 200 specialized coastal resorts, beachfront metropolises and beachfront cities. What is interesting about their findings is that the LPP is divided into two distinct sections and neither is sustainable. The southern LPP is similar to a developed country LPP, which is long established and displays an organic growth pattern. The northern LPP is a hybrid of organic and induced growth patterns with planned growth and low density. Despite a multiregional tourism initiative to facilitate economic and social equity by targeted tourism investment, the northern LPP resembles classical Third World LPPs where coastal development and investors displace locals. In both the north and the south LLP, unsustainable outcomes illustrate there is minimal progress to what Weaver (2014) proposed as enlightened mass tourism. While some argue that sun, sea and sand tourism is entering a decline as a result of environmental damage and a shift in

Plate 4.1 *Cuba, Varadero: beach resort*

Source: Photo by D. Telfer

Plate 4.2 *South Africa, Sun City Resort: tourists swimming at the man-made beach at the resort in a water-scarce area*

Source: Photo by D. Telfer

Plate 4.3 *Tunisia, Monastir: luxury beach resort, Amir Palace Hotel (note the architectural design)*

Source: Photo by D. Telfer

consumer demand to post-tourists or 'new tourists', Aguiló *et al.* (2005) argue that is not the case in the Balearic Islands. Following restructuring and quality improvements, it is still a competitive destination. The authors, however, argue that the sun, sea and sand model of tourism needs to be adapted to the framework of sustainable development.

While these resorts bring in large numbers of tourists, there is lively debate over their overall benefit to the destination. Criticism has been aimed at these resorts, especially if they are enclave resorts, cut off from the local community. They are often controlled from abroad, generate high rates of leakages, and the multinationals are more interested in profit than establishing strong links to the local community. As noted in Chapter 3, Britton (1982) developed an enclave model of tourism, and is particularly critical of package tours. Tourists are transported directly to resort complexes where they typically reside for the duration of the trip. Travel in the destination is only between resort clusters and to the airport to return home. The transportation, organization of the tour and accommodation are largely within the formal sector. Criticisms have also been directed to the trend of all-inclusive packages at these resorts, which include transportation, accommodation and meals. After arrival at the resort, tourists need to be enticed to venture out of the complex. All meals have been paid for in advance so tourists may be hesitant to venture out to a local restaurant to pay for another meal, thereby reducing the potential multiplier effect. Tips are often included in the package, thereby removing this opportunity for hotel staff. In an examination of three resort enclaves in Indonesia, including Nusa Dua on Bali, Kuta on Lombok and Bintan Beach Resort on Bintan, Shaw and Shaw (1999) argue that the notion of enclavity is inherently unsustainable, marginalizes local entrepreneurs, and widens the social, economic and cultural gaps that already exist between hosts and guests. Similarly, in the Okavango Delta of Botswana, Mbaiwa (2005) found the enclave tourism was dominated by foreign companies. While the tourism here is in much smaller numbers since they are following a policy of high cost and low volume for safari tours, the concerns over enclave developments have surfaced. This type of development has led to repatriation of tourism revenue, domination of management positions by expatriates and lower salaries for citizens. Mbaiwa (2005) argues that tourism has failed to significantly contribute to rural poverty alleviation in the Okavango region.

Integrative development and alternative tourism

While criticisms have been aimed at large-scale and enclave resorts, others have called for more integrative tourism development that promotes linkages to local communities. At one end of the spectrum of integrated tourism are larger tourism developments, and at the other end are small-scale or alternative tourism destinations where locals are provided with a greater opportunity to participate. Telfer (1996) and Telfer and Wall (1996, 2000) examined the efforts of both large- and small-scale hotels in Lombok and Yogyakarta, Indonesia, that established linkages to the local agricultural sector (see Plates 4.4 and 4.5). In both locations, hotels established direct and indirect links to local suppliers, traditional markets and farmers for the purchase of food products for the hotels. While these initiatives are promising and generate increased multipliers, there are challenges to maintaining these relationships in the long term, especially when five-star hotels place increasing demands on small producers and suppliers at certain times of the year. Another example of alternative tourism that involves locals directly is village-based tourism. In the case of the Solomon Islands, village tourism takes the form of tourists travelling by motorized canoe between villages situated on the edge of lagoons (Lipscomb 1998). Tourists undertake bushwalking, snorkelling, historical and nature tours, cultural displays, and arts and crafts in each village, as well as staying overnight in the village to get a greater understanding of village life. Lipscomb (1998) suggests that the potential for village-based tourism to persist at an early or exploration stage of development for a long period of time, producing a small income for the village, is in itself a compelling argument to develop village tourism. There are, however, challenges with this type of tourism, including marketing, planning and cultural impediments, and the fact that there is a danger that the local elites may primarily be the ones who benefit. The emphasis on integrated or community-based tourism as a form of sustainable tourism, where the control and benefits remain in the local community, including pro-poor tourism designed specifically to address poverty issues by enhancing opportunities for those in greatest need, will be further addressed in Chapter 5.

Plate 4.4 *Indonesia, Lombok: fisherman turned supplier in black leather jacket purchases fish in a local fish market, which in turn will be sold to an international hotel (see Telfer and Wall 1996)*

Source: Photo by D. Telfer

Plate 4.5 *Indonesia, Lombok: small, local fruit and vegetable supplier makes a delivery to the Sheraton Hotel in Sengiggi Beach, the products were purchased at local markets*

Source: Photo by D. Telfer

Ecotourism

Ecotourism continues to receive attention, since it theoretically represents a win-win scenario through environmental protection and improved local livelihoods (Cater 2004). For many developing countries, it is promoted as a means of reconciling economic growth and environmentally sustainable development (Duffy 2006). It is often seen on the opposite end of the spectrum from mass tourism as a form of alternative tourism. Ecotourism strategies have been endorsed by tourism businesses and destinations, but also advocated and promoted by a wide range of agencies such as the UN World Tourism Organization, United Nations Environment Programme (UNEP), international lending agencies (for example, World Bank), indigenous rights groups, development non-governmental organizations (NGOs), environmental NGOs (for example, World Wide Fund for Nature and Conservation International), the International Ecotourism Society, and bilateral development agencies (Cater 2004; Duffy 2006; Jamal *et al.* 2006). While there have been debates as to precise definitions of ecotourism, Diamantis (2004: 5) puts forward as a guiding conceptual principal that ecotourism 'occurs in natural settings (protected and non-protected) with an attempt to increase benefits to the economy, society and environment through sustainable educational practices from locals to tourists and vice versa'. It is particularly promoted in the undeveloped world, as it is their 'underdevelopment' or 'lack of modernisation' that makes their environments attractive to tourists from the 'developed' world (Duffy 2006). Ecotourism is promoted as a means for poorer communities to generate income and for many communities living adjacent to national parks or reserves and it is presented as a beneficial return 'for relinquishing rights over using the plant and animal resources within those reserves for subsistence purposes' (Duffy 2006). The United Nations declared 2002 as the International Year of Ecotourism. During that year, the World Ecotourism Summit was held and resulted in the Quebec Declaration on Ecotourism. The declaration recognizes that ecotourism embraces the principles of sustainable tourism, with respect to the social and environmental impacts of tourism, while embracing the following concepts that distinguish it from the wider concept of sustainable development (UNEP/WTO 2002, in Cater 2004: 485):

- contributes actively to the conservation of natural and cultural heritage;

- includes local and indigenous communities in its planning, development and operation, and contributes to their well-being;
- interprets the natural and cultural heritage of the destination to visitors; and
- lends itself better to independent travellers, as well as to organized tours for small-size groups.

It is interesting to note that southern NGOs expressed concerns over the UN declaring 2002 as the International Year of Ecotourism as it would open the floodgates to eco-opportunist Western exploitation (Cater 2006). As a follow-up to the Quebec declaration, the Global Ecotourism Conference released the Oslo Statement on Ecotourism in 2007, which further assessed the state of global ecotourism and evaluated challenges for ecotourism.

Duffy (2006: 2) suggests that the politics of ecotourism is revealed in debates over definition and raises the question: can it be 'provided by global tour operators and luxury nature based resorts, or is genuine ecotourism found in small scale local community run projects and campsites?' Efforts have also been made to develop certification programmes, and one example of note from a developing country is the Costa Rican Sustainable Tourism Certification (Jamal *et al.* 2006). The programme evaluates companies on five aspects: physical-biological parameters; infrastructure and services (for example, energy savings, handling of waste); service management; external client; and socio-economic environment (for example, interaction with local communities) (CST 2015). Weaver (2004) outlines the structural dimensions of ecotourism, which range from hard to soft ecotourism. Hard ecotourism traces its origins to the 1980s as a form of nature-based, small-scale alternative tourism that emerged as a reaction against the perceived environmental, sociocultural and economic excesses of large-scale development. Traits of hard ecotourism include a high level of environmental commitment among specialist ecotourist participants involved in specialized, often physically and mentally challenging experiences over longer periods of time in smaller numbers. At the other end of the spectrum is soft ecotourism, which has traits of a moderate or 'veneer' commitment to environmental issues, and participants are conventional tourists experiencing ecotourism in larger numbers as one part of a diversified experience. They have extensive reliance on services and an emphasis on package travel through travel agencies and tour operators. A cruise ship, for example, may offer a tour to a nature garden or reserve. Weaver (2004) points out while hard ecotourism purists would not agree with soft ecotourism, it is perhaps soft ecotourism that can generate the financial objectives to help enhance the natural environment and can be managed into site-hardened intensive zones.

The diversity of ecotourism is evident in South Africa where there are National Parks with no accommodation facilities (Knysna National Park) while others, such as Kruger National Park, have luxury accommodation, extensive roads networks, 4x4 routes, wilderness trails, guided safari drives, swimming pools, golf courses and banking. In South Africa, steps have been taken to commercialize the ecotourism industry. Many protected areas have management plans that employ zoning to protect some areas while allowing tourism to be developed in other areas. The commercialization of Kruger National Park is now being undertaken by the private sector, with companies submitting proposals that are evaluated in terms of financial aspects, environmental management, social objectives and empowerment. The linking of tourism with environmental protection and empowerment for human development illustrates the potential benefits that can occur if nature-based tourism is properly planned and managed (Spenceley 2004). Elsewhere, ecotourism lodges, such as those investigated by Johansson and Diamantis (2004) in Thailand and Kenya, demonstrate how the private sector can also contribute towards providing benefits to host communities.

While the various forms of ecotourism continue to be a source of debate, there are also debates surrounding its contribution to conservation and the involvement of local communities. Investigating community-based ecotourism in northern Vietnam from a gender and development perspective, for example, Tran and Walter (2014) found while there was a more equitable division of labour, increased income, self-confidence, community involvement and new leadership roles for women, issues of inequalities in social class, childcare and violence against women need to be addressed. Community-based ecotourism in Botswana has also had mixed results in terms of biodiversity conservation and community livelihoods as a result of multiple stakeholders in the design, planning and implementation of ecotourism projects (Stone 2015). Caution also needs to be raised over the possibility that increased income from ecotourism may result in social inequalities in remote communities (Gurung and Seeland 2011). In an investigation of the contribution of ecotourism to nature conservation and livelihood improvements around a nature reserve in Tanzania, it was found that few benefits accrued to local residents (Shoo and Songorowa 2013). As Cater (2004) points out, the ideals of ecotourism have met with harsh market realities resulting in a considerable divergence between theory and practice, and ecotourism has frequently been misinterpreted, misappropriated and misdirected. Products are marketed as ecotourism when they do not meet the basic criteria of the concept. Writing 10 years earlier, Cater (1994) warned that ecotourism may share many of the same characteristics as mass tourism in terms of leakages, as much tourist expenditure is not made in the destination. Investigations in Panama suggest there are opportunities for cruise ship companies to partner with local ecotourism operators (Thurau et al. 2015), thereby raising the question whether this may lead to some form of mass ecotourism and to what extent the environment and communities can be protected if numbers increase drastically. Concerns with ecotourism in Ghana also relate to marginalization of local communities, dependence on international forces, neo-crisis representations (for example, locals unable to manage resources) and marginalization of local ecological knowledge (Eshun and Tagoe-Darko 2015). Citing the case of Belize, Cater (1994) notes there is a high degree of foreign investment in ecotourism, which has created inflationary pressures in the local economy. As sites become popular, the concentration of visitors leads to degradation and there is debate about whether ecotourists themselves are 'an environmentally sensitive breed' (Cater 1994: 76; Sharpley 2006).

Duffy (2006) raises a number of important issues in examining the politics of ecotourism in developing countries. In addition to issues of definition, there are concerns over how destinations that are marketed as undeveloped regions perpetuate a particular image of the developing world. Ecotourism resorts also rely on global networks of travel and global tour operators, raising issues of sustainability and power and control. Power issues are also evident in the fact that funding from international agencies may also come with strings attached and, at a local level, not all residents will benefit equally from an ecotourism resort (Duffy 2006). Duffy (2006) argues that ecotourism links to the promotion of neo-liberalism as the path to development. Jamal et al. (2006) argue that ecotourism has been developed along a modernistic and commodified paradigm and it needs to be reoriented to a social-cultural paradigm based on participatory democracy and meaningful relationships with the biophysical world. Community participation in tourism will be further explored in Chapter 5. With the shift towards pro-poor tourism, ecotourism has also received attention for its potential role in poverty reduction. While ecotourism employment has been found to contribute to local socio-economic development, increased household incomes, welfare and opportunities, the level of contribution varies between areas, different ecotourism operations and different communities (Snyman 2014). More generally, Sharpley and Naidoo (2010) argue that while tourism may generate immediate economic benefits to the poor, it may not present a long-term solution for poverty reduction.

A final trade-off considered here with respect to ecotourism is the financial aspect. While this sector is meant to protect and enhance the natural environment, it also dictates that lower numbers be attracted to the destination in order to protect the initial resource sought out by tourists. Lower numbers, unless they are high-paying tourists, result in smaller overall profits at the destination level compared to more conventional forms of tourism. On an individual basis, those directly involved in ecotourism will see an improvement in income; however, it may be difficult to ascertain how this contributes to overall levels of development for the entire community. If the destination becomes well known, not only tourists, but also those seeking a job in the industry, will visit the destination, causing a variety of impacts if not properly controlled.

Culture and heritage-based tourism

Cultures and historic monuments are part of the attraction in many developing countries (see Plate 4.6). Cultural attractions generate tourism flows and thereby present opportunities for locals to interact with and generate income from the tourists. Festivals, souvenirs, traditional dances, religion and local food are all part of the attraction. Smith (2007) suggests that cultural tourism is changing, and while there will still be opportunities to enjoy unique cultural experiences, cultural excitement is more likely to exist in cosmopolitan locations (for example, world cities) than in small villages. Whether cultures are associated with a state, a region or a specific ethnic group, it is important to consider how those cultures are portrayed, who has ownership over how the culture is presented, and whether or not locals are being exploited as culture is incorporated into tourism. Wood (1997) explores the relationship between the state, identity and tourism, and notes two examples of the state having a major influence in culturally related tourism.

Plate 4.6 Thailand, Bangkok: tourists visiting the Grand Palace Complex

Source: Photo by D. Telfer

In China, the government officially recognizes 55 ethnic minority 'nationalities' that mostly reside in officially designated autonomous regions ranging in size from a village to a province. The Chinese government designates official tourist sites and then determines which areas should be opened up to foreigners. In Singapore, the state and the tourism industry treat the country as having four categories of ethnicity (Chinese, Malay, Indian and Other). Wood (1997) comments that these four categories do not accord with the self-identity and lived experiences of Singaporeans and they do not reflect ethnic tradition. However, reinforcing these labels for the tourism industry raises questions of authenticity.

Authenticity has been widely discussed in the tourism literature (Hashimoto and Telfer 2007). In the context of tourism, 'authentic' has been used to describe products such as works of art, cuisine, dress, language, festivals, rituals, architecture, or everything that is part of a country's culture (Sharpley 2008). However, authors such as Timothy (2005) and Sharpley (2008) suggest that authenticity is a subjective notion that can vary from place to place, culture to culture and from person to person. Hughes (1995) illustrates political influences as follows: 'authenticity in tourism is held to have been produced by a variety of entrepreneurs, marketing agents, interpretive guides, animators, institutional mediators and the like' (Hughes 1995: 781). Tourism behaviour will be explored in greater detail in Chapter 6; however, tourists do have different motivations and some are more concerned with experiencing 'authentic' cultures than others.

Wall and Xie (2005) examined the Li dancers of Hainan, China, who are members of an ethnic minority employed to represent their culture to tourists. Their following study focused on authenticity and utilized five themes, which, for the purpose of this chapter, can be broadened to reflect on the role of cultural tourism and development. The five themes or continua include:

- spontaneity versus commercialism;
- economic development versus cultural preservation;
- cultural evolution versus museumification (the freezing of culture);
- ethnic autonomy versus state regulations; and
- mass tourism versus sustainable cultural development.

The authors found that there is a desire to celebrate and portray ethnicity through commodification, and that mass tourism is desirable as it leads to job security and hence economic prosperity. The commodification of culture is viewed as a positive mechanism in the pursuit of sustainable development, as it is seen as inseparable from economic development, which is desirable. However, the authors caution that the control of the commodification of culture and the associated benefits are not in the hands of the minority people. While cultures tend to evolve with or without tourism, how culture is used and who benefits will influence the overall contribution to development broadly defined. In 2011, for example, the government of Peru spent US$13 million on heritage mega-projects at Chan Chan, Chiclayo and Humachuco in an effort to attract more foreign visitors to the northeast coast (Herrera 2013). However, Herrera (2013) argues that open dialogue and cross-cultural consultation with people in the local communities in Peru is rare and the impact of outside institutions driving the tourism agenda can exacerbate existing intra-community tensions. A further discussion of the social and cultural impacts of tourism is found in Chapter 7.

Living cultures are often linked to historic monuments, which are increasingly being utilized as tourist attractions (Timothy and Boyd 2003). UNESCO designates World Heritage Sites for countries that are participating members (see Plate 4.7), as well as maintaining a list of World Heritage Sites that are 'in danger' due to a range of problems from site deterioration and management issues to natural disasters, climate change and armed conflict. World Heritage designation presents both opportunities and challenges for local communities. Designation may generate additional tourists but there is also the danger of attracting too many tourists, causing problems for sustainable development, as well as raising questions of control over the site. At the Taj Mahal in Agra, India, for example, Chakravarty and Irazábal (2011) found a set of paradoxes surrounding the site being designated a World Heritage Site, specifically identifying a restructuring of the tourism industry as a result of the World Heritage designation that did not lead to proportional advances in local community development. Evans (2005) examined the promotion of Mundo Maya linking the heritage sites of pre-Columbian civilizations with the all-inclusive Mayan Riviera resorts of Cancún and Cozumel in Mexico. Evans (2005) explored the power relations between the indigenous communities, the state and dominant groups, along with the marketing of the region. He found that there is an absence of genuine local community and cultural involvement in the heritage site management, and, in some cases, interpretation, all leading to a pattern of commodification. Evans (2005: 46) states it is perpetuating a 'system in which national authorities and power groups (including heritage intermediaries) effectively collude in international agency and corporate foundation programmes and development aid, using the dualistic heritage tourism and conservation rationale for intervention'. The World Heritage Site of Angkor, Cambodia, has a rapidly growing tourism industry, and Mackay and Palmer (2015) also stress the importance of the intangible heritage of this 'massive lived-in sacred space'. They argue the site provides an opportunity to examine the interaction between human rights, intellectual cultural property, heritage management and ethics. They advocate using a wide variety of approaches to ethics rather than those rooted in Western philosophy in the practice of decision-making in cultural heritage management. As with any tourism development, steps need to be made so that locals are incorporated in the planning of the site so they themselves do not become marginalized and only serve to be part of the tourism product and not the beneficiaries. It is important to note also the economic and political realities in these locations to understand the barriers and constraints to local participation.

Plate 4.7 *Tunisia, El Jem: Roman Colosseum, UNESCO World Heritage Site*

Source: Photo by D. Telfer

Tourism development outcomes: the need for increased linkages and participation

The final element in Figure 4.1 is the development outcomes as a result of tourism. What, then, can tourism do to contribute towards social, economic and environmental goals of development? Which of the above tourism options just discussed, for example, could a destination select as a viable development option? Page and Connell (2006) suggest that tourism development has the potential to assist with development in terms of poverty reduction and employment growth. However, if tourism is viewed from a broader political dimension, then tourism can transform an economy in the following ways (Page and Connell 2006). Tourism can be a way of obtaining hard currency to improve the balance of payments and indebtedness. It can be a catalyst of social change establishing greater contact between the indigenous community and the tourists. Tourism can also be seen as a symbol of freedom, whereby citizens are allowed to travel freely within and outside their country. The industry can be a mechanism for improving local infrastructure to serve both the tourists and locals. Tourism can be an integral part of economic restructuring through privatization, as well as exposure to transnational corporations and national and international markets. Finally, tourism can complement commercial development through the growth of business tourism and encouraging small-scale entrepreneurial activity (Page and Connell 2006). If planned and managed correctly, tourism can also contribute to sustainable development, underlying the need to preserve and protect resources for future generations.

What Page and Connell (2006) have outlined is perhaps an ideal that a developing country can aim for. Tourism development also comes at a cost, and so, in reality, achieving these goals is much more difficult. This chapter has outlined the tourism development process. The agents of tourism development, including the state, private corporations (both domestic and multinational), not-for-profit organizations, and related financial and technical planning organizations, as well as tourists themselves, all have their own values, ideology, goals, priorities, resources and strategy for tourism development. The state provides the framework for tourism development to occur in the destination by establishing policy, creating plans for the destination and outlining any specific regulations governing tourism development. The policy and planning environment is a very political environment where values of the decision-makers will greatly influence the type of tourism that actually goes forward for construction. Will the state support casinos or beach resorts or ecotourism? Will it choose to diversify and build tourism based on a number of different product options? The political structure in the destination, and degree to which the state can act in a more unilateral fashion and whether there is a more democratic process involved allowing various groups and individual citizens to voice their support or opposition, will also influence development. Tourism plans created by the state cover different geographic scales (local to national) to guide development; however, there can be competing interests between various levels of the state, as well as between the state and private interests. As Reid (2003: 134) suggests:

> often, local communities are assumed to be represented by their local governments, but, as we have seen, national governments have their own agenda, usually having to do with earning foreign exchange in order to pay down debt held by foreign banks, but not necessarily related to issues like regional development and income generation at the local level.

In this age of globalization, private corporations will look to the destination that offers the most support and incentives and the fewest barriers. Tourism firms will conduct their own corporate planning, and will, in most cases, locate their product where they can get the best value for their investment dollar.

Once a tourism project winds its way through the filter of policy, planning and politics, construction occurs (ranging from large-scale to small-scale, enclave to integrated) and its potential to contribute to development is largely based on the level of linkages to the destination economy. The more the development project is linked to the local economy though purchasing products locally, hiring local people, utilizing local services, involving area citizens in the planning process, and reinvesting in the area through infrastructure, the greater the net benefit will be. Many destinations choose to diversify their tourism offerings in the hope of attracting more tourists and to potentially expand the linkages to local communities. Part of the challenge is to what extent the local economy can be integrated into the tourism sector and to what degree locals can participate. If local products cannot be obtained in sufficient quality and quantity, the challenges become greater. If, however, there is potential for increased backward linkages, efforts need to be made (or in some cases regulated) to enhance these partnerships. In the case of the hotel industry in China, Zhang *et al.* (2005) identified that there is significant potential for outsourcing, yet the market is still immature due to structural barriers in place. Referring to a study of hotel development in Shanghai, the authors indicated that outsourcing was occurring in the areas of a shopping arcade, recreation centre, flower kiosk, beauty salon, restaurants, dance hall, karaoke parlour, sauna, as well as some housekeeping and gardening departments. If the local community is able to successfully fulfil the outsourcing contracts, the backward linkages will continue to be enhanced. As economic growth from tourism occurs, there is potential for it to be funnelled into other areas of the destination society to help promote the broader social and environmental developmental goals. If, however, there are limited connections to the local economy and profits are repatriated out of the country, there will be limited potential for tourism to contribute to the development process. Telfer and Wall (1996) summarized from the literature the difficulties in locally sourcing agricultural products in developing countries for hotels. The main challenges with supply were with respect to quality and quantity of product, yet once linkages are established, the other challenge becomes making these partnerships last. Other barriers, such as tourism being controlled by the local elite or the government using the funds from tourism to only service international debt, will also have a negative effect on development outcomes.

Not only do the economic linkages need to be made; there is also a need for greater involvement of locals in the tourism planning process. However, as identified by authors such as Liu and Wall (2006), without adequate human resource management locals may not have the skills and knowledge to participate. They also suggest that 'many, perhaps most communities in the developing world may require an outside catalyst to stimulate interest in tourism development and external expertise to take full advantage of their opportunities' (Liu and Wall 2006: 160).

Planning and regulating for development

As illustrated in the previous chapter, there are competing arguments over the role of the state in the economy and in planning. Through setting policies, plans and regulations, the state can direct tourism to promote development. The direction of tourism planning has changed over time and sustainable development has come to the foreground. According to Costa (2006), sustainability, competitiveness, profitability, self-sustained development, governance, public–private coordination, partnerships, and local involvement are becoming priorities in tourism planning. Added to this is the shift towards utilizing tourism planning to reduce poverty and the beginnings of the recognition that climate change is going to have a major impact in some developing destinations (see Chapter 8). The strategies selected to develop tourism can range from facilitating the development of large-scale growth poles to encouraging local entrepreneurial development or placing restrictions on development in certain areas to protect fragile resources. Tourism can be developed as part of a regional development strategy or for revitalizing an urban or rural area. Hall and Jenkins (1998) outlined a range of tourism development policy instruments for rural tourism; however, as illustrated through some of the examples added below, they can be adapted to other locations and scales. The first category is regulatory instruments, which involve laws, regulations, permits and licenses. Laws, for example, can be passed to zone an area suitable for cruise ship development or a resort complex. The second category is voluntary instruments, which include providing technical assistance, or assisting volunteer associations and non-governmental organizations. Governments may support a regional tourist organization or a heritage conservation group. The third category is expenditure that includes spending money on specific activities, operating a public enterprise, public–private partnerships, monitoring and evaluation and promotion. A government may have to spend money on improving an airport or want to enter into an agreement with a private company to develop tourism in an area. The fourth category is financial incentives, including pricing, taxes and charges, grants and loans, subsidies and tax incentives, rebates and rewards and vouchers. As illustrated in Box 4.1, Mexico offers a variety of incentives to attract tourism developers. Economic incentives will attract developers. The drawback, however, is potentially giving up too much. The final category is non-intervention, whereby a state may choose not to get involved in order to achieve their objective. Recent developments in planning point to governments becoming more entrepreneurial, focusing on partnering with the private and non-profit sector to try to successfully brand their destination.

Within the options open to the state, regulating practices to promote backward economic linkages and improving or overcoming obstacles for local participation in the planning process are both options to promote development. Integrating tourism development with the local economy not only stimulates the formal tourism sector, but also stimulates the informal sector, which in some developing countries may be very important to overall development. Similarly, establishing programmes to enhance local entrepreneurs will help enhance the potential for them to participate in the economy. By enhancing opportunities for locals to participate in the planning process, new opportunities may open up, creating the prospect that more locals may participate in the benefits of tourism. The inclusion of the community in tourism will be further explored in the next chapter. The decision as to how much regulation to enforce is difficult; investment incentives may attract developers, while regulations may chase them away. Who ultimately controls the industry will determine, to a large extent, where the benefits go. As illustrated in the examples in this chapter, there is no one form (or forms) of tourism development that is suitable to all destinations, and no one set of plans, policies and instruments that can ensure success. It is also important to recognize that even with good plans, there are often problems in implementation (Lai *et al.* 2006). There is increasing awareness, however, that tourism cannot be planned in isolation, but needs to be integrated as part of broader development strategies within the context of sustainable development.

Discussion questions

1 How does government policy impact tourism development?
2 What are the advantages and disadvantages of different types of tourism developments?
3 What planning approaches can be used to maximize benefits to the local population?
4 What are the difficulties in implementing a sustainable development policy for a government and for a tourism corporation?

Further reading

Hall, C. M. (2008) *Tourism Planning: Policies, Processes and Relationships*, London: Prentice Hall.

The book provides a good overview of tourism planning and the linkage to sustainability. It explores planning at various geographic levels, as well as the importance of cooperative structures for planning.

Dredge, D. and Jenkins, J. (eds) (2011) *Stories of Practice: Tourism Policy and Planning*, Farnham, Surrey: Ashgate.

This book covers a range of international tourism planning and policy cases, highlighting a variety of issues. The book uses a storytelling approach and challenges readers to question existing research, wisdom and paradigms in understanding the influences of power.

Websites

www.fonatur.gob.mx/en/index.asp
The website for Fonatur in Mexico outlines their various tourism development projects along with investment incentives.

www.watg.com
This is the site of Wimberly, Allison, Tong & Goo, which is an international architecture, design, planning and engineering company with a focus on tourism. The web page has numerous case studies of tourism developments the company has been involved with.

5 Community response to tourism

Learning objectives

When you finish reading this chapter, you should be able to:

- understand the nature of communities
- recognize how communities and individuals respond to the introduction of tourism development
- be familiar with the links between community involvement and sustainable development
- evaluate different community-based tourism initiatives

There is a wide range of perspectives that can be taken on communities in the context of tourism. For some, the community can be considered as the main attraction and the gatekeeper to local knowledge, while for others the community is simply the setting where tourism occurs (Mowforth and Munt 1998). For others still, a community may, in fact, stand in the way of other potential tourism developments and therefore should be moved. Communities are increasingly being drawn into tourism not only from the demand side, as tourists actively seek out new destinations and communities to experience, but also from the supply side, as communities are becoming aware of the potential of the products they can offer to tourists and the economic gains that can be made. An important question to consider is who controls community-based tourism and whether the benefits from tourism go to the local people or whether they are controlled by the local elite or external tourism development agents, thereby exploiting the local community. Communities are not homogeneous and not all residents support the integration into tourism. Locals must not only deal with the constant attention of tourists, but also potential commodification of their culture (Boissevan 1996). With the shift towards sustainable development, there are increased calls for tourism to contribute in a positive way to the host community. Not only should tourism development protect and enhance local cultural heritage and environment, but also local residents need to have input in the tourism planning process and have greater participation in the benefits of tourism. With greater local control and by integrating tourism into the local economy through the use of local labour, products and resources, there is enhanced potential for tourism to contribute to the broader notions of development such as empowerment and greater self-reliance. In the context of indigenous peoples, Ryan (2005: 4) argues that:

> tourism is increasingly viewed not simply as a force for the creation of a stereotypical image of marginalised people, but a means by which these peoples aspire to economic and political power for self advancement, and as a place of dialogue between and within differing worldviews.

The purpose of this chapter is to explore the various relationships between tourism and local communities, focusing on how communities respond to tourism and some of the various concepts, programmes and agents that may promote development though the use of tourism. After examining the nature of communities and social capital and how communities respond to tourism, the chapter then focuses on the linkages between sustainable development and community-based tourism. Related issues such as participation and empowerment are considered. The power of communities in the face of tourism will be explored with the concepts of vulnerability and resilience and examples ranging from relocation to resistance. Finally, the chapter will focus on several current issues in tourism as they relate to communities, including the role of NGOs, fair trade, pro-poor tourism, volunteer tourism, and gender and community development. While all of these initiatives hold the promise of increased opportunities and benefits for local communities, they face numerous difficulties and challenges explored throughout the chapter.

Nature of communities

For some tourists, experiencing the local culture, language, traditions, lifestyles and natural environment are essential components of the trip. Increasingly, these elements are packaged and sold both by local inhabitants and by external agents under various labels such as cultural tourism, heritage tourism, indigenous tourism, village tourism and community-based ecotourism. This process is often referred to as commodification. However, debates continue as to whether the changes afforded by this type of tourism will forever change the host community in a negative way or whether in fact it may revive it. Modern advances in travel and the impacts of globalization have opened distant communities to outsiders, which in the past may have had little contact with the 'outside world'. Harrison and Price (1996) go so far as to argue that no community in the world now exists in isolation and few ever did. 'Willingly or not, all communities are part of nation-states with policies for all aspects for their inhabitants' lives; for education, health care, communications and security (social and otherwise)' (Harrison and Price 1996: 2). The implementation of these policies in part establishes the framework for tourism (Harrison and Price 1996). Not all destinations have effective governments, and in some locations national priorities to open remote areas will take precedence over local communities' wishes.

If communities in developing countries are increasingly becoming the focus of attention for tourism, what are the central attributes of the term 'community'? The concept of community is complex, and there have been various ways that communities have been contextualized. Rothman *et al.* (1995) conceive of community as the territorial organization of people, goods and services, and commitments that are important subsystems of society where locally relevant functions occur. Urry (1995) built upon the work of Bell and Newby (1976), who identified three main concepts of community. The first is a topographical sense, which can refer to the boundaries of a community. The second is a sense of community as a social system implying a degree of local social interconnection of local people and institutions. The third is a sense of 'communion', a human association implying personal ties and a sense of belonging and warmth. To these three concepts, Urry (1995) added the concept of ideology, which can often hide the power relations that inevitably underlie communities (Richards and Hall 2000a). With community linked to a geographic area, it is interesting to consider the diversity of destinations visited ranging from the rural hill tribe people in northern Thailand to the controversial urban slum tours of Kibera (the largest slum in East Africa near Nairobi, Kenya).

Richards and Hall (2000a) state that community has permeated the sustainability literature, and there are few sustainable tourism policies that do not refer to the importance of long-term benefits for the host community. An important point with respect to the nature of communities is made by Hall (2000), who notes that the emphasis on local, bottom-up approaches, often associated with sustainable development, appears to derive its legitimacy from an implicit assumption of the cohesion of local communities. However, communities are not homogeneous. They are made up of individuals and organizations who may well have different values, aims and objectives, and who may or may not adhere, in varying degrees, to the dominant traditions of the community. These differing opinions can lead to conflict and power struggles. The political realities in the destination may well reflect that it is the local elite that controls tourism in a destination rather than a truly grass-roots community-based initiative.

How individuals and communities respond to tourism will play a part in how receptive the area is to tourists. The geographic setting and the strength of the local culture have roles to play. The arrival of a busload of tourists into a remote village will have potentially greater impact than the arrival of the same tourists into an urban area that is used to seeing tourists on a regular basis. At a broad level, Wall and Mathieson (2006) suggest that tourism can be viewed as the interaction of three types of cultures: the destination culture, the cultures of the visitors' origins and the tourist culture. The tourist culture reflects that although tourists may have different backgrounds, they often use the same facilities, visit similar sites and exhibit common behaviours. The three different cultures are not all homogeneous and they interact in a diversity of ways (Wall and Mathieson 2006). The diversity of the tourist culture will be explored further in Chapter 6.

Social capital and communities

Social capital has become the focus of an increasing number of studies examining the nature of communities in relation to tourism development. Drawing on Hegel's work from the 1820s, Ottaway (2005) defines civil society as comprising the entire realm of voluntary associations between the family and the state, and these groups may try to provide services the state is not able to deliver. These voluntary organizations have many different forms and foci, ranging from small informal local groups focused on a single issue (for example, raising concerns over an environmental issue) to large bureaucratic international organizations with large budgets covering a range of concerns. NGOs, covered in more detail later in this chapter, have evolved into international NGOs (for example, CARE International) raising questions in terms of the global development paradigm as to whether there is a global social capital.

High levels of social capital indicate strong bonds and networking within a community. Putnam (1993) defines social capital as 'networks, norms, and trust that facilitate coordination and cooperation for mutual benefit'. From a development perspective, social capital consists of trust, reciprocity and cooperation (Flora 1998), and when these elements are strong in a community, the residents are more likely able to take advantage of economic community-building and capacity-enhancement opportunities (Park *et al.* 2012). Reviewing the literature on social capital, Liu and Lee (2015) found a number of different perspectives, and, in the context of community, social capital contributes social exchange, compensation and cooperation either collectively or independently (Park *et al.* 2012). In the tourism industry, it places emphasis on values, interactions with visitors or customers, and the creation of shared norms with cooperative partners, along with mutual trust and reciprocity (Jones 2005). Social capital then can be viewed as a resource derived from social relationships (Payne *et al.* 2011, cited in Liu and Lee 2015).

Strengthening social capital is important for development. In an era when public services are often being reduced, the ability of communities to support one another is critical. In some communities, the level of social capital is high and may be tied to strong traditions such as village life through cultural norms and expectations. While there may be strong community linkages within the community, as tourism evolves there will be a need for the community or entrepreneurs within the community to build social capital with those in the tourism industry outside of the community. Rastogi *et al.* (2014) studied social capital in two villages bordering the Corbett Tiger Reserve in India and to what extent they supported or opposed tiger conservation. They found that specific components of social capital, including solidarity, reciprocity and cooperation, networks and mutual support, were critical to understanding potential community action. A key finding was that there was potential for the tourism industry to disrupt social capital. The development of tourism around the tiger reserves resulted in an unequal distribution of benefits, undermining social capital. The authors argue that as the villages around the tiger reserve develop, the nature of social capital will transform, and there may be the need to formalize institutions to reinforce mutual trust to ensure the continuation of community action. Another criticism related to social capital is that it diverts the responsibility for development away from government to the voluntary sector (Mohan and Mohan 2002). As indicated in previous chapters, there is new emphasis on studying governance, and if the state does not support community development or abdicates its responsibilities, hoping the resource of social capital will offset a reduction in government services, then long-term development goals may suffer. In the context of developing communities, it is important to understand the level of social capital within a community and the potential for change as tourism develops.

Interactions between tourism and communities

Upon arrival in a community, tourists will encounter and interact with local residents; however, the nature of their trip and the type of accommodation and transportation they use will often determine the level of contact. The locals they encounter may originally be from the destination or may have migrated to the area in search of employment. Those that are employed in tourism or benefit indirectly from tourism may be positive towards the industry, while others may view tourism as an inconvenience or major problem to the community. Interaction may occur in the 'front stage', where a more formal service, performance or demonstration is given (see Plate 5.1), or the interaction may occur in the 'backstage', whereby the tourist is allowed a view into the 'real life' of the residents, which may or may not be anything like the formal performance. The increasing demand for tourists to engage in experiential activities is intensifying the interactions between tourists and locals. Smith (1977 and subsequent editions) explored the nature of the complex interactions that occur between 'hosts' and 'guests', and the resulting impacts. These impacts are addressed in greater detail in Chapter 7. In a review of the literature, Burns and Holden (1995) summarized some of the main concerns relating to the interaction of the international tourism industry with developing countries. Table 5.1 presents those concerns with corresponding concerns for communities.

Plate 5.1 *Argentina, Estancia Santa Susana near Buenos Aires: tour guide at a historic ranch presents traditional implements; refreshments to be served to the tourists are on the side table*

Source: Photo by Tom and Hazel Telfer

Table 5.1 *Concerns over tourism interaction in developing countries and implications for communities*

Concerns over tourism	Potential implications for communities
Tourism development creates 'islands of affluence' in the midst of poverty	Local communities cut off from potential linkages to the tourism industry; possible resentment; migration may occur to communities near tourism sites for employment altering community structure; unequal sharing of benefits
Scarce national resources used for the enjoyment of wealthy foreign tourists	Loss of local resources taken for tourists such as water, land; communities may need to adjust to new resource availability
Demonstration effect	Some members of the community may adopt tourists' behaviour patterns, turning them away from traditional patterns of behaviour; local community traditions may be under threat
Economic multipliers, the main tool for measuring economic impacts, are controversial and unreliable	Economic benefits may not be as large as first expected; distribution of financial gains may not benefit entire community
Commercialization of culture and lifestyles	Community change to focus on high consumption lifestyle; culture and cultural artefacts are turned into commodities for sale; may revitalize culture
Benefits are likely going to foreign companies or local elite	Communities are controlled by local elites or foreign companies; few benefits to the local communities; challenges for communities to benefit from the industry
Control of international tourism is external to the destination and defined by transnational tourism corporations	Loss of control; limited opportunities for communities to participate in the tourism planning and development process; if controlled locally, the community-based tourism still has to interact with the international corporations who supply tourists

Source: Left column, 'Concerns over tourism', after Burns and Holden (1995)

Although a number of concerns are outlined in Table 5.1, it is also important to note that tourism can also have positive impacts on communities, such as reviving culture, creating a sense of empowerment and generating income. In the face of globalization, there has also been a resurgence of regional or local identity, as mentioned in Chapter 4. Citing the work of Ray (1998), Richards and Hall (2000a: 4) state that 'regions on the periphery of the global economy are asserting their identity as a means of preserving their cultural identity and developing their socio-economic potential'. The selling of regional or local identity has become part of the tourism product. For example, Gerritsen (2014) found that for indigenous women in rural western Mexico, tourism offers the potential of an alternative livelihood strategy that allows for the re-localization of rural life threatened by globalization. Hall (2002) suggests there is an increasing recognition of the intellectual property dimension of tourism associated with regional characteristics and, as a consequence, destinations and communities are using strategies of 'place branding'. Local narratives provide insights into communities and are incorporated as part of culture into place branding (Lichrou *et al.* 2010).

Response to tourism by communities

Responses to tourism in a community can come from a variety of sources, including individuals, community groups, business operators, non-governmental organizations (NGOs), environmental groups and the government, to name a few. Individuals may oppose tourism development while other community groups may band together to launch a community-based tourism project. In 1982, Mathieson and Wall published one of the first tourism texts, entitled *Tourism: Economic, Physical and Social Impacts*. In 2006, they published an updated edition of the book and added the terms *Change* and *Opportunities* to the title in order to highlight the fact that not only are communities impacted by tourism development, but they also respond to the changes that tourism can bring. The term *consequence* is adopted in the updated book to refer to the changes that occur as a result of tourism since it has fewer negative connotations than the term *impacts*. Communities often seek to attract tourists and tourism developers, and so development is often sought rather than imposed (Wall and Mathieson 2006).

In terms of the nature of communities, Wall and Mathieson (2006) stress that tourism takes many forms, and communities that embrace tourism have diverse characteristics. As a result, the consequences of tourism are highly contingent, reflecting the specific forms and locations. The authors conclude, therefore, that making generalizations about the impacts in a community is very difficult since both the type of tourism and characteristics of the community need to be taken into account before speculating on the consequences of tourism (Wall and Mathieson 2006). Murphy (1985) suggests that how a community responds to the opportunities and challenges of tourism depends, to a large degree, on its attitudes to the industry. He notes that attitudes are personal and complex; however, in terms of community attitudes, there are three main determinants. The first is the type of contact that exists between resident and visitor. The second is the relative importance of the industry to the individual and the community, and the third is a tolerance threshold. This is the resident receptiveness expected in relation to the volume of business a specific destination can handle. Increasingly, attention is being drawn to how communities and individuals cope with tourism as well as community well-being. Jordan *et al.* (2015), for example, explored how residents in Jamaica responded to the development of a new cruise ship port. They argue that large-scale tourism developments do cause stress for residents, but this may be reduced if there is input from stakeholders and if individuals utilize more problem-coping responses through established communication channels as opposed to less effective emotion-focused coping responses.

Models of attitudes towards tourism

The attitudes of residents towards tourism is one of the most systematic and well-studied areas of tourism (McGehee and Andereck 2004; Sharpley 2014). Early studies adopted the perspective that communities were relatively homogeneous places, while later research recognized that communities are heterogeneous with a great variety of attitudes. Reaction to any tourism development will typically range across a continuum from acceptance to rejection and, according to attitude, certain behaviours may occur. Over time, a number of models have been developed to articulate the responses of communities to tourism. Doxey (1976) suggests that attitudes towards tourism in communities go through a series of stages, which include euphoria, apathy, irritation, antagonism, and a final stage when a community is undermined and what attracted tourists initially no longer has the same attraction. The interaction between 'hosts' and 'guests' gradually becomes more formalized over time. This model, however, has received criticism on two fronts. The first is the inevitability of attitudes moving from positive to negative when the opposite may be true, and, second, it may be misleading for communities to have dominant attitudes (Wall and Mathieson 2006). Butler (1975) adopted the attitude framework on cultural interaction from Bjorklund and Philbrick (1972) to tourism, which indicates that the attitudes and behaviours of groups or individuals will be either positive or negative and active or passive (Wall and Mathieson 2006). As an example, entrepreneurs who are financially involved in tourism may aggressively promote the industry, while a small but highly vocal group uninvolved in tourism may lead aggressive opposition (Wall and Mathieson 2006). A key consideration here is to what extent opposition is permitted in differing countries. Dogan (1989) conducted a cross-cultural study of European tourists vacationing in Turkey and examined the response to tourism ranging from active resistance to adoption of Western culture. Dogan (1989) proposed the following categories: resistance, retreatism, boundary maintenance and adoption. In a similar fashion, Ap and Crompton (1993) examined interactions of tourists and locals from the same culture, and they devised the embracement–withdrawal continuum, where responses fell into one of the following four strategies: embracement, tolerance, adjustment and withdrawal. These categories are further explored in Chapter 7 on the impacts of tourism (see Figure 7.6). The theoretical base for many studies has been social exchange theory, which suggests that people evaluate an exchange based on costs and benefits incurred as a result of the exchange (McGehee and Andereck 2004). Therefore, those residents who perceive themselves as benefiting from tourism will most likely view it positively, while residents who perceive themselves as incurring costs will be negative towards tourism. McGehee and Andereck (2004) do, however, note that there has been mixed support for social exchange theory in the studies of residents' attitudes.

Entrepreneurial response to tourism and capacity building

It is often those residents or communities who see the potential economic benefits of tourism that will take initiative and start a tourism enterprise. Entrepreneurship can be found at all scales, from an individual operating in the informal sector as a beach vendor, to a community binding together to offer village tourism, to the development of a new domestic hotel chain or to government actions, such as those in Dubai reorienting the economy around tourism development. It is important to remember that the majority of tourism enterprises are small to medium-sized enterprises (SMEs). Brewer and Gibson (2014) refer to many of these individuals as 'necessity entrepreneurs' who have no choice for generating income other than to start a small enterprise. In examining night market operators in Taiwan, Liu and Lee (2015) found that they exhibited the entrepreneurial orientations of risk-taking, innovativeness and opportunity discovery, and that social capital among the operators was a valuable asset to success. They found that 'networking relationships, reciprocity, trust, and social norms' could be used as marketing tools, which together can strengthen competitiveness (Liu and Lee 2015).

Reflecting the shift towards poverty reduction, governments and development organizations have started to focus their development policies towards indigenous small enterprises in an effort to promote economic independence, capacity building and community empowerment (Zhao 2009). Building community capacity has become a key element in tourism planning and development (Moscardo 2008). In Ghana, the government is focusing on developing the tourism industry and, specifically, on female entrepreneurship for the country's socio-economic advancement (Elijah-Mensah 2009). Also researching tourism in Ghana, Koutra and Edwards (2012) argue that capacity building can make a significant contribution towards developing viable and socially responsible economic activities. As individual entrepreneurs or more community-based entrepreneurial enterprises gain more experience, there is potential for local capacity building. There are a variety of different definitions and elements to capacity building, and these relate to strengthening human resources, individual and organizational capacity, developing appropriate facilities and training for tourism and evaluating the impacts of tourism (Bushell and Eagles, cited in Aref 2011). Koutra and Edwards (2012) refined a 2007 UNDP definition of capacity building to include investments in social, human, physical and financial capital, and it is the outcome of the interactions between the interests of individuals, businesses, networks, organizations and policy institutions at national and supranational levels.

Capacity building has been explored in the context of community-based tourism in South Africa, and it is suggested that the process should be expanded beyond developing tourism-specific technical skills. That is, capacity building should focus more broadly on individual and community empowerment, thereby encouraging more holistic community development (Giampiccoli *et al.* 2014). Government programmes supporting entrepreneurial activities and capacity building, such as microcredit loans, can be critical in helping small operators to start and gain access to regional and international markets. NGOs can also play an important role in capacity building, such as helping a village become known for handicraft production. Exploring IT capacity building in the tourism sector in the Maldives, Adam and Urquhart (2007) argue that a lack of information technology skill and human capital are significant obstacles to successfully implementing IT projects in developing countries. If local entrepreneurs and communities can bridge the digital divide and gain access to the Internet, the creation of their own websites promoting their tourism products can create direct links to international and domestic markets. These Internet sites increase the chance local operators can avoid the traditional tourism distribution system, thereby reducing economic leakages. There is, however, a number of barriers in tourism to community capacity building that relate to communities as a whole, as well as individual entrepreneurs. These barriers include:

● barriers at the individual level, including lack of skill, knowledge and lack of sense of community among individuals in the tourism industry;
● barriers at the organizational level associated with community organization, lack of external support, lack of resource mobilization and lack of community leaders;
● barriers at the community level, usually related to community factors, which include limited local participation, lack of community power in tourism decision-making and lack of appropriate community structure (Aref 2011: 350).

Other challenges for family-owned tourism micro business in the mountain destinations of Nigeria cover a range of issues, including access to start-up funds, seasonality, concerns over livelihood survival and who will take over the business in the future (Banki and Ismail 2015). Banki and Ismail (2015) contend that there needs to be greater government involvement in ensuring adequate funding to microfinance banks, as well as information provision on what funds are available. Overcoming the challenges in building community or individual capacity is a critical challenge for tourism and entrepreneurship in developing countries.

Sustainable development and community involvement in tourism

In a review of the literature, Choi and Sirakaya (2006: 1275) state: 'sustainable development for community tourism should aim to improve the residents' quality of life by optimising local economic benefits, by protecting the natural and built environment and provide a high quality experience for visitors'. As indicated in Chapters 2 and 4, there have been a number of calls within tourism for approaches that incorporate the local community in the planning and development process (for example, Jamal and Getz 1995; Dredge and Jenkins 2011b). Murphy (1985) argues that the industry holds great potential for social and economic benefits if the planning is redirected from a pure business and development approach to a more open and community-oriented approach that views tourism as a local resource. Jamal and Jamrozy (2006) proposed an integrated destination management framework that brings in the destination's ecological-human communities as equitable and integrated members of goal setting and planning-marketing. In the context of ecotourism, Jamal *et al.* (2006) call for a shift towards more of a social-cultural paradigm based on participatory democracy.

The importance of partnerships and the involvement of stakeholders have also been explored in the context of sustainable tourism (Jamal and Getz 1995; de Araujo and Bramwell 1999; Bramwell and Lane 2000). Partnerships are especially important in community-based tourism in developing countries where funds may be limited. Different organizations need to work together to effectively develop, promote and operate their product. De Araujo and Bramwell (1999) examined stakeholder assessment in tourism planning in 10 municipalities in Alagos State in the north-east part of Brazil. The area is economically poor and is part of a larger tourism plan that is seeking to use tourism as a regional development tool. The authors argue that with inadequate involvement of the affected parties, the potential for conflict and reinforcement of inequalities can increase. They also comment that acceptance of a plan may be enhanced even by those affected by it if they have been involved in the planning process. The challenge with any stakeholder involvement, as will be illustrated later, is managing a potentially time-consuming process.

Agenda 21 is the action plan adopted for sustainable development at the United Nations Conference on Environment and Development in 1992 held in Rio de Janeiro. This was a move, in part, to take the next step from the 1987 World Commission on Environment and Development, which produced *Our Common Future* and generated one of the most cited definitions of sustainable development. Chapter 28 of Agenda 21, which became known as Local Agenda 21, focuses on the role that local political authorities can take in introducing a comprehensive planning process aimed at promoting sustainable development in their locality (Baker 2006). Local authorities have specific and significant environmental management functions in the following areas:

- developing and maintaining local, economic, social and environmental infrastructure;
- overseeing planning and regulations;
- implementing national environmental policies and regulations; and
- establishing local environmental policies and regulations (Baker 2006: 106).

The framework of Local Agenda 21 establishes a possible path that promotes sustainability and encourages local involvement in the planning process. Government can act as a facilitator in the development of community-based tourism through the provision of start-up funds or training programmes. Subsequent international environmental conferences have re-emphasized the importance of Local Agenda 21. At the World Summit on Sustainable Development in Johannesburg in 2002, for example, a new toolbox of quantifiable actions was established to follow Local Agenda 21, titled Local Action 21 (Baker 2006). As extensive public involvement is an integral part of the Local Agenda 21 process, it helps establish a sense of purpose resulting in the community having new confidence in its ability to shape its future (Baker 2006). In resource management, there are increased calls for cooperative management, which involves sharing rights between government and civil society (Plummer and Fitzgibbon 2004). Jackson and Morpeth (1999) argue that Local Agenda 21 may be used as a potential mechanism for implementing sustainable tourism. The World Tourism Organization, along with the World Travel and Tourism Council (WTO 1996) also focused on the importance of local communities in their document *Agenda 21 for the Travel and Tourism Industry*. Jackson and Morpeth (1999: 33) suggest that 'sustainable tourism developments do offer genuine hope for marginalized communities, and we can find evidence of where those communities are being enabled to play a meaningful role in delivering tourism initiatives which are appropriate to those specific localities and regions'.

The importance of community has remained a key element of sustainable development with the shift towards poverty reduction in wider development agendas (see Chapter 1). This is evident in a range of subsequent international programmes and conferences, including the UN Millennium Development Goals and the UNWTO's ST-EP (Sustainable Tourism for Eliminating Poverty) programme. At the Rio+20 Conference in 2012, tourism was noted in the outcome document, *The Future We Want*, for its potential to contribute to sustainable development. One of the major themes for the 2012 conference was a 'green economy in the context of sustainable development and poverty eradication' (UNCSD 2012b). The subsequent Green Economy Report includes a chapter on tourism containing a number of recommendations for green tourism as the industry has a potential to create employment, support livelihoods and enable sustainable development (UNEP 2011). The UNWTO has also developed reports on tourism in small island developing states (SIDS), on how tourism can contribute to the UN Millennium Development Goals in a sustainable manner for the benefit of the host population (UNWTO 2012).

The goals of these various initiatives and their possible adoption in tourism are, in part, to promote community development. The United Nations tentatively defined community development as 'a process designated to create conditions of economic and social progress for the whole community with its active participation and the fullest possible reliance on the community's initiative (UN 1955: 6). Nozick (1993) developed a set of principles of sustainable community development, which include:

- economic self-reliance;
- ecological sustainability;
- community control;
- meeting individual needs; and
- building a community culture.

Richards and Hall (2000a) effectively make the connection of the sustainability efforts of local communities with the global: 'Local communities become not only important in terms of actions taken to preserve their own immediate environment, but also form part of wider alliances to preserve the environment globally (act local, think global)'. The authors also highlight the role of NGOs and other pressure groups having like-minded members who are environmentally aware as they themselves can be viewed as communities. However, even though there are continued calls for local involvement, Liu and Wall (2006) emphasize the importance of moving beyond rhetoric and understanding the capabilities of locals to actually participate. They call for an increased recognition of human resource planning: 'If tourism is really to be a "passport to development" and a means to enhance the lives of destination residents then greater attention must be given in tourism plans to their needs and capabilities' (Liu and Wall 2006: 169). Milne and

Ewing (2004) explore community participation in tourism and suggest that in the Caribbean and elsewhere, regardless of political systems, evolving stakeholder relationships and access to information technology, participation will not occur in a sustainable fashion unless people have the will and interest to take control of their own destinies. Later in the chapter, the barriers in terms of participation in the planning process are addressed more directly in the context of tourism.

Community-based tourism

Community-based tourism is one type of tourism that incorporates high levels of community involvement under the sustainability umbrella. It is often viewed at the opposite end of the spectrum from large-scale, all-inclusive, mass tourism resorts owned by corporations that have limited economic linkages to communities with, perhaps, some residents of the local community being hired in low-skill and low pay jobs (Hatton 1999). Community-based tourism is local tourism developed in local communities in innovative ways by various individuals and groups, small business owners, entrepreneurs, local associations and governments (Hatton 1999). Funding can come from a variety of sources, including international donors. The Asian Development Bank, for example, is providing funds through its Mekong Tourism Development Programme to community-based and pro-poor initiatives and, through this programme, funds have been provided to provinces in Lao PDR (Harrison and Schipani 2007). Community-based tourism can also be linked to some forms of indigenous tourism. Indigenous tourism is defined as 'tourism activity in which indigenous people are directly involved either through control/and or having their culture serve as the essence of the attraction (Hinch and Butler 2007: 5).

There are several goals of community-based tourism (Hatton 1999). The first is that it is socially sustainable. Tourism activities are developed and operated for the most part by local community members and participation is encouraged. In addition, the revenues are directed towards the community through various potential ways such as co-ops, joint ventures, community associations, businesses that employ local people, or to a range of entrepreneurs starting or operating small and medium-sized enterprises. The second major goal for community-based tourism is respect for local culture, heritage and traditions. It has been suggested that community-based tourism can also reinforce or rescue local culture, heritage and traditions. In addition, respect is implied for natural heritage especially when the environment is part of the attraction. To the north of the city of Yogyakarta, Indonesia, is the village of Bangunkerto, the site of a community-based agritourism initiative (see Plate 5.2). With the help of local government, the community switched their crops to higher-quality salak fruit and opened an agritourism site for tourists. Based on interviews, it was found that all aspects of the site are controlled at the local level and the operation has strengthened local identity (Telfer 2000).

Based on a brief survey of community-based tourism in members of the Asian-Pacific Economic Cooperation (APEC), Hatton (1999) identified a number of recurring themes. The first theme is focused on why community-based tourism started in the various destinations. The common factor identified is the expectation of economic gain, which

Plate 5.2 *Indonesia, village of Bangunkerto: the site of community-based agritourism project based on tours through a salak plantation; these women are laying stones by hand on the roadbed in order to prepare it for paving; this is being done so that tour buses are better able to get through the community*

Source: Photo by D. Telfer

in some cases was directly related to need. The second theme is leadership linked to the initiative from one person, a small group or, in some cases, the government. Cultural heritage is often one of the most important aspects of community-based tourism, and in many cases is the attraction for the tourists, and so it was identified as a third theme. The natural environment is also a key theme for many communities where tourists are drawn to experience the environment. See Box 5.1 for a discussion on community-based ecotourism in Cuba (see Plates 5.3 and 5.4). The fifth theme is that community-based tourism is linked to the growth of employment opportunities, particularly for women, young people and indigenous peoples. Finally, in community-based tourism, there is an emerging theme where corporations and communities are starting to work together.

In the context of indigenous tourism, Hinch and Butler (1996) discuss the key aspect of control of the industry: 'Whoever has control can generally determine such critical factors as the scale, speed and nature of development'. The authors have established a framework for indigenous tourism that looks at the level of control the indigenous people have over tourism enterprises and also whether there is an indigenous theme featured in the attraction. Attractions, services and infrastructure controlled by indigenous people and developed around indigenous themes represent the strongest manifestation of indigenous tourism. In 2007, the authors revised their framework and highlighted the central role of culture in the indigenous tourism system. Smith (1996) identifies four interrelated elements in indigenous tourism:

- the geographic setting (habitat);
- the ethnographic traditions (heritage);
- the effects of acculturation (history); and
- the marketable handicrafts.

As the culture itself becomes the attraction, there are concerns raised over authenticity as the culture is opened up to tourists. Indigenous culture can be sold to tourists through cultural performances and souvenirs, which may change over time to suit the tourists' needs. Yamamura (2005) examined the culture of the *Naxi* people in the old town of Lijiang, China, where new tourism-related merchandise is being produced in accordance with the traditional handicraft techniques. Smith (1996) notes that although there are conflicts among some indigenous people over the desirability of tourism in

Box 5.1

Community-based ecotourism in Cuba

Traditionally, local participation in the tourism planning and development process in Cuba has been limited due to government controls. With the collapse of the Soviet bloc, Cuba turned to tourism to generate foreign exchange. The government allowed foreign capital to be invested in joint ventures. With a lack of a participatory process, the government has been able to 'fast-track' beach resort development in places such as Cayo Coco and Cayo Guillermo, as well as other locations. However, the government has been attempting to diversify the product away from mass tourism towards more niche products such as health, nature and cultural tourism. This brings tourists in more direct contact with locals and the pressure for increased levels of participation is growing (Milne and Ewing 2004). One example of nascent local participation is La Moka Ecolodge (26 rooms), built in 1994 next to the community of Las Terrazas in Sierra del Rosaria, a biosphere reserve. The community-managed hotel was the idea of the Minister of Tourism; however the local community was consulted about the project. Promoted as a nature and ecotourism destination, the Tourist Map for the Reserve states the following:

> Las Terrazas Tourist Complex is a rural experience of sustainable development that combines 5,000 ha of forest in the heart of Sierra del Rosario Reserve of the Biosphere and a working community of 890 inhabitants, offers a unique product in which the endemic Flora and Fauna, hotel services and social experiences will give you memories to be treasured.
>
> (Hecho En GeoCuba 1997)

The hotel is small-scale and was designed to fit into the surrounding environment. In places, trees were left intact, such as the one that grows up through the main lobby of the hotel and out through the roof. Excursions into the biosphere include guided nature walks, birdwatching, trekking, hiking, horseback riding and mountain bike tours, as well as visits to the partially restored remains of a coffee plantation. Residents of the community have the opportunity to work in the hotel, or sell crafts to the tourists while others work in the park (Telfer 2001). New community facilities were built close to the hotel and contain a small range of services for the residents such as shops and a health clinic. The community is scheduled to repay the government investment ($6 million) over a 15–20 year period and 40 per cent of the profits go into a community development fund overseen by the neighbourhood Committee for the Defence of the Revolution, with 10 per cent going to the community health clinic (Milne and Ewing 2004).

Source: Telfer (2001); Milne and Ewing (2004)

Plate 5.3 *Cuba, La Moka Ecolodge: visitors approach the main entrance to the hotel; note the architectural design of the roof*

Source: Photo by D. Telfer

their home communities, along with conflicts over the previously mentioned changes to handicrafts, there is an economic element to be considered as it can provide jobs and income. The shift towards the dynamics of active involvement in indigenous tourism is the focus of Hinch and Butler's (2007) volume. In addition, rural and indigenous people's environmental ethics may differ from more urban counterparts and they may view environmental controls and protection as contradictory, limiting their development to satisfy the desires of urban sophisticates (Butler 1993).

While community-based tourism has many positive points, these relatively small-scale tourism operations do face a number of challenges. Cleverdon and Kalisch (2000) suggest that one of the major challenges facing community-based tourism is the competition and threat posed by large-scale resorts in the vicinity. With their resources and marketing techniques, larger operations have the potential to take business away from small operators. The key challenge for the tourism industry and policymakers is to find a way for large and small firms to coexist and work together as part of an integrated local economic development policy (Cleverdon and Kalisch 2000). This has been a focus of pro-poor tourism for the tourism industry to build stronger ties to neighbouring communities. Other challenges relate to the long-term viability of a community-based project and whether it becomes controlled by just a few. In an analysis of power, Coles and Church (2007: 7) suggest that while 'equitable, fair and locally empowering forms of tourism production, governance and consumption remain the aspiration, inevitably they require interaction among human beings; in other words they are political processes and they are the subject of power relations among constituencies'. It is also important to remember that while sustainable community-based tourism may put some of the control of tourism in the hands of the local people, they must still interact with the external agents of the tourism industry, which is largely responsible for bringing in the tourists in the first place. Jamal and Dredge (2015) caution against the proponents of community-based tourism who do not challenge or question tourism's role in community development, assume homogeneous power bases and do

Plate 5.4 *Cuba, Las Terrazas: located next to La Moka Ecolodge, the community of Las Terrazas is home to some of the people that work in the resort*

Source: Photo by D. Telfer

not fully recognize external barriers and constraints. Finally, Blackstock (2005) states that from a community development perspective, community-based tourism has three major failings:

- it seeks to ensure the long-term survival of the tourism industry rather than social justice;
- it tends to treat the host community as a homogeneous bloc; and
- it neglects the structural constraints to local control of the tourism industry.

Participation

If there is to be participation by communities in the planning process, it is important to consider what the actual level of participation is. In an analysis of Local Agenda 21 initiatives in developing countries, Baker (2006) highlighted the structural difficulties for participation. There also may be difficulties for those in power to release control and allow the locals into the process. What, then, are the different levels or forms of participation? Arnstein (1969) developed a typology or ladder of citizen participation with eight rungs. The bottom two rungs are manipulation and therapy. These levels are described as non-participation and are more about allowing power holders to educate or cure the participants. The third and fourth rungs of the ladder progress to tokenism, allowing the have-nots to be heard and have a voice. These rungs are informing and consultation, and citizens lack the power to ensure that those in power will heed their views. The fifth rung, placation, is a higher level of tokenism, as citizens can advise but the power holders still make the decisions. The last three rungs are levels of citizen power and increasing levels of decision-making clout. The sixth level is partnership, which allows for negotiation and trade-offs with those in power. The seventh rung is delegated power and the last rung is citizen control, where the have-nots have the majority of decision-making seats or managerial control (Arnstein 1969).

In addressing the limitations of the typology, Arnstein (1969) acknowledges it does not address the roadblocks to achieving genuine levels of participation. On the power holder side, they include racism, paternalism and resistance to the redistribution of power. On the have-nots side, they include inadequacies of the poor community's socio-economic political infrastructure and knowledge base. In addition, there are difficulties of organizing a representative and accountable citizens group in the face of futility, alienation and distrust (Arnstein 1969).

Limits to participation in tourism

While, in theory, communities should be involved in the tourism planning process in developing countries, there can be significant barriers. Tosun (2000) indicates that the concept of the participatory development approach originated in the developed world. Concepts, theories or models created in the developed world will face challenges when applied to developing countries. Tosun (2000) identified operational, structural and cultural limits to community participation in the tourism development process in developing countries. These limits are often a reflection of the prevailing sociopolitical, economic and cultural structures in many developing countries. At the operational level, obstacles included the centralization of public administration of tourism development, lack of coordination between involved parties, and a lack of information made available to locals in the tourist destination. Structural limitations involve institutional barriers, power structures and legislative and economic systems. More specific items identified within the structural limitations include the problem of persuading professionals to adopt participatory tourism development, lack of expertise, elite domination, lack of appropriate legal system, lack of trained human resources, relatively high cost of community participation and lack of financial resources (Tosun 2000). Finally, Tosun (2000) identifies cultural limitations as limited capacity of poor people and apathy and low levels of awareness in the local communities. These constraints are similar to what Timothy (1999) found in Yogyakarta, Indonesia, and he grouped them under the following categories:

- cultural and political traditions;
- poor economic conditions;
- lack of expertise; and
- lack of understanding by residents.

It is also important to consider the existing political structure of the destination to see if a more open process of community participation is allowed or possible. In the case of both Cuba (Milne and Ewing 2004) and China (Li 2004), significant barriers to a democratic form of community-based tourism planning are apparent. In Yogyakarta, Indonesia, Timothy (1999) found that in Javanese culture, one of the most apparent traditions is that of authority and reverence towards the people in positions of power or high social standing. This concept permeates from high-level political jurisdictions down to the village level and to the level of family. In the case of the village head, for example, villagers view that person as authority, and few would bypass the village head for receiving and giving advice, as this behaviour would be offensive and cause the village head to lose face. In overcoming some of the barriers to community-based tourism in the rural multi-ethnic village of Viscri, Romania, Iorio and Corsale (2014) found that it was existence of a local leader in association with external actors that facilitated the start of the community-based initiative. The challenge will be how much control the external actors have.

Empowerment

The concept of empowerment is difficult to define, yet it is embraced by a wide range of organizations with diverse social aims, as it is attractive and viewed as politically correct (Scheyvens 2003). Arai (1996) outlined the following five concepts related to empowerment based on the literature:

- it involves a change in capacity or control, or an increase in power and the ability to use power;
- it has multiple dimensions, including psychological, economic, social and political change;
- it is a multilevel construct with changes occurring within an individual, group or community;
- it should be understood from a holistic perspective combining the above listed dimensions and levels; and
- it is a process or framework that describes changes as an individual, group or community mobilizes themselves towards increased citizen power.

Scheyvens (2003) used the same dimensions (psychological, economic, social and political) of empowerment as Arai (1996) in the context of tourism, which are explained below. If a community is empowered economically, it will need to ensure access to productive resources in the tourism area. Scheyvens stresses this is particularly important in the context of common property resources. If a park is established, the local community needs to be able to benefit from the park, but if indigenous rights to harvest from the park are undermined, then they will not be empowered economically. Social empowerment occurs when a community's sense of cohesion and integrity is confirmed or strengthened by being involved with tourism. Social empowerment is increased if money generated from tourism is used for social development projects such as improving water supply subsystems or health clinics. Social disempowerment occurs if there are some of the negative social impacts sometimes associated with tourism such as crime, displacement from traditional lands or prostitution. Psychological empowerment is a reflection of a community's confidence in its ability to participate equitably and effectively in tourism planning, development and management. Tourism that is sensitive to cultural norms and respects traditions can be empowering; however, tourism that undermines local culture will have a negative effect. Scheyvens (2003: 235–6) suggests that 'feelings of apathy, depression, disillusionment or confusion in the face of tourism development could suggest that psychological disempowerment has occurred'. Finally, Scheyvens (2003) explores political empowerment. If community members are politically empowered, their voices are heard and should guide the tourism development process. The importance of diverse interests groups in a community is raised, and all voices need to be heard through democratic processes or along more traditional lines of communication.

Power of communities in the face of tourism

An externally controlled tourism industry can be a powerful force for developing countries to deal with. The forces of globalization have multinationals searching for the lowest production costs, resulting in governments in developing countries offering various incentives to attract international developers. In developing countries, much of the planning is top-down, with governments controlling the framework for the industry. Butler's (1980) model of resort development illustrates the various stages a resort could go through over time following the pattern of a product life cycle, rising and falling and perhaps rising again. In the early stages of the model, where there is limited tourism, it is suggested that communities have more control over development. As the destination evolves and external companies come into the destination, control is passed from the local community to external agents. Various levels of governments may also intervene to try and use tourism in the area as part of a regional or national development plan, which may mean relocation of communities, as illustrated below.

As tourism grows, changes and opportunities occur. Tourism also has the potential to have a positive impact in terms of promoting a sense of place and identity for a local community. Richards and Hall (2000a) state that communities are not simply victims of the globalization process and commodification, but they can become centres for resistance, as is discussed below. James (2006) suggests that one of the dominant trends in the present period is the tension between globalism and localism, and interest in local differences is also what attracts tourists to locations. The promotion and marketing of local or regional identity in terms of tourism can generate employment opportunities, as is illustrated later.

Vulnerability and resilience to tourism

The complimentary concepts of vulnerability and resilience have come to the forefront in wider examinations of sustainability outside of tourism as ways to explore a more holistic perspective related to social and ecological domains (Larsen *et al.* 2011). Vulnerability explores the characteristics of exposure, susceptibility and coping capacity shaped by dynamic historical processes, differential entitlements, political economy and power relations as opposed to the direct outcome of a stress (Larsen *et al.* 2011). Resilience is the ability of a system to absorb shocks so as to avoid crossing a new threshold to an alternate state (which may be irreversible) and then to regenerate once a disturbance has occurred (Resilience Alliance 2009, cited in Larsen *et al.* 2011). Within the context of tourism in developing country communities, what is the vulnerability and resilience to the introduction to tourism? For example, in exploring tourism in the context of social inclusion/exclusion of the porters in the Peruvian Southern Andean communities along the Inca Trail to Machu Picchu, Arellano (2011) found that the sporadic employment opportunities could not justify the mistreatment of porters. The porters' enduring poor health, of which the tourism industry was well aware, raises fundamental problems about endemic inequalities and injustice associated with the porter communities. To what extent are porter communities or other communities vulnerable to the external shock of tourism or other sources of stress, and how can resilience be built? Tourism scholars have been slow to adopt the recent conceptual ideas on community resilience, although there has been some movement towards examining resilience centred on complex adaptive systems and disaster planning and relief (Lew 2014). Lew (2014) stressed the importance of understanding community resilience and proposed the SCR (scale, change and resilience) model that considers the rate of change (slow to fast) and scale of tourism (from individual entrepreneur to community), which result in distinct resilience issues, methodologies and measurements. Similarly, Calgaro et al. (2014: 341) developed a Destination Sustainability Framework that incorporates vulnerability and resilience in the context of Thailand's tourism-dependent coastal communities with the following components:

1   the shock(s) or stressor(s);
2   the interconnected dimensions of vulnerability – exposure, sensitivity and system adaptiveness;
3   the dynamic feedback loops that express the multiple outcomes of actions taken (or not);
4   the contextualized root causes that shape destinations and their characteristics;
5   the various spatial scales;
6   multiple time frames within which social-ecological change occurs.

Understanding vulnerability and resilience as well as building community resilience is important for development. In Dominica in the Caribbean, Holladay and Powell (2013) examined residents' perception of the social, institutional economic and ecological resilience of their community. To build resilience, the authors argue that communities should invest in

> strengthening social bonds, developing capacity in local institutions, in diversifying the tourism product and controlling infrastructure development. Indicators measuring trust, networks, local control, flexible governance, leakage prevention and controlled infrastructure development emerged as important in assessing social-ecological resilience and sustainability.
>
> (Holladay and Powell 2013: 1188)

Relocation

In some cases, entire local communities have been displaced to make way for new tourism developments, with some receiving compensation while others do not. Long (1993) explored the complex challenges and the differing reactions in the seaside community of Santa Cruz Huatulco in Mexico (population 735 with 250 households) to the relocation of their community 1 kilometre inland. This was done to make way for the Las Bahais de Huatulco, a master-planned mega-resort development project by Fonatur. Fonatur employed a number of social impact mitigation techniques; however, Long (1993) found that the original residents experienced changes in every aspect of their community. These included an influx of outsiders, a shift from an ocean environment to cement houses on dusty streets, development of new social classes, new occupations groups and new organizations. As with any controversial development project, Long (1993) also commented that there were those who supported the project. This finding reinforces the fact that communities are not homogeneous and it is not correct to portray communities as simply the victims of globalization. Wang and Wall (2005) explored the consequences of tourism-caused displacement of a minority community in Hainan, China, where tourism is being used as a regional development strategy and is being planned in a top-down manner. During the preparation period for the resettlement, the people had expressed great stress about the displacement and rejected additional risk-taking activities. The actual relocation distance was short and, along with the integrated resettlement pattern, resulted in no obvious external settling change in this relocation. After the move to the new village, stress and conservatism decreased, resulting in more innovative behaviours by the residents. Wang and Wall (2005) argue that if those being resettled are to get their share of benefits to tourism, improvements in planning are required, as well as training opportunities and greater access to jobs. The pressure group Tourism Concern has run campaigns to raise awareness of the displacement of locals due to aggressive tourism development, focusing on communities including the Maasai and tribal peoples of East Africa, coastal communities of Kerala and Tamil Nadu in India, and the 2004 tsunami-displaced coastal communities in Sri Lanka. After the 2004 tsunami, fishing communities in Sri Lanka were relocated as land was set aside as buffer zones against possible future tsunamis; however, the land was later sold for tourism development (Tourism Concern 2014a).

Resistance to tourism

Kousis (2000) explored resistance movements to tourism and examined local environmental mobilizations against tourism developments in Greece, Spain and Portugal, while Routledge (2001) examined local resistance to tourism in Goa, India. In Goa, the expansion of luxury and charter tourism has increasingly involved investments from transnational corporations. The emphasis has shifted away from backpackers and charter tourism to luxury tourism as it brings higher-spending tourists. Routledge (2001) states that the development of Goa as a tourist site has had serious consequences for the state's people, ecology and political economy, and has transformed the coastal areas into dispensable space. Economic growth is considered of prime importance, and environmental problems are secondary to growth (Routledge 2001). The situation led to the emergence of a number of Goan opposition movements, including citizens, activist village groups and environmentalist non-governmental organizations concerned about tourism development. Routledge (2001) examined two organizations, including the Goa Foundation and *Jagrut Goenkaranchi Fouz* (JGF), or Vigilant Goan Army, and identified a number of resistance examples that ranged from information dissemination to village mobilizations and protests, demonstrations and roadblocks. In China, as elsewhere, tourism is being used to promote economic and cultural development in remote rural and ethnic regions, which has affected ethnic autonomy and increasing involvement of higher levels of government in local issues (Cornet 2015). Cornet (2015) found that the Dong villagers of Zhaoxing in Guizhou province expelled the Han-managed tourism company from their community. By expelling the company and defending how they wanted the Dong culture to be represented to tourists, the villagers reaffirmed their distinctiveness from the Han and also reaffirmed their desire for economic development on their own terms. Cornet (2015) calls for more sophisticated, ethnically rooted and locally infused understanding of social change incorporating agency in terms of resistance and everyday politics. Boissevan (1996) highlights that local residents have devised a range of strategies to prevent tourists from penetrating their back regions, which include:

- covert resistance – low-key resistance of sulking, grumbling, obstruction, gossip;
- hiding – withdrawal from tourists to celebrate;
- fencing – fence off private areas/rituals;
- ritual – resurgence in celebrations helps protect against impacts of tourism and helps cope with change;
- organized protest;
- aggression.

A final example here of protest is the 2015 clashes by those in Hong Kong against shopping tourists from mainland China. The protestors argue that mainlanders are exploiting visa rules in order to buy high-quality goods at lower prices (BBC 2015).

Tourism and the informal sector

Part of the interesting dilemma of tourism development that has been highlighted in previous chapters is the debate over resort enclaves targeting mass tourists. While these types of developments have been criticized for being isolated, in some cases it is the existence of these resort complexes that enables small-scale community-based projects to open, taking advantage of the tourists already in the destination. The informal sector plays a large role in developing countries. In the context of urban areas, Pal (1994: 79–80) suggests the potential for:

> small-scale informal sectors to provide employment and generate income for countless urban dwellers in burgeoning cities of the developing world is considerable. Indeed, given the surplus of labour and high levels of poverty in urban areas, the encouragement of these sectors should be a cornerstone of any policy promoting sustainable development.

For tourism in rural or urban areas, the response of the informal sector will be important for the overall economic impact of the industry. In studying tourism employment in Bali, Cukier (2002) found evidence of migration to the tourism area of Bali from other islands and those in the informal sector saw their current employment as the means to acquire the skills to move to the formal sector. Citing studies in Yogakarta, Indonesia, and elsewhere in South Asia, Shah (2000) comments that domestic or regional tourists tend to buy more from local vendors than Western tourists. These findings further illustrate the importance of domestic tourism in development plans.

How communities and individuals respond to tourism, whether through resistance or trying to take advantage of the possible employment opportunities, will, in part, dictate the role that tourism can play in development. The chapter now turns to examine several current issues in tourism as they relate to communities including the role of NGOs, fair trade, pro-poor tourism, volunteer tourism, and gender and community development. As mentioned in the introduction, some of these various concepts, programmes and agents may promote development through tourism yet they all face significant challenges.

Role of NGOs in community tourism

Non-governmental organizations (NGOs) are playing an increasing role in influencing tourism. Baker (2006: 9) defines NGOs as 'organisations operating at the national and increasingly, at the international level, which have administrative structures, budgets and formal members and which are non-profit-making'; however, they can also operate at a very local scale. As discussed above, a key resource of NGOs is social capital. NGOs have taken on a number of different roles within developing countries, including providing development relief, raising awareness over specific issues such as environmental concerns or sex tourism, lobbying governments, assisting local communities with projects, and assisting building community capacity to name just a few. Some NGOs also help coordinate volunteer tourism, which will be discussed below. NGOs are also involved in the movement towards 'new practices of environmental governance' (Baker 2006). Environmental concerns such as climate change, biodiversity loss and deforestation have resulted in issues that cross political borders, raising questions about traditional environmental governance practices. The new practices have NGOs alongside states and international organizations, as well as the utilization of a wide range of policy instruments (legal, voluntary and market instruments) and normative and governance principles to promote sustainable development (Baker 2006).

Tourism Concern has had projects focused on raising awareness over the poor treatment of employees in the tourism sector in developing countries, issues of human rights abuses and tourism, promoting ethical and responsible tourism, and campaigning against specific tourism developments for environmental concerns, to name a few. The Ecumenical Coalition on Tourism (ECOT), based in Hong Kong, has a number of campaigns against the abuses of the tourism industry. TEN (Tourism European/Ecumenical Network) is a member of ECOT and has 11 NGO members concerned over unjust practices in tourism in developing countries. Equations, based in India, is a research, campaign and advocacy organization that focuses on the question: 'Who Really Benefits from Tourism?' In the context of communities, Burns (1999a) suggests that NGOs can often act as a bridge that promotes cooperation within communities, establishing initial links with the local and regional government tourism sector to form partnerships. While some NGOs work across international borders, many are very small and work in specific countries or communities. NGOs that traditionally focus on a specific issue are increasingly cooperating with each other to achieve a common goal. Strauß (2015) documented the alliance between various Balinese NGOs to block a major tourism development on the island. Barkin and Bouchez (2002) documented the work of the Centre for Ecological Support (CSE) on the Pacific coast of the Mexican state of Oaxaca. The area is the location of a mega-resort developed by Fonatur, the Mexican Tourism Development Fund. Approximately 30 km of the coast was expropriated by Fonatur. The authors state that the tourism development and accompanying infrastructure integrated the previously isolated community into the international market, sparking a self-reinforcing cycle of speculation and investment. This accelerated process of social and spatial polarization impoverished the native populations and created tensions. The community was also hit by Hurricane Paulina in October 1997. The CSE was originally created in 1993, after the first large hotels were inaugurated in the new resort, to promote regional development. Work had begun on a resource management plan for sustainable development, and the NGO worked with native communities to regenerate smaller river basins to promote community welfare, including replanting trees with commercial and cultural value. The reforestation was complemented by a variety of initiatives, including ecotourism. By creating a favourable environment, they would possibly attract tourists. The ecotourism project was designed to be owned and managed by indigenous communities participating in the programme and sensitive to the natural heritage they were rescuing and protecting.

One of the main questions for this book is whether tourism can contribute to development broadly defined. Archabald and Naughton-Treves (2001) examined tourism-revenue sharing around national parks in Western Uganda with neighbouring local communities. In the case of Bwindi Impenetrable National Park, two international NGOs (International Gorilla Conservation Programme (IGCP) and Cooperative for Assistance and Relief Everywhere (CARE)) worked with the Uganda Wildlife Authority in the initial stages of the tourism-revenue programme on things such as strategic planning assistance and workshops, as well as technical, logistical and monetary support for community training. Mountain gorilla ecotourism was part of the park's offerings. The tourism revenue-sharing programme funded community development projects in 19 of the 21 parishes bordering the park, and each parish received approximately US$4,000, with the money going to building primary schools, health clinics or roads. While the authors note the programme has been suspended, it does highlight the possibilities of tourism revenue-sharing programmes for some of the broader aspects of development.

There are a variety of other international NGOs, such as the WWF, IUCN and Conservation International, that have been involved in a number of environmental and conservation projects, including tourism and ecotourism. The African Wildlife Foundation is an example of an NGO that has evolved over time, illustrating how nature conservation has transitioned. The focus has shifted from 'conservation logic' that focused on nature protection to 'development logic' focusing on community consultation and participation, and finally to 'market logic', which promotes private sector involvement (Van Wijk *et al.* 2015). On a cautionary note, however, Mowforth and Munt (1998) raise the concerns that through coordinating links to the World Bank, some international NGOs have adopted an approach advocating corporate schemes for environmental projects, giving 'total management control' to the private or NGO sector. The rise of NGOs also illustrates the changing nature of governance, as states are not only dealing with multinational corporations, but also with, in some cases, international NGOs.

Fair trade in tourism

Local communities face challenges trying to interact with the highly competitive and international nature of tourism. The practice of fair trade has often been associated with the production of foodstuffs (coffee, for example) and handicrafts, and is often linked to aid programmes in developing countries (Evans and Cleverdon 2000). The idea is that local producers in developing countries receive not only support, but also a fair price for the products they produce. If this concept can be translated into tourism, local communities will benefit in a number of ways. Evans and Cleverdon (2000), citing the work of Barratt Brown (1993), outline the main concepts of fair trade as:

● improving working conditions;
● improving production and marketing of goods;

- implementing premium pricing, training and investment;
- minimizing economic leakage;
- widening distribution of economic benefits;
- guaranteeing price stability; and
- guaranteeing more sustained income.

Fair trade can also be certified around business operations, including labour standards, skill development, employment equity, community involvement and environmental management (Vatikiotis *et al.* 2011). Fair Trade Tourism (FTT), based in South Africa, promotes responsible tourism in southern Africa and offers a fair trade certification programme (FTT 2015). Fair trade is distinguished from free trade as it fights against poverty in developing countries and brings the consumer in contact with the producer. This is especially so for tourism that occurs in local communities in developing countries. It is also different from free trade as it targets small-scale producers and the trading method is designed to strengthen the partner's position in developing countries. This includes assistance from developed countries in product development, control over image and representation and access to marketing techniques (Cleverdon and Kalisch 2000). It is important to keep in mind that the tourism industry consists of predominantly small and medium-sized enterprises (SMEs) (Richards and Hall 2000b). These may range from a family opening their home to tourists in village tourism or an individual waiting at a historical monument for tourists to arrive to act as their guide. The tourism industry relies on innovation, and so, as noted above, entrepreneurs are an important part of the industry and can represent a crucial alternative to transnational companies (Richards and Hall 2000b). If these small firms receive the main benefits from the fair trade process, which include 'transparency of trading operations, commitment to long-term relationships and the payment of prices that reflect an equitable return for the input provided' (Cleverdon 2001: 348), they will be better able to compete in the tourism market.

Although there is a great deal of opportunity in this area, there are a number of differences between trade in primary products, such as coffee, and tourism that will present challenges to implementing fair trade in tourism. Some of these differences are listed below:

- Fair trade organizations are non-profit while tourism organizations operate in a highly competitive market seeking profits.

- The tourism product is intangible and invisible, and involves multi-sector activity.
- The success of tourism depends on low, flexible prices while primary products such as coffee have a 'world price'.
- Developing countries are used to exporting primary products; however, in many grass-roots communities, there is little experience using tourism as an export item.
- Tourism is typically based on competitive entrepreneurship, and the collective organization of small-scale tourism providers is a new concept.
- Tourism is consumed in the destination, bringing social and cultural intrusion (Cleverdon and Kalish 2000).

While these differences may provide challenges for larger tour operators to meet, there may be opportunities for smaller operators to create a strategic advantage by incorporating fair trade principles in their business operations (Chandler 1999, in Cleverdon 2001).

Pro-poor tourism

Pro-poor tourism, introduced in Chapter 2, is an approach to poverty reduction through tourism that targets the benefits of the industry to the poor rather than a specific type of tourism, as all forms of tourism can be made to be more pro-poor. Pro-poor tourism is revisited here for its potential links to community development. Since the 1992 Earth Summit, sustainability efforts had been heavily biased towards environmental considerations with tourism being seen primarily as a form of sustainable use of natural resources and a way to enhance conservation (Rogerson 2006). Rogerson (2006) argues that what was missing or downplayed was concerted attention to the impacts of tourism upon the poor and, until recently, few governments in developing countries considered linking tourism development directly to poverty reduction. It is not surprising, then, with poverty alleviation strongly evident in programmes such as the United Nations Millennium Development Goals and the shift to the Human Development paradigm, that tourism should be viewed as strategy to alleviate poverty. The concept of pro-poor tourism has received increased attention (for example, Holden 2013; Scheyvens 2015) with the UNWTO launching the ST-EP programme (Sustainable Tourism – Eliminating Poverty) at the World Summit on Sustainable Development in Johannesburg in 2002. The journal

Current Issues in Tourism has dedicated a special issue to the topic (see volume 10, numbers 2/3, 2007).

Pro-poor tourism 'focuses on how tourism affects the livelihoods of the poor and how positive impacts can be enhanced through sets of interventions or strategies for pro-poor tourism' (Rogerson 2006). The concept has similarities to fair trade in tourism for its focus on poverty and it has the potential to be implemented in a variety of types of tourism whether they are large-scale or small-scale, urban or rural. In the context of this chapter, more specific types of tourism, such as community-based tourism or ecotourism, have an emphasis on enhancing local participation in the benefits of tourism, and so have affinities with pro-poor tourism, but would need to also have specific strategies to assist the poor. There are a number of advantages inherent in tourism that makes it attractive for promoting pro-poor growth, as identified by Rogerson (2006). There is a wide scope for participation in tourism, including the informal sector. The customers come to the product creating opportunities for linkages to other sectors. Tourism is highly dependent upon natural capital (wildlife and scenery) and culture and the poor have some of these assets. Tourism is able to be more labour-intensive than manufacturing. Finally, compared to many other economic sectors, a greater proportion of benefits from tourism in jobs and entrepreneurship opportunities accrue to women.

Community tourism is thought of as the main avenue for the poor to participate through community-run lodges, campsites or craft centres, often supported by NGOs, yet poor individuals participate in all types of tourism through self-employment, such as hawking and casual labour (Ashley *et al.* 2000). In the introduction to this chapter, it was pointed out that communities are not homogeneous. Ashley *et al.* (2000) also stress the fact that the poor are not homogeneous and the impacts of tourism (positive and negative) will inevitably be distributed unevenly among poor groups, reflecting differing patterns of assets, activities, opportunities and choices. This concern is evident in a study on pro-poor tourism in a village in Thailand where King and Dinkoksung (2014) found that villagers with entrepreneurial skill benefited more. There are also a number of ethical dilemmas linked to communities and pro-poor tourism that need consideration. If poor communities are to benefit, is it 'tourism of the poor', whereby villagers are the objects of the tourists' gaze, or is it 'tourism for the poor', with benefits flowing to the local communities instead of tour operators? Finally, is it 'tourism by the poor', whereby communities are empowered and take the place of the tour operators (King and Dinkoksung 2014)? All of these questions relate to who wins and who loses (King and Dinkoksung 2014).

With the focus on livelihoods, the question turns to what policies or programmes would contribute to pro-poor tourism. Ashley *et al.* (2000) outline the following strategies to enhance economic participation of the poor in tourism:

● Target education and training at the poor (particularly women) to take up employment and self-employment.
● Expand access to microfinance.
● Recognize and support organizations of poor producers.
● Develop core tourism assets and infrastructures in relatively poor areas where commercially viable product exists.
● Strengthen local tenure rights over land, wildlife, cultural heritage, access to scenic destinations and other tourism resources.
● Use planning to encourage investors to develop strategies to assist poor.
● Minimize red tape that excludes least skilled.
● Enhance vendors' access to tourists.
● Provide business support to improve quality, reliability of supply, transport links.
● Incorporate domestic/regional tourism and independent tourism into the planning process.
● Avoid excessive focus on international all-inclusives.
● Recognize the importance of the informal sector in the planning process.

One of the key items above is the expansion of microfinance to help the poor who do not have sufficient capital to launch new initiatives (see Vargas 2000). The importance of this was demonstrated in 2006 when the Noble Peace Prize was awarded to Muhammad Yunus and Grameen Bank in Bangladesh for the work on microcredits as an instrument against poverty. It is interesting that the Peace Prize went to an economic initiative. An important challenge associated with microcredit schemes, especially those funded from international agencies, is that in some communities in developing countries, the nature of development may be defined differently from that of the West, and so macroeconomic indices used by international funding agencies to assess the microcredit schemes may not work (Horan

2002). Horan (2002) found that a group of women in Tonga who received funding through a microcredit scheme for textiles that were to be sold to tourists were engaged in their concept of development known as *fakalakalaka*, which is a broader notion of development than can be measured by economic indicators. It involves all areas of the individual (physical, mental and spiritual) and extends to personal relationships, family dynamics, and the development of the community. The women used the money to make ceremonial textiles for use in their local culture rather than textiles for tourists. Although there was a low default rate on the loans, the international lenders did not recognize that development was occurring. The projects funded by microcredits need to fit with local conditions.

In the democratic era of South Africa, tourism has become an essential sector for national reconstruction. The government has developed a wide range of interventions to involve poor communities in tourism. Through a variety of different funds and programmes, the government has encouraged small-scale entrepreneurs to develop new products as well as the tourism industry to reach out to poor local communities and incorporate their workers, products and services into the tourism industry (Rogerson 2006). See Box 5.2 for a more detailed discussion of pro-poor tourism initiatives in South Africa (see also Plate 5.5).

While pro-poor tourism represents a step towards involving more of the local community, Mowforth and Munt (1998: 272) argue that given the small size, it is not 'a tool for eliminating nor necessarily alleviating absolute poverty, but rather is principally a measure for making some sections of poorer communities "better-off" and of reducing vulnerability of poorer groups to shocks (such as hunger)'. It has also been pointed out that it can place considerable demands on tourism stakeholders who operate in a market environment to adopt strategies that maximize benefits for the local poor, which may include higher wages that need to be paid regardless of business cycles (Chok *et al.* 2007). These authors go on to point out that successful pro-poor tourism relies, to a great extent, on the altruism of non-poor tourism stakeholders to move the industry towards increasing benefits along with reducing the costs for the poor. Hall (2007) summarizes the critics of pro-poor tourism as suggesting it is 'another form of neo-liberalism that fails to address the structural reasons for the north south divide, as well as internal divides within developing countries'. Schilcher (2007) examines the ideological aspects of pro-poor strategies from neo-liberalism to protectionism.

Box 5.2 Pro-poor tourism in South Africa

In the democratic era in South Africa, tourism has been identified as a development tool. Rogerson (2006) has identified a number of government programmes and pro-poor tourism pilot projects. Beginning with the release of a White Paper on the Development and Promotion of Tourism in 1996, there has been a focus on responsible and sustainable tourism, and an important recognition of the need for communities to be involved and benefit from tourism. There is 'a strong emphasis in tourism planning upon job creation and enterprise development in support of the country's neglected black communities' (Rogerson 2006). The government has instituted a number of different programmes to support the involvement of poor communities in tourism. These initiatives include funding for infrastructure and new product development, such as cultural tourism and handicrafts in rural areas and township tourism in urban areas. Funds have also been set up for assisting tourism entrepreneurs and training programmes. A Fair Trade in Tourism South Africa brand and trademark have also been established. Guides for responsible tourism and guides of good practice have been available to operators. Some of the recommendations in these guides include prioritizing opportunities for local communities, tourism businesses buying locally made goods and using locally available services, and local people should be hired. Some ecotourism operators, for example, have been expanding benefits to local communities. Rogerson (2006) identifies township tourism in Alexandra Township on the edge of Johannesburg as part of the pro-poor initiatives. As a result of the apartheid government, the township is one of the most impoverished and densely populated in South Africa. It has been the site of township tourism bringing tourists to sites of significance to the anti-apartheid movement generating understanding on issues of poverty and historic oppression. As well as being part of a campaign against urban poverty, Alexandra Township tourism has been part of a pro-poor pilot project linking it to the Southern Sun Hotels Group. A series of existing programmes and proposals to link the hotel group to the community were developed/proposed, including outsourcing from the township, marketing tours to the township through the hotels, recycling guest amenities to township entrepreneurs to make into crafts, and the sale of locally produced goods (souvenirs) to hotel guests. Rogerson (2006) reports difficulties were encountered in the project, highlighting the challenges in maintaining partnerships. Challenges were also found in strengthening pro-poor linkages by leveraging tourism's backward linkages from the accommodation sector to the local agriculture sector in KwaZulu Natal. Pillay and Rogerson (2013) argue that to strengthen the pro-poor impacts of tourism requires integrated policy interventions addressing demand-side, production-related and intermediary factors that constrain poor farmers from participating in the food supply chain to the tourism industry.

Source: Rogerson (2006); Pillay and Rogerson (2013)

Plate 5.5 *South Africa, township near Pretoria: tourists visit a township*

Source: Photo by D. Telfer

She argues that for strategies to be pro-poor, growth must deliver disproportionate benefits to the poor to reduce inequalities that have limited the potential for poverty alleviation. She argues that there is a need to shift policy from growth to equity; however, that would involve a shift away from the neo-liberal approach, and therefore a policy based on equity is more likely to remain as rhetoric rather than be implemented on a large-scale basis.

Volunteer tourism

A recent type of tourism that has links to pro-poor tourism and has particular relevance to communities is volunteer tourism. Scheyvens (2002) describes volunteer tourism as an element of 'justice tourism', with individuals paying to travel to developing countries to assist with a development or conservation project. There is growth in the volunteer tourism market for students doing a 'gap year' (before starting university), particularly in Britain, Australia and New Zealand, although the demographics of the 'gap year' are expanding to people taking 'career breaks' (Simpson 2004). Wearing (2001) suggests that volunteer tourism can be viewed as a development strategy leading to sustainable development. The first type of volunteer projects are related to conservation work, and it can occur in a variety of destinations, such as Africa, Asia, Central and South America, and can occur in places such as rainforests, cloud forests, conservation areas and biological reserves (Wearing 2001). An example of this is volunteer tourists working for sea turtle conservation in Tortuguero, Costa Rica. The volunteering is organized by the Caribbean Conservation Corporation (CCC), an NGO headquartered in the United States but with a national office in Costa Rica, and a year-round field station in Tortuguero (Campbell and Smith 2006). The second main type of volunteer tourism is linked to development work by tourists. These types of projects can involve offering medical assistance, projects linked to heritage and cultural restoration, and other types of social and economic development initiatives (Wearing 2001). In the case of South Africa, Stoddart and Rogerson (2004) examined volunteer tourists volunteering with Habitat for Humanity International's Global Village Work Camp Programme. Volunteers work with local community members to construct affordable housing.

While volunteer tourism has the potential to assist communities, concerns have also been raised. Simpson (2004: 690) states that 'currently, the gap year industry promotes an image of a "third world other" that is dominated by simplistic binaries of "us and them", and is expressed through essentialist clichés, where the public face of development is one dominated by the value of western "good intentions"'. Other concerns raised include whether the volunteers are actually making a difference or causing more difficulties than providing benefits, and whether the gap year is a new form of colonialism (Brown 2006). There has been limited research on how volunteer programs impact host communities (Lupoli et al. 2014) and there are concerns over whether volunteer tourism is being commodified by large tour operators, whereby local desires are neglected, volunteer tourists promote their own values and the cycle of dependency is promoted (Guttentag 2009, in Lupoli et al. 2014). Other possible negative impacts on communities include the hindering of work progress and completion of unsatisfactory work, disruption of local economies, reinforcement of the conceptualization of the 'other', rationalization of poverty and instigation of cultural changes (Guttentag 2009). In an investigation in a cluster of communities that have hosted NGO volunteer tourism for over 20 years, Zahara and McGehee (2013) found that the volunteer tourists do have a positive impact and, specifically, the volunteer tourists exert bridging social capital, which, in turn, has impacts in all aspects of community capital. As in many sectors in the tourism industry, indicators, codes of conduct and certifications are being developed for volunteer tourism. Volunteer tourism is further explored in Chapter 6 in the context of consumption.

Gender and community development

A theme in this chapter is that communities are made up of individuals and groups; they do not all have the same values and goals and they do not face the same challenges or have the same opportunities. Within the United Nations Millennium Development Goals, both women and children are specifically mentioned. In India, for example, a law has been introduced banning children under 14 years of age from working as servants in homes, or in restaurants, tea shops, hotels and spas (Associated Press 2006). Over time, there has been a variety of different theoretical perspectives in examining women in relation to development, from 'women in development', to 'woman and development' and then to 'gender and development', with the shift towards gender equality and the social construction of gender (Tucker and Boonabaana 2012). In 2012, the World Development Report from the World Bank was titled *Gender Equality and Development* (World Bank 2012). Moreover, the women and environment discourse highlights the links between the social and economic position of women in developing countries and environmental degradation (Baker 2006: 167):

> The position of women makes them more vulnerable to the negative effects of environmental degradation than their male counterparts. They are more marginalised, usually work harder, especially if engaged in agricultural labour, have a less adequate diet and are often denied a voice in the political, economic and social spheres.

The shift within the women and environment debate towards sustainable development saw early discussions focused on women as passive victims of environmental degradation stemming from global processes to more of an emphasis on women's positive roles as efficient environmental resource managers in developing countries (Baker 2006). Women can act as promoters of sustainable development in many ways, including:

- With their domestic, agricultural and cultural roles, women are key agents in promoting sustainable development.
- Women hold knowledge of their local environment and they are the key to developing appropriate biodiversity strategies.
- Promoting sustainable livelihoods at the community level can be accelerated by giving women the right to inherit land as well as to have access to credit and resources.
- A strong connection exists between promoting human rights, especially for women, and promoting sustainable development as it is based on equity and partnership.
- The promotion of democratic environmental governance has a gender dimension as participation based on gender equality of access is more democratic, legitimate and effective (Baker 2006: 168).

The above roles are of much significance in the context of tourism. In community-based tourism, for example, women have the potential to help promote sustainable development in their interactions with the tourists and the natural environment. In the tourism impact literature, the relationship between women and tourism has been discussed from a number of different perspectives, including tourism generating greater independence and more income for women (Momsen 2004; Hashimoto 2015). The tourism industry has created new opportunities for women, particularly in developing countries and rural areas (Wall and Mathieson 2006), with a variety of positions in both the formal (for example, hotels and running guest houses) and the informal sector (for example, guides and street/beach vendors) (see Plate 5.6). However, studies have also identified managerial positions are often taken by men, with women more likely to have lower pay, part-time and seasonal positions (Wall and Mathieson 2006), and the exploitation of women by the tourism industry such as in the sex tourism industry (Hashimoto 2015; Nkyi and Hashimoto 2015). Moving beyond the tourism impact literature and examining gender relations, women are often denied employment and related benefits as a result of 'culturally defined, normative notions of gender identity, roles and relations' and, as a result, some pro-poor tourism development projects have targeted women's empowerment (Tucker and Boonabaana 2012: 437). However, pro-poor tourism projects that promise quick fixes in terms of empowerment may be naïve as they do not take into account the complexities of the sociocultural norms of gender roles and relations in the specific location and may not recognize women's agency in negotiations of power relations

Plate 5.6 *Indonesia, Lombok: young women present traditional Sasak weaving*

Source: Photo by D. Telfer

(Tucker and Boonabaana 2012). Tucker and Boonabaana (2012) critically examine the complexity of the gender and development approach to tourism and two issues they identify that demand future research are as follows. First, the fact that many women in tourism destinations often suffer the 'double burden' of having to look after the home and children and also work in the tourism industry demands attention; and, second, it is necessary to explore what happens in a community when, as a result of working in tourism, women make more money than men, when the reverse was true before the introduction of tourism. Tucker and Boonabaana (2012) go on to state that cultural specifics on gender relations need to be considered when examining the relationships between gender, tourism and poverty reduction, and that poverty reduction initiatives can have complex and seemingly contradictory outcomes in terms of gender.

Conclusion

Preston (1996) suggests that there are three main paradigms in terms of characterizing and securing development. The first paradigm focuses on state intervention, the second on the role of the free market, and the third relates to the power of the political community. This chapter largely focuses on the third aspect: power of the political community and the development of tourism at the local level. As Preston (1996) suggests, this approach is oriented to securing formal and substantive democracy and has an institutional vehicle through NGOs, charities, and dissenting social movements. In the context of tourism, it is important to recognize to what degree local communities interact with the state and market forces. We need to consider how vulnerable or how resilient a community is to tourism. A key issue with respect to the relationship between tourism and communities is the power relationships that exist within the community and between communities and the tourism sector. Does the local community have a chance to participate in a meaningful way? As Milne and Ewing (2004: 215) state, the critical issue becomes 'how do we ensure not only that local involvement and participation occurred, but that it can be sustained in such a way that it leads to effective development outcomes?' The development benefits of tourism need to spread out to as many as possible in the community and even pro-poor approaches may not enhance equity. Critics focusing on the power of the political community have highlighted the alleged impractical idealism of the approach along with its potential to generate conflict between powerful groups in the peripheries and metropolitan centres, which can worsen the situation of the poor (Preston 1996). Others have suggested problems of consensus building, barriers to participation, lack of accountability, weak institutions and a lack of integration with international funding organizations are barriers for indigenous or community development (Wiarda 1983; Brinkerhoff and Ingle 1989).

Communities are complex, and are increasingly recognized as important resources for tourism. With the shift to sustainability and poverty alleviation, the participation of local communities and capacity building have become central to many tourism plans. Communities are not only impacted by tourism, but they also respond and take advantage of the opportunities from tourism (Wall and Mathieson 2006). This chapter has referred to several types of tourism, as well as several strategies that can be adopted to promote additional benefits to disadvantaged communities. Caution, however, needs to be used when comparing different types of tourism. In the tourism literature, comparisons have been made between large-scale mass tourism and small-scale community-based tourism. As Butler (1993: 34) suggests 'making simplistic and idealised comparisons of hard and soft, or mass and green tourism, such that one is obviously undesirable and the other close to perfection, is not only inadequate, it is grossly misleading'. He goes on to state that mass tourism does not have to be uncontrolled, unplanned, short term or unstable; equally, green tourism, or, in the case of this chapter, community-based tourism, is not always inevitably considerate, optimizing, controlled, planned and under local control. As we saw in the context of tourism and gender issues, community-based tourism may well generate conflict, inequities and resource exploitation. The economic and political realities of the marketplace with respect to tourism, and the nature of communities where tourism occurs, need to be considered. A highly fragmented and extremely competitive tourism industry in the public and private sectors often militates against self or internal control (Butler 1993). Having examined the nature of 'host' communities and various strategies of community-based tourism, the next chapter will focus on the 'guest' by exploring the consumption of tourism.

Discussion questions

1   Distinguish the strengths and weaknesses of both the informal and formal tourism employment sectors.
2   Can community-based tourism development help promote community and individual empowerment?
3   Does hosting international tourists reinforce notions of dependency for communities?
4   Does community-based tourism represent a form of sustainable tourism?
5   What role should NGOs take in community-based tourism?

Further reading

Beeton, S. (2006) *Community Development Through Tourism*, Collingwood: Landlinks.

This book contains eight chapters covering a range of issues, including the nature of communities, the relevance of tourism theories, planning, strategy, marketing, rural communities, and dealing with crisis.

Singh, S., Timothy, D. and Dowling, R. (eds) (2003) *Tourism in Destination Communities*, Wallingford: CABI.

This edited volume contains 14 chapters divided into three sections that explore the relationships between tourism and the destination community, the impacts of tourism on destination communities, and the challenges and opportunities for destination communities.

Richards, G. and Hall, D. (eds) (2000) *Tourism and Sustainable Community Development*, London: Routledge.

This edited volume of 20 chapters has a range of international case studies and explores what local communities can do to contribute towards sustainable tourism and what sustainability is able to offer local communities.

Websites

www.un.org/en/civilsociety/
This website illustrates how the United Nations works with civil society organizations.

www.un.org/womenwatch/
This United Nations web page has links to information and resources on gender equality and empowerment of women.

www.fairtrade.travel/
This is the website for Fair Trade Tourism, a non-profit promoting organization responsible tourism in southern Africa.

6 The consumption of tourism

Learning objectives

When you have finished reading this chapter, you should be able to:

- appreciate the factors that influence tourist-consumer behaviour
- understand the changing nature of tourism demand
- identify the extent of emerging new, 'responsible' tourism markets
- identify ways in which the tourism industry can influence tourist behaviour
- appreciate the importance of domestic tourism in development

Tourism is, essentially, a social activity. Certainly, it is big business and one of the world's largest economic sectors, providing a significant source of income and foreign exchange earnings for many countries; certainly, it is a vast and diverse industry, providing employment for up to 10 per cent of the global workforce; and certainly, as this book explores, it is a potentially effective agent of economic growth and development. However, it is important not to lose sight of the fact that, first and foremost, tourism is about people – it is about millions of individuals travelling within their own countries or overseas, visiting places and attractions, staying in destinations and engaging in various activities during their stay.

More specifically, tourism is about people, tourists, who interact with and impact upon other people and other places. In other words, a fundamental characteristic of tourism is that the product is 'consumed' on site; whether in their home country or overseas, tourists must travel to the destination to enjoy or participate in tourism. In fact, as with all services, the tourism product – normally thought of as tourist experiences – cannot actually be produced without the input of tourists into the production process. Although hotels, restaurants, shops, transport operators, attractions and other businesses within the tourism sector offer potential services, these services (or experiences) are not provided until tourists actually purchase and consume them (Smith 1994). Thus, tourism development inevitably results in the presence of tourists in the destination; equally, the presence of tourists inevitably brings about consequences (both positive and negative) on the destination's environment, its economy and local communities.

The consequences or impacts of tourism development are explored in more detail in the next chapter. However, it has long been recognized that the nature of these impacts cannot be divorced from the nature of the consumption of tourism itself. Although the emergence of mass international tourism in the 1960s was greeted with some degree of optimism, the potential economic benefits of tourism being seen as a panacea to the developmental challenges facing destinations and their communities, it was not long before concern was being expressed about the adverse effects of mass tourism consumption on local environments and cultures. One commentator at that time, for example, observed that, as mass tourism develops, 'local life and industry shrivel, hospitality vanishes, and indigenous populations drift into a quasi-parasitic way of life catering with contemptuous servility to the unsophisticated multitude' (Mishan 1969: 142).

During the 1970s, books such as *Tourism: Blessing or Blight?* (Young 1973) and *The Golden Hordes: International Tourism and the Pleasure Periphery* (Turner and Ash 1975) adopted a more balanced perspective, yet the overall tone was that mass tourism may have consequences for the destination that potentially outweigh the developmental benefits of tourism. Indeed, by the 1990s, most, if not all, of the problems associated with tourism development were being blamed on mass tourism and tourists:

> the crisis of the tourism industry is a crisis of mass tourism; for it is mass tourism that has brought social, cultural, economic and environmental havoc in its wake, and it is mass tourism practices that must be radically changed to bring in the new.
>
> (Poon 1993: 3)

The 'new' proposed by Poon refers to alternative (to mass), sustainable forms of tourism development, the achievement of which, as noted in Chapter 2, is dependent upon the adoption of a new 'social paradigm' with respect to the consumption of tourism. In other words, fundamental to the successful development of more appropriate, sustainable forms of tourism is the need for tourists to act more sustainably, to become 'good' or 'responsible' tourists. Poon also suggested that by the early 1990s, there was in fact evidence of the emergence of 'new' tourists – that is, tourists who are more environmentally aware, more quality conscious, more adventurous and more ready to reject the passive, structured, mass-produced package holiday in favour of more individualistic, authentic experiences. Since then, there is clear evidence that, with increasingly widespread use of the Internet, many tourists now book and travel independently. In the UK, for example, almost 60 per cent of holidays are booked independently, although this includes domestic holidays so the proportion of overseas trips booked independently will be lower (Taylor 2014). At the same time, not only has it become widely assumed that tourists are becoming 'greener' or increasingly disposed towards consuming tourism in more responsible or environmentally appropriate ways, but also that assumption has frequently been utilized as the justification for developing or promoting sustainable forms of tourism.

Importantly, however, there is little evidence to support this claim that tourists are in fact becoming more environmentally conscious or responsible, and indeed little evidence of who these tourists are (Dolnicar *et al.* 2008). For example, it is often suggested that the growth in demand for ecotourism is a sign that tourists are increasingly seeking environmentally appropriate tourism experiences. It is certainly true that ecotourism is becoming increasingly popular; research has suggested that the number of tourists taking ecotourism holidays is growing three times faster than those choosing 'mainstream' holidays, and that, by 2024, ecotourism will represent 5 per cent of the global holiday market (Starmer-Smith 2004). However, significant doubt exists about the extent to which 'ecotourists' are motivated by genuine environmental concerns (Sharpley 2006; Beaumont 2011). Similarly, while surveys have long suggested that many tourists would be willing to pay more for environmentally friendly holidays (for example, Tearfund 2000), there is limited evidence that this is manifested in practice.

The purpose of this chapter, therefore, is to explore the nature of the consumption of tourism and the consequential implications for sustainable tourism development. The first questions to address briefly, however, are: What is the tourism demand/consumption process? What are the factors that influence that process? And, how has the consumption of tourism changed over time?

The tourism demand process

Tourism demand or consumption is a complex process, described by one commentator as 'discretionary, episodic, future oriented, dynamic, socially influenced and evolving' (Pearce 1992: 114). In other words, the demand for tourism involves making choices about how to spend specific periods of leisure time, choices that may be influenced by a variety of factors and which may change over time. Moreover, it is not only about how and why people decide to participate in tourism, but also about how they behave as tourists, why they choose particular types of tourism, what tourism means to them, and why their 'tastes' in tourism may change.

Despite this complexity, however, tourism demand is typically, though somewhat simplistically, seen as a sequential set of stages that may be summarized as follows:

Stage 1: Problem identification/felt need
Stage 2: Information search and evaluation
Stage 3: Purchase (travel) decision
Stage 4: Travel experience
Stage 5: Experience evaluation

Each stage in the demand process may be influenced by personal and external variables, such as time and money constraints, social stimuli, media influences, images/perceptions of the destination, or marketing, while each consumption experience feeds into subsequent decision-making processes (Figure 6.1).

At the same time, of course, tourism demand is not a 'one-off' event. People consume tourism over a lifetime, during which tourists may progress up or climb a 'travel career ladder' (Pearce 1992) as they become more experienced tourists. As a result, travel needs and expectations may change and evolve, but these may also be framed and influenced by evolving social relationships, lifestyle factors and

Figure 6.1 *The tourism demand process*

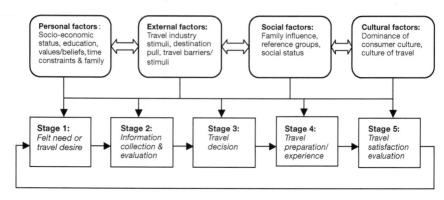

constraints, and emerging values and attitudes. This latter point is of particular relevance to the theme of this chapter, and will be returned to shortly, but fundamental to understanding the consumption of tourism is an appreciation of the factors that may influence the demand process. These can be categorized under four headings (see Cooper *et al.* 2005: 53):

1   *Energizers of demand*: These are the forces and influences (or personal 'push' factors) that collectively create the motivation to travel or go on holiday, or initiate the demand process.
2   *Effectors of demand*: The information search/evaluation process and subsequent purchase decision are influenced by the tourist's knowledge and perceptions of particular places, destinations or experiences. These are sometimes referred to as destinational 'pull' factors that lead the tourist to making particular travel choices.
3   *Filterers/determinants of demand*: A variety of economic, social and demographic factors determine particular choices or 'filter out' inappropriate products. These include: mobility; employment and income; paid holiday entitlement; education levels; and age, gender, race, stage in family life cycle. In addition, choice may be determined by intangible, psychographic variables, such as attitudes, values and lifestyle.
4   *Roles*: Holiday/travel choices are also influenced by roles within the purchasing 'unit' (for example, the different roles adopted by family members in choosing a holiday) and as tourists.

Evidently, then, an almost infinite combination of variables may influence how, when and where tourism is consumed by individual tourists, to the extent that predicting tourist behaviour may be seen as a difficult, if not impossible, task. In fact, according to Krippendorf (1987), many tourists themselves are unable to explain precisely why they participate in particular types of tourism, while others, such as Ryan (1997), suggest that tourism is an irrational (and, hence, inexplicable) form of behaviour! Nevertheless, for the purposes of this chapter, it is important to consider three issues or factors that influence the tourism demand process, namely tourist motivation, the influence of values on consumption, and tourism and consumer culture.

Tourist motivation

Academics and researchers have long been concerned with the questions of why and how people consume or participate in tourism (Hsu and Huang 2008). In particular, attention has been (and continues to be) focused on the first stage in the tourism demand process, or what is broadly referred to as tourist motivation. The reason for this is, perhaps, self-evident. Not only is motivation the 'trigger that sets off all events in travel' (Parinello 1993), the process that translates a felt need into goal-oriented behaviour (that is, tourism) directed at satisfying that need, but also the nature of the felt need influences the consequential behaviour of the tourists and their potential impact on destinations.

Given the complexity of the subject, it is not surprising that tourist motivation is explored from a variety of perspectives (Sharpley 2008). The key themes that emerge from these are summarized below, but, first, it is useful to consider so-called 'tourist typologies' as descriptors of distinctive forms of tourist-consumer behaviour.

Tourist typologies

Tourist typologies are, in essence, lists or categorizations of tourists based upon a particular theoretical or conceptual foundation. As such, they tend to be descriptive as opposed to predictive, yet they do reflect, if not explain, different motivations, interests and styles of travel on the part of tourists. One of the first such typologies was proposed by Gray (1970), who coined the terms 'sunlust' and 'wanderlust', where sunlust tourism is essentially resort-based and motivated by the desire for the three S's – sun, sea and sand.

Conversely, wanderlust tourism is typified by the desire to travel and to experience different places, peoples and cultures. Implicit in each term are the characteristics of the different forms of travel and the potential destinational impacts of each.

The distinction between these two types of tourism was expanded upon by Cohen (1972) in his widely cited tourist typology based upon a 'familiarity-strangerhood' continuum. In other words, Cohen suggested that tourists are more or less willing to seek out different or novel places and experiences; some travel within an 'environmental bubble' of familiarity – they seek out the normal/familiar (food, language, accommodation, fellow tourists) and are unwilling to risk something new or different – whereas others seek out different or unusual experiences. This, in turn, determines how different tourists travel. Some are 'institutionalized', inasmuch as

Figure 6.2 *Cohen's typology of tourists*

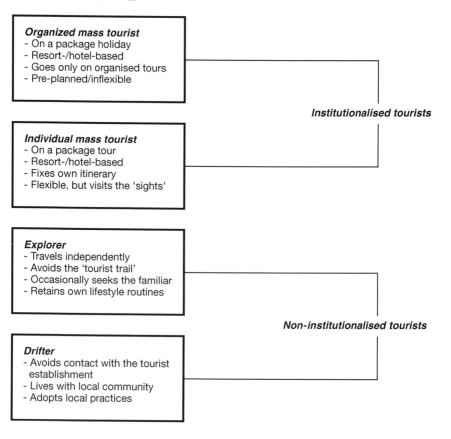

they depend upon the tourism industry to provide familiar, predictable, organized and packaged holidays; others, conversely, are 'non-institutionalized', travelling independently and requiring little contact with the tourism establishment. His typology proposes four types of tourist, from the organized mass tourist to the drifter (Figure 6.2).

The extent to which this typology remains relevant to contemporary international tourism is debatable. Not only has the practice of tourism matured over the last four decades – as discussed later in this chapter, the demand for tourism in terms of places, experiences and frequency has changed significantly, as has the way in which people actually organize and purchase their trips or holidays – but also the phenomenon of globalization has transformed the world within which tourism occurs. In particular, global communications and information systems have made it virtually impossible (for explorers or drifters) to escape the realities of their 'home' world, while the practice of independent travel has, in effect, become institutionalized. Nevertheless, it provides a useful framework for thinking about how different types of tourists behave and interact with the destination.

The same may be said for a number of other typologies. Cohen (1979), for example, subsequently constructed a 'phenomenology of tourist experiences', based again on the familiarity-strangerhood continuum. However, the focus was on the extent to which tourists either enjoy a sense of belonging or feel like strangers in their normal, home environment. Those who benefit from a satisfying, fulfilling home life are, according to Cohen, likely to seek only trivial, recreational experiences on holiday, while, at the other extreme, 'alienated' individuals are likely to seek out more meaningful or authentic experiences while on holiday. Alternatively, Plog's (1977) widely cited (though frequently criticized) typology attempted to link tourists' personality with destination choice; risk-averse (psychocentric) tourists travel to familiar, closer destinations, while more adventurous risk-takers (allocentrics) choose more distant exotic destinations. Again, however, transformations in the supply of tourism, particularly the expansion of the 'packaged' pleasure periphery, limit the contemporary relevance of this typology.

Smith's (1989) typology deserves mention in that it explicitly links tourist types and numbers to destinational impacts. At one extreme, small numbers of 'explorers' travel to undiscovered places and, Smith suggests, attempt to blend into the destination; at the other extreme, large numbers of charter tourists care little for the destination and have maximum negative impact. Within these two extremes, different types/numbers of tourists travel with different attitudes towards the destination. However, as with most typologies, a causal link between tourist type, tourist behaviour and subsequent destinational impact is, often mistakenly, assumed. For example, in many destinations in the less-developed world, mass-institutionalized tourism exists in the form of all-inclusive resorts, yet, in some cases, these may bring substantial economic benefits but only limited environmental impact (see Box 6.1).

Yiannakis and Gibson (1992) developed a typology that identified 15 tourist roles, from 'sun lover' to 'educational tourist', each determined by a preferred balance between familiarity–strangeness, stimulation–tranquillity and structure–independence. Their work was subsequently 'tested' in an Australian context, confirming that tourists choose destinations according to the extent that they will be able to enact their preferred roles as tourists (Foo *et al.* 2004). In recent years, research into tourist typologies has remained popular, although more contemporary studies have tended to focus on subgroups of tourists. Uriely (2009), for example, deconstructs the 'backpacker' typology into four subtypes, while Mehmetoglu (2007) identifies three clusters of nature tourists.

Generally, however, tourist typologies tend to imply that larger groups of mass, 'institutionlized' tourists on organized holidays and seeking more traditional sun-sea-sand holidays have a greater negative impact on destination environments and societies than smaller numbers of independently minded, environmentally aware tourists seeking more individualistic and cultural experiences. However, as the following section on tourist motivation suggests, care must be taken in making such a broad assumption.

Tourist motivation: key themes

As noted above, tourist motivation is fundamental to the consumption of tourism. Not only does the motivational stage 'kick-start' the entire tourism demand process, but also the manner in which people behave as tourists (the experiences they seek as well as their behaviour in a more specific sense) is very much determined by what motivates them in the first place. Therefore, understanding tourist motivation is fundamental to understanding the overall consumption of tourism.

Box 6.1

All-inclusive resorts: the case of Sandals Resorts, Jamaica

The concept of all-inclusive international holiday resorts can be traced back to the development of Club Méditerranée (Club Med) resorts in the 1950s. However, the opening of the first Sandals resort at Montego Bay, Jamaica, in 1981 marked the beginning of a new era in all-inclusive, club-based holidays. Not only was the resort truly all-inclusive – virtually every aspect of the holiday was pre-paid – but also it provided a high-quality experience for couples only. Since then, Sandals has expanded its operations across the Caribbean, with a total of 15 resorts under the Sandals name in Jamaica, St Lucia and Antigua, Grenada and The Bahamas, as well as three Beaches Resorts, two Grand Pineapple Resorts and the private island resort of Fowl Key, all providing luxury holidays for a variety of niche markets. In fact, the Caribbean has become widely associated with all-inclusive holidays, which are now the most popular form of international tourism in the region; in addition to Sandals, SuperClubs and Allegro Resorts also own a large number of resorts around the Caribbean, and it has been claimed that 48 of the world's top 100 all-inclusive resorts are located in the Caribbean.

All-inclusive resorts have long been criticized for limiting the 'trickle-down' economic benefits of tourism, for a high level of imports, for restricting employment and promotion potential and, generally, for providing little developmental benefit to the destination (Freitag 1984; Patullo 1996). However, in addition to meeting substantial demand for all-inclusive holidays (that is, tourists who seek safe, luxurious holidays in an exotic location), the resorts in Jamaica collectively generate the largest contribution to GDP on the island, while, in terms of employment, they create relatively more jobs than conventional hotels. For example, research has shown that five-star all-inclusives create between 1.5 and 2 jobs per room compared with 1 job per room in a conventional hotel. At Sandals in particular, staff are relatively well paid and benefit from free meals and transport; it is claimed that they can save up to one-third of their monthly salary. All line staff receive at least 120 hours of training each year and, in 2003, Sandals Montego Bay initiated a skills training centre for local school leavers. Additionally, Sandals established a Farmers Programme in 1996, under which local farmers were encouraged to supply produce to the resorts. By 2004, 80 farmers were supplying hotels across the island and total annual sales had risen to US$3.3 million. Local craft producers are also able to sell their products to tourists within the resorts.

In 2009, the Sandals Foundation was established as the philanthropic arm of the Sandals organization, its purpose being to 'fulfill the promise of the Caribbean community through investment in sustainable projects in education; environment and community which improve people's lives and preserve our natural surroundings'.

Through harnessing the resources of Sandals and its partners, the Foundation supports projects across the Caribbean, such as ensuring as many people as possible have the educational opportunity to fulfil their potential, developing community support programmes to address community and social issues and, of course, supporting environmental programmes. Hence, Sandals is directly committed to the sustainable development of the Caribbean through its corporate social responsibility activities.

Source: Issa and Jayawardena (2003); Ashley *et al.* (2006); Sandals Foundation (2015)

There is enormous diversity in the treatment of tourist motivation within the tourism literature, reflecting both the variety of ways in which it is possible to approach the subject and also the fact that a generally accepted theory or understanding of tourist motivation has yet to emerge (Sharpley 2008). Nevertheless, a number of key points or themes are in evidence, the first of which is the fact that motivation should not be equated with demand; that is, tourism demand or consumption is the outcome of motivation. To put it another way, tourist motivation can be seen as a 'meaningful state of mind which adequately disposes an actor or group of actors to travel, and which is subsequently interpretable by others as a valid explanation for such a decision' (Dann 1981: 205). Therefore, what we are concerned with essentially are the factors or influences that create this 'meaningful state of mind'. Such influences may be either intrinsic or extrinsic to the individual tourist.

(1) Intrinsic motivation

The study of motivation has traditionally been concerned with people's innate needs and desires; it is an individual's intrinsic and deep-rooted needs that lead to motivated, goal-oriented behaviour. In other words, every individual has unique personal needs, the satisfaction of which has long been considered the primary arousal factor in motivated behaviour. In this context, one of the best known and most widely applied theories is Maslow's hierarchy of needs; in fact, many tourism texts refer to Maslow's model, linking specific needs with identified goal-oriented tourist behaviour, while others explicitly adapt it, as, for example, in the case of the travel career concept mentioned above. Similarly, Crompton (1979) suggested that tourist motivation emanates from the need to restore an individual's psychological equilibrium, which may become unbalanced as a result of unsatisfied personal needs. However, most commentators accept that tourist motivation is not simply a function of intrinsic, psychological needs – extrinsic or social factors are also significant influences on why and how people consume tourism.

(2) Extrinsic motivation

From an extrinsic perspective, tourist motivation is structured by the nature and characteristics of the society to which the tourist belongs. To put it another way, there are a variety of forces and pressures in an individual's social and cultural environment that may influence that individual's needs and motivations, or desire to consume tourism. At a basic level, for example, the motivation for many people to take a holiday is to relax, to rest, to have a change and to get away from the routine; in order to survive in modern society, an individual must periodically escape from it. Indeed, it is somewhat ironic that modern society has created both the need for and means of mass international travel. That is, technology, economic growth and socially sanctioned free time have brought freedom and mobility to many in the developed world, yet, in modern society, tourism has become an essential 'social therapy, a safety valve keeping the everyday world in good working order' (Krippendorf 1986).

Importantly, it is not simply the desire to get away that may motivate people. Tourism also offers the opportunity to travel beyond not only the physical, geographic boundaries of normal, home existence, but also the social boundaries. In other words, tourism releases people from the constraints or social 'norms' of their home existence, providing them with the opportunity to indulge in so-called 'ludic' behaviour (that is, behaviour that can be described as play), manifested in, for example, excessive drinking or sexual activity, or more generally 'letting their hair down'. Conversely, tourists may seek meaning or 'authenticity' through tourism, compensating for what may be regarded as the lack of meaning or 'reality' in modern society (Cohen 2007). Indeed, MacCannell (1989) views the tourist's search for meaning and authenticity as a modern form of pilgrimage, the motivation for tourism thus becoming a form of secular spiritual quest (Sharpley 2009a).

Beyond the more general relationship between the nature of modern social life and tourist motivation, a number of other social influences can also be identified. Some commentators explore the relationship between work and leisure/tourism experiences (Ryan 1991a), the nature of work having a potential impact on the desired from of tourism consumption. For example, it is suggested that those in demanding, stressful jobs may seek compensatory (relaxing, restful) holidays, and vice versa. Others refer more generally to cultural and social factors, including social class, reference groups and family roles, as dominant social influences on tourist motivation and behaviour.

Collectively, this diversity of intrinsic and extrinsic motivating factors suggest that identifying specific or dominant determinant factors may be a difficult, if not impossible, task, particularly given the fact that tourists themselves may be unwilling or unable to express their real travel motives. Nevertheless, there are two generally accepted features of tourist motivation:

- *Tourism as escape*: Whether implicitly or explicitly, tourists are motivated primarily by the desire to escape, to travel away from rather than to travel to somewhere or something. Reviews of the relevant literature consistently find that the escape motive predominates.
- *Tourism as 'ego-enhancement'*: Tourists are motivated by the potential rewards of participating in tourism. Such rewards may be personal, interpersonal, psychological or physical and, collectively described as 'ego-enhancement', they compensate for the deficiencies or pressures and strains of everyday life. At the same time, tourists' motivations are markedly self-oriented or egocentric, focusing on personal wants and needs. In other words, tourism represents a form of self-reward or self-indulgence.

The implication of this is, of course, that tourism is essentially an egocentric, 'selfish', escapist activity. It is, for the most part, about relaxation, fun and entertainment, and hence tourists are most likely to give priority to satisfying their personal needs rather than demonstrating and responding to a positive concern for the consequences of their actions – their focus will be inwards, on the satisfaction of personal needs and wants, rather than on the external tourism environment. This is not to say that the 'green' tourist does not exist in reality, but the study of tourist motivation certainly confirms McKercher's assertion, referred to in Chapter 2, that tourists are simply consumers, not anthropologists wishing to blend into and learn about local communities and cultures. This debate is addressed in more detail later in this chapter, but, as suggested earlier, both values and consumer culture must also be considered as influences on tourism consumption.

Values and tourism consumption

Psychographic variables, such as values, attitudes and opinions, are important determinants in the tourism decision-making process. In particular, values, which serve as standards or criteria for all forms of personally or socially preferable behaviour, have been found to be potential predictors of travel behaviour. In other words, how people consume tourism is likely to be influenced strongly by their personal values. For example, if family security is highly valued, then safe and predictable family holidays may be the preferred tourism experience. Conversely, a backpacker or independent traveller may have an exciting life as a dominant value.

Importantly, values are usually organized into a value system or hierarchy. That is, individuals typically hold a number of values that may be more or less influential in different contexts, while different people attach varying values to similar objects or behaviour; 'an individual relies on his/her value system to maintain self-esteem or consistency in those situations where one or more conflicting values are activated' (Madrigal and Kahle 1994). In the tourism context, the likelihood of value conflict is high, particularly between personal values, such as pleasure, freedom or happiness, and social values that serve as guidelines for socially acceptable behaviour. Implicitly, therefore, only individuals who hold strong or dominant environmental values are likely to adopt a 'responsible' approach to the consumption of tourism, particularly given the motivating factors identified in the preceding section (Hedlund 2011).

Tourism and consumer culture

The discussion so far has focused specifically on tourism. However, tourism is just one of a whole host of goods and services that people in modern societies consume, and, as Solomon (1994: 536) observes, 'consumption choices simply cannot be understood without considering the cultural context in which they are made'. Therefore, it is important to consider briefly how the consumption of tourism is framed by a wider consumer culture.

In modern tourism-generating societies, consumption has become a culturally significant activity. For modern consumers, goods, services and experiences, such as tourism, have assumed a meaning and significance well beyond their simple, utilitarian values and the practice of consumption plays an important role in contemporary social life (Lury 2011). In other words, all goods and services satisfy basic needs – food to satisfy hunger, clothes to keep warm, holidays for rest and relaxation, and so on – but their consumption also fulfils a wider role.

In particular, consumption is seen as an effective means of social classification, creating self-identity or social status. This is not, of course, a new phenomenon – so-called 'conspicuous consumption' has, in general, long been a marker of wealth and social status, while tourism in particular has always used as an indicator of status, wealth or taste. However, in modern societies, consumption has become a dominant and widespread element of social life, with people increasingly seeking identity or distinction through the products they consume. As a result, producers in general have had to become increasingly responsive to the diverse and rapidly changing demands of consumers; in tourism, this has been manifested in, for example, the development of specialized, niche products or styles of tourism that, though relatively affordable, have the aura or status of luxury. For example, ecotourism, which tends to be exclusive (expensive) tourism based in exotic or distant places, is often referred to as 'ego-tourism' (Wheeller 1992).

In addition to classification purposes, products and experiences may be consumed in other ways. It has been suggested, for example, that three further modes of consumption exist (Holt 1995):

- *Consumption as play*: Consumption that focuses on interaction with other consumers rather than on the object of consumption, such as the shared experience of 'white-knuckle' rides at a theme park.
- *Consumption as integration*: The integration of the self into the object of consumption, or vice versa, whereby tourists may adapt themselves to the destination environment and culture (that is, become 'good' tourists), or participate in activities that reflect their self-image, such as trekking holidays if they consider themselves to be adventurous.
- *Consumption as experience*: Consumption that is framed by a social world that provides meaning to the object of consumption. For tourists, the social meaning of tourism (for example, opportunities for hedonistic pleasure) defines their tourism consumption.

Quite evidently, then, the demand for tourism is a highly complex issue, the study of which embraces a variety of themes and perspectives, which this brief overview has only been able to touch on. Nevertheless, it has demonstrated that the consumption of tourism, or why and how people participate in tourism, is influenced by an almost infinite variety of personal (intrinsic) and sociocultural (extrinsic) factors. At the same time, not only do people become more experienced tourists as they progress through their 'travel career' and adapt their behaviour as their values, attitudes and personal circumstances change, but also the overall practice of tourism consumption is constantly evolving. Therefore, it is a difficult, if not impossible, task to predict individual tourist behaviour, although, as the next section shows, a number of distinctive trends exist in the demand for tourism that are relevant to the management and development of tourism.

Tourism demand: trends and changes

As noted in Chapter 1, there are two principal ways of exploring trends and changes in the demand for tourism: historical and contemporary flows in international tourism (that is, spatial transformations evidenced by statistical data); and transformations in the nature of tourism demand (that is, changes in the style of tourism demand). Both perspectives are important in the context of this book. First, contemporary trends in the volume, value and direction of international tourism flows provide an indication of tourism's increasing contribution to development in emerging destination areas. More specifically, international tourist flows have, traditionally, been polarized and regionalized between and within the wealthier, developed regions of the world. However, in recent years, many less-developed countries have been claiming a greater share of global tourist arrivals and receipts, and, consequently, have benefited from tourism-related economic growth and development. Regional flows of tourism between developing countries are also on the rise. Second, changes in the nature of demand point to both opportunities and challenges in managing tourism development effectively. For example, niche products, such as adventure tourism (see Plate 6.1), ecotourism or heritage tourism, represent opportunities for tourism development in countries with a rich natural and cultural heritage; equally, the increasing popularity of all-inclusive holidays has underpinned the successful development of tourism in a number of

Plate 6.1 *Tunisia, near Matmata: these camels are used to providing rides to tourists*

Source: Photo by D. Telfer

less-developed countries. For example, the rapid and successful development of tourism in Cuba since the late 1980s (tourism is now the island's principal source of hard-currency earnings) has been based primarily on all-inclusive resorts (Cerviño and Cubillo 2005; Sharpley and Knight 2009). With the easing of relations between Cuba and the US in 2015, some US airlines are now planning to operate direct flights to Cuba.

International tourism flows are considered in some detail in Chapter 1. The two key trends highlighted are as follows:

● *Global growth*: Over the last 60 years, worldwide tourist arrivals and receipts have demonstrated consistent overall growth, pointing to tourism's resilience to external 'shocks', such as economic recession, wars, health scares and so on. Such growth continues unabated; despite a variety of events during the first decade of the new millennium, including the Indian Ocean tsunami, oil price rises, health scares, terrorism and, not least, the global economic crisis of 2008, which resulted in temporary falls in worldwide arrivals, the overall trend remains one of average annual growth of around 4.5 per cent. Thus, international arrivals grew from 687 million in 2000 to surpass the one billion mark in 2012, reaching 1.087 billion in 2013. Moreover, it is forecast that by 2020, this figure will have reached 1.6 billion. As also noted in Chapter 1, it is also important not to overlook domestic tourism, which, globally, is thought to be between six and 10 times greater in terms of volume/trips. Accurate data do not exist, but for some less-developed nations, domestic tourism undoubtedly remains far more significant than international tourism. In India, for example, international arrivals amounting to 2.6 million in 2000 were dwarfed by an estimated 145 million domestic trips, while contemporary domestic tourism in China alone is, in terms of trips, more than double the volume of total international tourism activity (see Plate 6.2).

- *Global spread*: While Europe has maintained its dominant share of international arrivals, that share has been declining. Conversely, between 1995 and 2010, the highest average annual

Plate 6.2 *China, Beijing: tourists visiting the Forbidden City*

Source: Photo by R. Sharpley

growth rate in arrivals (10.5 per cent) has been experienced by the Middle East region, followed by Africa (6.7 per cent) and Asia and Pacific (6.3 per cent). Over the same period, Europe experienced just 3 per cent annual growth in arrivals. Most recently, Asia and Pacific has enjoyed the fastest growth, and this region is forecast to experience the strongest annual growth in arrivals (4.9 per cent) through to 2030 (UNWTO 2014a: 14). This is evidence of both tourists from the main generating countries in Europe and North America travelling to more distant or exotic destinations, and, more particularly, of increased intra-regional travel – that is, much of the growth in tourism in areas such as the Middle East and Asia has been generated within those regions, fuelled by rapid economic growth. Such an increase in intra-regional travel is likely to continue apace, particularly given the rapid economic development in China and India.

Changes in the nature of tourism demand

Although the extent to which the 'new' tourist, referred to earlier in this chapter, exists in reality (or, more precisely, the extent to which tourists are adopting a more responsible attitude towards their consumption of tourism) is debatable, there is no doubt that styles of tourism consumption have undergone significant transformations over the last half-century, and particularly over the last two decades. Such transformations relate to: types of holidays, with a rapid proliferation in differentiated, niche products; how such holidays are organized and purchased; and an increasingly fuzzy relationship between tourist markets and tourism products. Collectively, they represent a shift from the standardized two-week, short-haul, sun-sea-sand package holiday to more individualistic, active/participatory forms of tourism that provide a wider variety of physical, cultural or educational experiences. Nevertheless, the traditional 'package' remains popular. In 2012, for example, some 48 per cent of overseas holidays taken by UK residents were booked on an inclusive/package basis (ABTA 2013).

It is not possible here to explore all these changes in detail. Not only are there innumerable categories of new tourism products or experiences that are frequently subcategorized (for example, adventure tourism is subdivided into 'hard' and 'soft' adventure tourism), but also a lack of definitional distinction exists between them. For instance, terms such as wildlife tourism, wilderness tourism, nature tourism and even ecotourism are frequently used interchangeably. As a result, accurate data regarding participation in particular forms of tourism do not always exist. Nevertheless, for our purposes here, it is useful to highlight the general trends occurring under the three interrelated headings referred to above.

Types of tourism/holidays

The principal trend in the demand for tourism over the last two decades has been the dramatic increase in the demand for specialized, niche products as an alternative to the traditional summer-sun holiday. Of course, other forms of tourism have long been offered by the tourism industry, such as winter sports tourism or winter sun tourism, both of which were initially upmarket forms of tourism but soon became more mainstream products, or cultural/educational tourism. However, there has been a rapid increase in the supply of niche tourism products designed to meet the needs of tourists with special interests who wish to do something different, or who perhaps want to be seen doing something different, exclusive or exotic. Such niche products are frequently labelled under generic headings such as adventure tourism, cultural tourism, nature tourism, heritage tourism, sport tourism, health/wellness tourism, polar tourism, wilderness tourism, ecotourism and so on, although, as already noted, the distinction between these may not always be clear. Importantly, however, they are frequently marketed on the basis of being alternatives to mass tourism. Equally, specific forms of tourism reflect particular interests on the part of tourists – golf tourism, wine tourism or battlefield tourism are just three examples. The main point is, of course, that the demand for tourism is becoming increasingly diverse, providing destinations with opportunities to exploit specific resources or attractions. For example, and as mentioned above, over the last 25 years, Cuba has successfully developed resort-based, all-inclusive tourism. However, the country's capital, Havana, has also become an important heritage destination in its own right, with tourism underpinning the redevelopment of the city's colonial Spanish quarter (Colantonio and Potter 2006) (see also Plate 6.3). Equally, destinations also face the challenges of managing tourism in often-fragile environments, as well as remaining distinctive within an increasingly crowded global tourism market.

At the same time, traditional forms of tourism are now being consumed in more distant or exotic places. Europeans, for example, now travel to places as diverse as the Caribbean (see Plate 6.4), the Middle East, India, South East Asia, the Indian Ocean or the South Pacific to enjoy sun-beach holidays. Their 'pleasure periphery' has expended well beyond the Mediterranean, potentially bringing the alleged problems of mass tourism to places less able to withstand its impacts. One such destination is Dubai, which, though developing a successful tourism sector, faces potential problems in the future (Box 6.2).

Organizing and purchasing tourism

In addition to transformations in *what* tourists are consuming, changes are also occurring in *how* they are organizing and purchasing tourism experiences. On the one hand, packaged holidays remain popular with tour operators organizing and selling both 'traditional' holidays and new or niche tourism products, although it

Plate 6.3 *Cuba, Havana: horse and buggy rides for tourists can be contrasted with the local citizen form of transportation in the background where large trucks have been converted into buses*

Source: Photo by D. Telfer

Plate 6.4 *The Bahamas, Nassau: tourists walk through the Prince George Wharf area, which is the main shopping district for tourists arriving by cruise ship*

Source: Photo by D. Telfer

is mainly smaller, specialist operators that cater for specific, specialist markets. However, as a result of widespread Internet usage, there is an increasing trend towards tourists organizing and purchasing their holidays independently and online, while, of particular note, online reviews by tourists on websites such as TripAdvisor have become important in independent trip planning. There is also an increasing tendency for people to take more frequent holidays, supplementing or even replacing the two- or three-week holiday in the summer. A particular phenomenon has been the dramatic increase in short-break holidays facilitated by the growth in low-cost airline operations. Of less relevance to those countries that lie beyond the effective operational range of low-cost flights (although a small number of low-cost airlines operate long-haul flights, such as Norwegian Air Shuttle, which commenced flights from Scandinavia to New York and Bangkok in 2013), such short-break tourism has nevertheless brought significant economic benefits, for example, to many transitional economies in the former Eastern Europe. Low-cost carriers have also been operating in developing countries. SpiceJet is one of the leading low-cost carriers in India,

Box 6.2

Tourism development in Dubai

Over the last 30 years, Dubai, one of the seven sheikhdoms making up the United Arab Emirates, has rapidly evolved into one of the principal international tourist destinations in the Middle East. In the early 1980s, Dubai was reliant on oil production, which, at that time, accounted for two-thirds of GDP. However, dwindling oil stocks led to the development of tourism and, since then, the emirate's tourism sector has grown dramatically. Tourist arrivals increased from 374,000 in 1982 to 5,420,000 in 2004, and over the last decade have continued to grow remarkably. In 2012, almost 10 million international arrivals were recorded, and in 2014 the figure rose to 11.95 million. Tourist receipts have grown commensurately; in 2003, they exceeded US$1 billion, surpassing oil revenues, and in 2012 reached US$12 billion, representing some 30 per cent of Dubai's GDP. Current plans, as set out in Dubai's Vision for Tourism 2020, seek to increase arrivals to 20 million and receipts to US$36 billion by 2020, an ambitious target that will require an annual increase in arrivals of 9 per cent, more than double the projected global rate.

Underpinning this remarkable growth in tourism has been the equally dramatic increase in the supply of accommodation. By the end of the 1990s, there were some 20,000 hotel rooms on offer. By 2010, there were 566 hotels in the emirate offering a total of more than 67,000 rooms. In 2013, more than 17,000 additional rooms were under construction. Included in the supply of accommodation are a number of flagship properties, including the Burj al Arab, often claimed to be the world's only seven-star hotel, and the 1,500 room Atlantis Hotel on the Palm. At the same time, tourism to Dubai is boosted by both high-profile events, including the annual shopping festival and various sporting events, and the development of attractions and infrastructural projects. These include the now well-known Palm, two man-made islands that have increased Dubai's coastline by 120 km, and the controversial World project, a collection of islands shaped like countries, both of which will support accommodation and leisure services. In addition, attractions such as the Ski Dubai centre (a 400 m indoor ski slope), the Burj Khalifa (the world's tallest building) and the Dubai Mall, which is home to the Dubai Aquarium, are major attractions, while planned developments include a new artificial island, Bluewaters Island, on which the world's largest ferris wheel will be constructed. It should also be noted that in 2014, Dubai airport achieved the status of being the world's busiest airport, with more than 52 million passengers passing through its terminals. This number is forecast to grow to a staggering 103 million by 2020.

The majority of visitors to Dubai are from within the region, the most important market being Saudi Arabia, although China and Russia are also growing markets. The emirate has also become an increasingly popular (though relatively expensive) winter-sun destination for European tourists. Average length of stay, however, is relatively short, reflecting Dubai's status as a major air travel hub.

Despite its continuing success in developing tourism, Dubai faces a number of challenges, not least in maintaining sufficient arrivals to fill its burgeoning supply of accommodation. However, current occupancy levels remain surprisingly healthy, averaging around 78 per cent for all hotels. At the same time, the development of new attractions will undoubtedly attract tourists and new markets will both diversify and expand the number of arrivals, yet not only are projected growth figures highly ambitious, but also the environmental consequences of infrastructural developments may be significant. For example, controversy surrounds the development of the Palm and the World, with many claiming that these projects have damaged the local marine ecosystem. Moreover, the more that Dubai develops and, indeed, diversifies its visitor base, the more it will lose its Middle Eastern 'flavour', while the potential for conflicts between international tourists and local society/culture may intensify.

Source: Dubai Attractions (2015)

with over 210 daily flights to 34 Indian and eight international cities, including in Nepal, Sri Lanka and the UAE (SpiceJet 2015).

Tourism markets – tourism products

As the supply of tourism products and experiences has become increasingly diverse, the markets for such products have become less distinct. In other words, whereas once it was possible to suggest with reasonable certainty that particular forms of tourism would be consumed by particular groups of tourists, this is no longer the case. On the one hand, experiences that were once the preserve of the better off, such as cruise holidays or long-haul trips, are now available to the mass market as the industry has recognized the potential of mass marketing 'exclusive' holiday experiences. Equally, the tourism industry is becoming increasingly responsive to the demands of new markets for their products. For example, adventure tour companies that have traditionally organized overland trips for groups of young 'travellers' now sell adventure holidays for families seeking more active or challenging tourism experiences. On the other hand, tourists themselves are consuming tourism in more unexpected ways – one notable example is the fact that, of the half a million or so people in the UK taking so-called 'gap year' holidays each year (that is, a long study or career break), over half are over the age of 55 and are referred to by some as 'denture-venturers'!

One form of 'new' tourism that deserves further mention in the context of this chapter is so-called volunteer tourism. The concept of travelling to undertake volunteer activities is neither new nor something that has not long been facilitated by relevant organizations – in the UK, for example, Voluntary Service Overseas (VSO) has for more than 50 years been providing opportunities for people to engage in voluntary development and aid work overseas (www.vso.org), while the British Trust for Conservation Volunteers (BTCV) has been organizing voluntary conservation work and holidays since 1959 (www.tcv.org.uk). However, over the last two decades, participation in what has come to be referred to as volunteer tourism (or by some as 'voluntourism') has become increasingly widespread, with one study suggesting that some 1.6 million people are involved in volunteer tourism projects each year (see Wearing and McGehee 2013). Considered by many to be a distinctive form of alternative tourism – that is, a form of tourism in which tourists are not only able to give something to the destination, but also to have more meaningful experiences – it has also attracted increasing academic attention, not least because what appears at first sight to be an effective means of supporting destination communities, whether through conservation, engaging on development projects or, as is common, working in schools or orphanages, is in fact a highly controversial form of tourism that some consider to be more destructive than developmental (Guttentag 2009; Sin 2009).

Broadly speaking, volunteer tourism may be defined as a form of tourism in which tourists 'for various reasons, volunteer in an organized way to undertake holidays that might involve aiding or alleviating the material poverty of some groups in society, the restoration of certain environments, or research into aspects of society or environment' (Wearing 2001: 1). For some, volunteering may be the sole purpose of the trip, whereas for others it may be a relatively short 'add-on' to a longer trip. This immediately points to the first area of controversy, namely the motives of volunteer tourists themselves. That is, it is argued that many volunteer tourists are motivated not by altruism, but by self-interest, although others suggest most volunteers fall somewhere between the two (Tomazos and Butler 2010). Either way, although many tourists undoubtedly participate in volunteering for genuinely altruistic reasons, nevertheless the implication is that the potential exists for local communities to be exploited by volunteer tourists for the experience they gain, irrespective of the work they might undertake. Hence, Tourism Concern, for example, is running a campaign to stop volunteer placements in orphanages.

The second area of controversy surrounds the role the organizers of volunteer tourism. Whereas early volunteer tourism experiences were offered by charitable organizations with established links to development partners 'on the ground', the increasing demand for volunteer tourism has led to a growth in commercial operators whose primary interest is making a profit. As a consequence, debate surrounds the extent to which commercial operators focus meeting the experiential demands of their clients (the tourist) rather than the developmental needs of the local community and, indeed, the extent to which they are able to construct appropriate volunteer programmes. And third, controversy also surrounds the benefits of volunteer tourism to the destination community, with some commentators raising concerns about the commodified relationship between the 'host' and 'guest', the potential for local communities to become dependent on volunteer organizations and the extent to which volunteer activities reflect external, volunteer tourism organizers' agendas.

Volunteer tourism, then, is an increasingly popular yet contested form of alternative tourism (for a detailed review, see Wearing and McGehee 2013), which, irrespective of the criticisms discussed above, reflects a transformation in styles of travel and tourism more generally. That is, traditional, packaged and passive forms of tourism are being rejected in favour of more individualistic, specialist and active experiences. At the same time, the controversy surrounding volunteer tourism also reflects the wider issue of the extent to which attitudes towards tourism have also changed. In other words, are tourists still motivated by the needs of escapism and/or ego-enhancement, or are we witnessing the emergence of the 'responsible' tourist (Goodwin 2011)?

Is there a responsible tourist?

The development of sustainable forms of tourism has long been justified by the belief that tourists in general are becoming 'greener' or more responsible. In other words, reflecting the trends in demand outlined in the previous section, it is claimed by many that the traditional, mass-package tourist is being replaced by a more experienced, aware, quality-conscious and proactive tourist-consumer. That is, following a shift in general consumer attitudes, tourists 'want more leisure and not necessarily more income, more environmentally sustainable tourism and recreation and less wasteful consumption' (Mieczkowski 1995: 388). At the same time, since the early 1990s, innumerable publications have exhorted tourists to act more responsibly, to become 'good' tourists (for example, Wood and House 1991), though this has been criticized as the 'moralisation' of tourism (Butcher 2002). However, the extent to which tourists are becoming more responsible remains the subject of intense debate, with some commentators concluding that the holiday choices and behaviour of few tourists is actually determined by environmental concerns.

There are two arguments in favour of the emergence of the responsible tourist. First, the rapid growth in demand over the last decade for activities or types of holidays that may collectively be referred to as 'responsible', such as ecotourism or volunteer tourism (see below), as well as the success of organizations such as Responsible Travel that promote and sell 'responsible' holidays, is often cited as evidence that ever-increasing numbers of tourists, as a result of their heightened environmental awareness or concern, are seeking out more appropriate and, in a developmental sense, beneficial forms of tourism. For example, over two decades ago, Cater (1993) reported that the number of arrivals to three 'selected ecotourism destinations', namely Belize, Kenya and the Maldives, virtually doubled over a 10-year period from 1981. Similarly, some have claimed that participation in ecotourism has increased annually by between 20 and 50 per cent since the early 1980s and now accounts for up to 20 per cent of all international tourism arrivals (Fennell 1999: 163), although others make more conservative estimates.

Second, surveys consistently point to the alleged emergence of the 'green consumer' both generally and in the more specific context of tourism. In the early 1990s, for example, it was found that increasing numbers of people in the UK considered themselves to be either 'dark green' (that is, 'always or as far as possible buy environmentally friendly products') or 'pale green' (that is, 'buy if I see them') consumers (Mintel 1994). A more recent survey by the same organization found that green consumerism is increasing, although consumers are more likely to buy environmentally friendly products to feel good about themselves rather than for altruistic reasons. Indeed, another study found that the act of purchasing may make people feel that they have 'done their good deed' and thus may act less altruistically when presented with other ethical dilemmas (Mazar and Zhong 2010). This suggests that that the act of buying a responsible holiday may absolve tourists from any sense of responsibility while actually on holiday, a point that has long been made: 'Responsible tourism is a so-called solution that keeps everyone happy. It appeases the guilt of the "thinking tourist" while simultaneously providing the holiday experience they or we want' (Wheeller 1991: 96).

Other studies have also pointed to the growth in responsible consumerism. For example, Cowe and Williams (2000) found that one-third of consumers are seriously concerned with ethical issues when shopping, while, in the tourism context, Diamantis (1999) suggests that 64 per cent of UK tourists believe that tourism causes some degree of damage to the environment, and that, generally, UK consumers would be willing to pay more for an environmentally appropriate tourism product. A survey by the charity Tearfund (2000) came to similar conclusions; specifically, it found that 59 per cent of respondents would be happy to pay more for their holidays if the extra contributed to better local wages, environmental conservation and so on.

However, in practice, there is little evidence to support either of these arguments. While there is no doubt that the demand for ecotourism and other forms of environmentally aware travel is on the increase, research has consistently failed to demonstrate that tourists consuming such experiences are motivated or influenced by environmental values (Sharpley 2006). Indeed, other research has consistently demonstrated that tourism is relatively immune to environmental concerns (or that 'responsible' tourist behaviour is motivated by factors other than environmental concern). For example, in one poll, just 1 per cent of tourists stated that their carbon footprint was an important factor when deciding on a holiday purchase, whereas cost is the most important consideration for 43 per cent of tourists (Skidmore 2008). More specifically, studies into the motivation of ecotourists show that the majority seek wilderness scenery, undisturbed nature and the activities that such locations offer as the prime reasons for participating in ecotourism. In other words, it is the pull of particular destinations or holidays (and the anticipated enjoyment of such holidays) that determines participation rather than the influence of environmental values over the consumption of tourism in general. In fact, research into the behaviour of ecotourists in Belize demonstrates that many are motivated by factors other than environmental concern (Box 6.3).

Box 6.3

Ecotourists in Belize

The Central-American country of Belize was one of the first countries to recognize and embrace the concept of ecotourism as a means of achieving sustainable development through tourism. In the late 1980s, the country adopted an Integrated Tourism Policy to develop ecotourism, and since then it has become an increasingly popular destination for ecotourists, drawn by the country's natural and cultural (Mayan) heritage and, in particular, the opportunity to explore its famous coral reefs. The subsequent development of tourism in Belize has been widely criticized, specifically the extent to which both national elites and overseas organizations have exploited the country's natural resources, but one study in particular has revealed the actual motivations of some 'ecotourists' in Belize.

Contrary to the assumption that ever-increasing numbers of tourists are motivated by environmental concerns, research has shown that many visitors to Belize demonstrate behaviour that is stereotypical of the mass tourist. Thus, many tourists are attracted to the country by the opportunity to dive on coral reefs or other natural/cultural experiences, though their visits are framed by the desire for a generic 'Caribbean' experience (beaches, palm trees, reefs, brightly coloured fish), to collect certain sights and experiences, rather than to learn about, experience and contribute to Belize in particular. Few have genuine environmental knowledge and awareness – their stated desire to avoid damaging the coral reefs, for example, is to maintain them as tourist attractions rather than for their intrinsic value – and many act in an inappropriate fashion. Moreover, for many tourists, the attraction of Belize lies in the opportunities for hedonistic behaviour, including heavy drinking, drug taking and sexual encounters, either with locals or with other tourists. In short, they lack the environmental awareness and self-reflexivity that might be expected of an ecotourist.

Source: Duffy (2002)

More importantly, perhaps, the assumptions about widespread adherence to the principles of green consumption in general, and the consequential adoption of responsible tourism consumption practices in particular, also lack foundation. Although there is widespread stated support for green consumerism, it is evident that few consumers regularly buy/consume according to environmental or ethical values. In fact, research has shown that in the UK, less than 1 per cent of people consistently practise green consumerism. Moreover, such behaviour is unlikely to remain constant over time or be applied to all forms of consumption. In short, consumers address environmental issues in complex and ambivalent ways and, as a result, their consumer behaviour is frequently contradictory. That is, their environmental attitudes towards the consumption of different products are likely to vary. Indeed, despite the surveys identifying strong environmental credentials on the part of tourists, UK tourists' spending with responsible tour operators amounted to just £112 million in 2004 (Cooperative Bank 2005), compared with some £26 billion spent on all overseas travel. More specifically, a study at Stansted airport in the UK in 2009 found that only 7 per cent of flyers were funding green energy projects and purchasing carbon offsets for their flights (Web 2010). Thus, although the responsible tourist undoubtedly exists, the great majority of tourists are more likely not to be influenced by environmental values in their tourism consumption choices and behaviour (Sharpley 2012).

Influencing tourist behaviour: a destination perspective

The form that tourism development in the destination takes is directly related to the nature of tourism consumption, in terms of both products demanded and actual behaviour 'on site'. In other words, for the achievement of sustainable tourism development (as defined in Chapter 2), it is necessary that not only should tourism be developed in an appropriate way with regards to character, scale, degree of integration with the local economy and so on, but also that tourists themselves should behave in a responsible manner. As the example of Belize (Box 6.3) demonstrates, developing an ecotourism product does not guarantee sustainable tourism development.

As this chapter has suggested, however, the extent to which tourists are becoming more environmentally aware is unclear. That is, an understanding of the consumption of tourism suggests that responsible tourist behaviour, or purposefully consuming tourism experiences that benefit destination environments and societies, is unlikely to be a widespread phenomenon. How, then, can tourist behaviour be influenced in order to enhance the sustainability of tourism development? Four courses of action deserve attention here:

- codes of conduct;
- effective destination planning (according to broader development goals);
- destination marketing; and
- role of tour operators.

Codes of conduct

In recent years, codes of conduct have emerged as a popular and widely utilized visitor management tool. This is not to say that they are a new phenomenon – in the UK, for example, the Country Code, advising visitors to the countryside how to act appropriately – was first published in 1953. However, codes of conduct are now considered a vital means of promoting responsible behaviour on the part of tourists and, nowadays, numerous such codes exist. Some relate to tourism in general, such as an early 'Code of Ethics for Tourists' (O'Grady 1980), which focuses upon potential sociocultural and economic impacts (Figure 6.3), while others may be destination-specific (for example, the Himalayan Tourist Code) or activity-specific. Moreover, most are directed towards tourists themselves, although some are targeted at the tourism industry or local communities. The majority of codes are produced by non-governmental organizations or special interest groups. Conversely, the tourism industry and governmental bodies have typically produced few codes (Mason and Mowforth 1996), although one contemporary example is the 'code of ethics' provided by Last Frontiers, a specialist UK operator that organizes bespoke holidays to Latin America (Last Frontiers 2015), which, perhaps not surprisingly, bears a strong resemblance to Figure 6.3:

- Travel with a genuine desire to learn more about the people of your host country.
- Your pre-trip research can be continued by asking questions while you are there. Any attempt to learn even a few words of the relevant language will make this a more rewarding experience for all involved.
- Realize that often the people in the country you visit have time concepts and approaches different from your own.
- Don't treat people as part of the landscape; they may not want their picture taken.
- Do not make promises to people in your host country unless you can carry them through.
- Ensure that your behaviour has no impact on the natural environment. Avoid picking flowers, removing seeds, damaging coral, and even buying souvenirs such as shells and skins (see Plate 6.5).
- Try to put money into local people's hands: drink local beer or fruit juice rather than imported brands, and buy and eat locally

produced food. When you are shopping, even where bargaining is expected, do inject humour and remember a low price almost certainly means a lower wage for the maker.
- If you want to take gifts, make sure they are appropriate.
- If you really want your experience to be a 'home away from home', it may be foolish to waste money on travelling!

Collectively, however, there is an enormous and potentially confusing diversity of codes of conduct in tourism. For example, one study found that, globally, there are 58 separate codes of conduct relating to whale watching alone (Garrod and Fennell 2004).

The purpose of codes of conduct is, most usually, to raise awareness among tourists of the need for responsible behaviour. In essence, they are lists of rules for appropriate behaviour in particular contexts, although, as codes, they have no legal status. That is, they depend upon voluntary compliance on the part of tourists, and, as a result, their effectiveness may be limited, particularly as tourism is seen as a means of escaping from rules and regulation. Therefore, tourists are more likely to respond positively to codes if, in addition to how, it is explained why they should behave in particular situations.

Figure 6.3 *A code of ethics for tourists*

1. Travel in a spirit of humility and with a genuine desire to learn more about the people of your host country.	7. Instead of the Western practice of knowing all the answers, cultivate the habit of asking questions.
2. Be sensitively aware of the feelings of other people, thus preventing what might be offensive behaviour on your part. This applies very much to photography.	8. Remember that you are only one of the thousands of tourists visiting this country and do not expect special privileges.
3. Cultivate the habit of listening and observing, rather than merely hearing and seeing.	9. If you really want your experience to be 'a home away from home', it is foolish to waste money on travelling.
4. Realise that often the people in the country you visit have time concepts and thought patterns different from your own; this does not make them inferior, only different.	10. When you are shopping, remember that the 'bargain' you obtained was only possible because of the low wages paid to the maker.
5. Instead of looking for that 'beach paradise', discover the enrichment of seeing a different way of life, through other eyes.	11. Do not make promises to people in your host country unless you are certain to carry them through.
6. Acquaint yourself with local customs – people will be happy to help you.	12. Spend time reflecting on your daily experiences in an attempt to deepen your understanding. It has been said that what enriches you may rob and violate others.

Source: O'Grady (1980)

Plate 6.5 *South Africa, near Pretoria: shops selling souvenirs to tourists; note the taxidermy animals for sale*

Source: Photo by D. Telfer

Destination planning

The purpose of sustainable tourism is to optimize tourism's contribution to a destination's sustainable development. Traditionally, the achievement of this objective has been seen to lie in the development of small-scale, appropriately designed, locally controlled projects, the implication being that these types of tourism development will attract the 'right' kind of tourists. However, it is now accepted that the challenge is to make all forms of tourism sustainable. In other words, different types of tourism development may be more or less suited to the particular resources and developmental needs of different destinations and, therefore, destination planners may be able to influence the nature of tourism consumption through effective planning and promotion of their tourism product.

Tourism in Bhutan, considered in some detail in Chapter 2, is one example of the effective manipulation of tourism consumption by the destination. There, tightly controlled pricing and distribution of holidays in the country, along with strict regulations on tourism activities, ensures that tourists behave in a way that is appropriate to local conditions and needs. Elsewhere, the development of all-inclusive tourism has made a significant contribution to local development while restricting the negative consequences of tourism through what is, in effect, the zoning of tourism development. Importantly, such developments also meet tourists' needs. When the tourism authorities in the Gambia attempted to influence tourism consumption by banning all-inclusive holidays, thereby hoping to encourage expenditure beyond the confines of hotels, demand for holidays in the Gambia declined dramatically.

Equally, there are, of course, numerous examples of destinations or tourism developments that, at the local scale, successfully meet the more typical principles of sustainable tourism (Buckley 2003). The Global Sustainable Tourism Council's Sustainability Criteria for Destinations is an example of criteria that can be used to plan destinations in a sustainable fashion. The point is that tourism consumption cannot be divorced from local developmental needs, and thus destination planners can play an important role in ensuring that tourism consumption (even if it is mass tourism) is appropriate to those needs.

Destination marketing

As observed earlier in this chapter, a significant influence on the tourism demand process (and the subsequent behaviour and expectations of tourists) are so-called 'effectors' of demand; that is, destination choice is, to a great extent, influenced by the knowledge, images and perceptions that tourists have of particular places or experiences. The impact of these images and perceptions upon tourist-buyer behaviour has been explored widely in the literature (Pike 2002); however, it is generally accepted that not only are destinations in the position to augment or adapt tourists' images and perceptions, but also that they have the opportunity to influence the nature of demand for their product. As Buhalis (2000: 97) notes, 'destination marketing facilitates the achievement of tourism policy, which should be coordinated with the regional development strategic plan'.

Despite this potentially influential role of destination marketing in contributing to the appropriate development and consumption of destination experiences, there has been a tendency, particularly in less-developed countries, to market or promote places in order to verify, rather than adapt, tourists' images and perceptions (Silver 1993). For example, it has been suggested that much tourism destination marketing in developing countries enhances stereotypical (though not necessarily accurate) images of traditional, undeveloped cultures, and colonial power relations (Echtner and Prasad 2003). As a consequence, tourists may be attracted by marketed representations of the destination that reflect their preconceptions rather than the reality of the place, influencing both their experience and their behaviour.

Nevertheless, destination marketing can play a more positive and proactive role in sustainable destination development. In general, marketing should be resource- rather than demand-based; that is, it should focus on promoting particular attributes of the destination rather than on meeting tourists' perceptions. More specifically, a variety of techniques may be employed, including (see Buhalis 2000):

- *Enhancing and differentiating the product*: Tourists' needs for unique or authentic experiences, for which they may be willing to pay a premium, will be met by a quality, differentiated experience.
- *Appropriate pricing policies*: Pricing is an effective tool in market segmentation (see below) and in ensuring the full cost to the destination is reflected in the price tourists pay.
- *Embracing the needs of all stakeholders*: The needs of destination-based stakeholders should, in particular, be taken into account in destinational marketing.
- *Developing effective public–private sector partnerships*: Such partnerships may facilitate the meeting of broader developmental objectives in the destination.
- *Effective segmentation to target appropriate markets (Leisen 2001)*: A closer matching of tourists needs/expectations with destinational resources reduces the potential for negative consequences.

Role of tour operators

Tour operators play a central role in tourism development. They have been described as the 'gatekeepers' of the tourism industry, being able to influence the scale and scope of tourism development as well as the volume and direction of tourist flows. As a result, tour operators are often seen (somewhat inaccurately) as epitomizing mass tourism development, providing cheap holidays to mass markets with little regard for the impacts on destination environments and societies.

In the context of this chapter, there is an evident link between tour operators and the consumption of tourism. Not only do they create and satisfy the demand for tourism experiences, from mass sun-sea-sand holidays to more specialized, niche products, but through the promotion and pricing of their products they are able to dictate the type of tourists travelling to any particular destination. Thus, the success of so-called 18–30s holidays (and the much-criticized behaviour of young tourists on such holidays) is very much a result of the manner in which they are promoted.

Increasingly, tour operators are adopting a more responsible approach to their activities, one notable example being the former Tour Operators' Initiative, which long sought to encourage sustainable practices among its members, and which in 2014 became part of the Global Sustainable Tourism Council (see Chapter 2). According to the Guide for Responsible Tour Operations (UNEP 2005), tour operators should focus on five areas in improving the sustainability of their operations, one of which (customer relations) is concerned with influencing tourist behaviour. Under this heading, a number of principles are proposed:

- *Raise clients' awareness of sustainability issues*: Provide positive messages in brochures and other information as to how tourists can contribute to the social and environmental well-being of destinations.
- *Develop or adopt a responsible tourism code of conduct*: Tour operators may adopt an existing code or develop their own, communicated to customers in a pre-departure information pack.
- *Communicate sustainability messages throughout the holiday cycle*: Information on sustainability can be provided at all stages of the holiday process, from booking through departure and travel (in-flight messages) to destination experience and post-holiday questionnaires.
- *Reinforce the message between holidays*: Regular mailings to existing customers, such as in-house magazines or promotional leaflets, provide a medium for reinforcing sustainability messages.

The success of any of these methods depends, of course, on tourists' responsiveness to these messages; this, in turn, may depend upon the extent to which a tourist holds strong environmental values. As considered earlier in this chapter, the tourism consumption experience may be dominated by other, more ego-centric values, in which case tour operators may be able to contribute to destination sustainability through more direct measures, such as influencing planning, working with local stakeholders or supporting local voluntary/charitable organizations.

Domestic tourism

Not surprisingly, perhaps, most attention in the context of tourism consumption and sustainable tourism development is focused upon international tourism, particularly with respect to tourism in the less-developed world. On the one hand, international tourism represents, for many countries, a significant export industry and a vital source of foreign exchange earnings. On the other hand, the potential for negative consequences is seen to be greater, and therefore the need for effective planning and management is more pressing.

Frequently overlooked, however, is the contribution that domestic tourism – that is, people consuming tourism in their own countries – can make to sustainable development (Ghimire 2001). In other words, domestic tourism provides many of the benefits of international tourism, such as employment, income, new business development and economic diversification; at the same time, it is likely to benefit locally owned and controlled businesses. Therefore, it is important to consider briefly the role of domestic tourism consumption in sustainable development.

Domestic tourism occurs in most, if not all, countries. It is also, in terms of total numbers of trips, significantly greater than international tourism activity; it is estimated that, globally, domestic tourism annually accounts for between six and 10 times more trips than international tourism. The reasons for this are self-evident. In wealthier, developed countries, residents frequently make a number of domestic tourist trips each year in addition to overseas holidays, and many only take domestic holidays. For example, research in the early 2000s revealed that that in the United States, residents collectively made 990 million domestic trips each year compared with some 60 million overseas trips (Bigano et al. 2004). In less-developed countries, domestic tourism is, for financial reasons, the only option for most residents, although in some the number of outbound trips is on the increase. However, the rise in low-cost carriers in countries such as India, China, Thailand and Indonesia, to mention but a few, illustrates the growing wealth among some in developing countries leading to more travel by air, both domestically and abroad. More importantly, in most less-developed countries, 'the number of nationals traveling for leisure is considerably higher than the number of international tourist arrivals' (Ghimire 2001: 2). The number of domestic trips is, of course, directly related to population size; according to Bigano *et al.* (2004), China, India, Brazil and Indonesia are the four largest domestic tourism markets in the less-developed world. However, even in smaller countries, such as Turkey, domestic tourism is a growing sector and a potential contributor to regional development (Seckelman 2002), with recent figures indicating that more than 68 million domestic tourist trips were made by Turkish residents in 2013 (Turkstat 2015). Interestingly, in South Africa, the National Department of Tourism has made domestic tourism a priority, seeking to increase its contribution as a percentage of the overall tourism contribution to gross domestic product from 54.8 per cent in 2009 to 60 per cent by 2020 (Kruger and Douglas 2015).

Singh (2009a) notes that the monetary exchanges with domestic tourism take place directly at the grass-roots level with those providing the product locally, while there are a number of other potential benefits of promoting domestic tourism consumption:

- It can act as a catalyst of local or regional economic growth, particularly in peripheral rural areas; this has been a particular driver of domestic tourism in China, for example.
- It may spread wealth from richer, urban areas as emerging middle classes engage in domestic leisure travel.
- In some instances, the potential for sociocultural impacts may be less than with international tourism development, while opportunities for community-based tourism may be greater.
- Domestic tourism may provide the basis for developing greater social cohesion, national cultural identity or, as in the case of South Africa, encouraging national reconciliation (Koch and Massyn 2001).

At the same time, of course, domestic tourism development faces many similar challenges to international tourism development, including environmental degradation, social pressures and cultural commodification, while local tourism resources (businesses, land, attractions, etc.) may come to be owned or controlled by urban elites. The key to success lies in effective planning and management, as has been adopted in countries such as Brazil and Mexico. Generally, however, there is still a lack of knowledge and understanding about the benefits and costs of domestic tourism in the less-developed world.

Discussion questions

1 What are the emerging trends in tourism demand?
2 Are tourists really interested in becoming green?
3 What are the advantages and disadvantages of targeting a specific sociodemographic tourist segment or tourist type such as cultural tourists or adventure tourists?
4 How can marketing be used to control tourist behaviour?
5 What are the advantages of promoting domestic tourism?

Further reading

Butcher, J. (2002) *The Moralisation of Tourism: Sun, Sand . . . and Saving the World?*, London: Routledge.

This book sets out to challenge the view that not only are tourists becoming 'greener', but also that tourists and the tourism industry should adopt a more responsible approach to the consumption and development of tourism. It provides a fascinating yet controversial critique of contemporary, responsible approaches to tourism.

Mann, M. (2000) *The Community Tourism Guide*, London: Earthscan.

Essentially a guide to so-called 'real' holidays, or responsible/sustainable community-based tourism products, this book sets out the reasons why tourists should behave responsibly and provides numerous examples of responsible tourism in practice.

Sharpley, R. (2008) *Tourism, Tourists and Society*, 4th edn, Huntingdon: Elm Publications.

This book explores the relationship between tourists and society from two perspectives: the influence on society of tourists/tourist behaviour, and the influence (impacts) of tourists on society. It provides an in-depth analysis of tourism demand and behaviour, and particular attention is paid to the nature of, and influences on, tourism consumption within a postmodern context.

Swarbooke, J. and Horner, S. (2007) *Consumer Behaviour in Tourism*, 2nd edn, Oxford: Butterworth-Heinemann.

This provides a thorough and broad introduction to tourism consumption. Introducing the main theories and concepts relevant to the consumption of tourism, it considers principal contemporary issues and debates and refers to a wide range of literature.

Websites

www.ecotourism.org

This is the website of the International Ecotourism Society (TIES), a global network of individuals, institutions and tourism industry representatives that promotes and supports responsible practices in tourism.

www.ecpat.org.uk

ECPAT is the leading children's rights organization that seeks to prevent the exploitation of children, including through tourism.

www.responsibletravel.com

Responsible Travel is an online travel agency specializing in 'responsible' travel and holidays.

www.slowmovement.com

The Slow Movement promotes a 'slower' approach to life that enables people to make connections with their world. This includes so-called slow travel, an approach to tourism that provides an opportunity to become part of local life and to connect to a place and its people.

www.tourismconcern.org.uk

Tourism Concern is a UK-based group that seeks to raise awareness of, and solutions to, the negative social, cultural, environmental and economic consequences of tourism.

7 Assessing the impacts of tourism

Learning objectives

When you have finished reading this chapter, you should be able to:

- appreciate the wider social, political and economic contexts within which tourism's impacts occur
- identify the main sociocultural, environmental and economic impacts of tourism
- understand how local residents react to the impacts of tourism
- recognize measures to help minimize the negative impacts of tourism

The central theme of this book is that tourist destinations, particularly those in the developing world, face a dilemma. On the one hand, tourism is widely seen as an effective means of achieving development; it represents a potentially valuable source of income and employment and a driver of broader economic, infrastructural and sociocultural development. On the other hand, such development cannot be achieved without cost. Not only does tourism, as a resource-based industry, inevitably exploit or 'use up' resources, be they natural, man-made or human (McKercher 1993), but also the activities of tourists themselves may have significant impacts on the destination, on its environment and local communities. If not controlled or managed, these costs may outweigh the benefits of tourism development and, in the longer term, reduce the attraction of the destination to tourists.

The dilemma facing destinations, therefore, is how to meet the broader developmental objectives of the destination by optimizing the contribution of tourism while at the same time keeping the costs or negative consequences of tourism development to a minimum. In other words, the challenge for destinations is to achieve sustainability of the tourism sector, fundamental to which is the need to balance the positive and negative impacts of tourism within the context of broader development goals.

The overall purpose of this chapter is to consider the nature of these impacts and to explore the ways in which they may be planned and managed effectively within a sustainable developmental context. First, though, it is important to emphasize that the impacts of tourism cannot, or should not, be viewed in isolation. In other words, the impacts of tourism or, as Wall and Mathieson (2006) suggest, the 'consequences' of tourism – the term 'impacts' is often seen as having negative connotations – are frequently considered generically in the tourism literature, with typical impacts simply listed and described. However, the way in which such impacts are perceived, the extent to which they are felt and the manner in which they are responded to varies enormously from one destination to another. In other words, tourism impacts arise from a complex interrelationship between the destination, the tourism industry and tourists, and should be viewed within the broader and dynamic economic, social and political contexts within which they occur. Therefore, the first section of this chapter provides a framework for assessing the impacts of tourism.

Tourism impacts: a framework

The impacts of tourism are felt across the tourism system; moreover, such impacts may be both positive and negative. In tourism-generating regions, for example, outbound tourism is an important source of employment in travel retailers, tour operators, airports, transport operators, and so on. Conversely, air transport – the 'transit region' (Leiper 1979) – is seen as a major contributor of greenhouses gases and, hence, a significant negative impact (see Chapter 8). Nevertheless, the impacts of tourism are usually considered in the context of the destination where tourism development occurs, where tourists come into contact with local people and environment and where there is arguably the greatest need to identify, measure and manage such impacts, both positive and negative.

In assessing the impacts of tourism, a useful starting point is to consider the destination as an overall tourism environment. Frequently, the 'environment' is thought of simply in terms of the physical attributes (natural and built) of the destination. However, tourists seek out attractive, distinctive or authentic tourism environments or destinations, which, as Holden (2000: 24) observes, possess 'social, cultural, economic and political dimensions, besides a physical one'. In other words, the tourism environment may be defined as:

> that vast array of factors which represent external (dis-)economies of a tourism resort: natural . . . anthropological, economic, social, cultural, historical, architectural and infrastructural factors which represent a habitat onto which tourism activities are grafted and which is thereby exploited and changed by the exercise of tourism business.
>
> (EC 1993: 4)

Thus, from a tourist's perspective, the social or cultural aspects of a destination are usually inseparable from its physical aspects in terms of their overall experience of that destination. The important point, however, is that perceptions of the tourism environment may vary significantly. That is, different groups will perceive or value the tourism environment in different ways. As we saw in Chapter 6, tourists' attitudes towards the destination may vary enormously, as might their subsequent behaviour while on holiday. Equally, there is likely to be a distinction between the ways in which local communities and tourists perceive the destination environment, reflecting local cultural/environmental values, economic need and so on. For example, while tourists may value a pristine or undeveloped environment, locals may view it as a legitimate resource for exploitation as they seek to enhance their lifestyle or achieve social and economic development. Conversely, local communities may strongly value or hold sacred particular places or environments and seek to protect them from tourism exploitation or development. Uluru (Ayers Rock) in Australia is a well-known example of the latter (Brown 1999; Digence 2003), while the authorities in Bhutan have long sought to protect sites of cultural or environmental significance from tourism exploitation (Dorji 2001; Nyaupane and Timothy 2010). At the same time, it has been found that grass-roots activist groups in southern Europe, specifically in Greece, Spain and Portugal, have in the past engaged in demonstrations and other forms of protest against environmentally damaging tourism developments (Kousis 2000). Elsewhere, in Kerela, India, fisherwomen protested against World Tourism Day celebrations in 2007 (which had the theme 'Tourism Opens Doors for Women') demanding that the government stop acquiring further coastal land for the development of tourism (Kerala Tourism Watch n.d.).

In other words, when assessing the impacts of tourism, it is essential to consider the broader social, political and economic context of the destination and to recognize that planning and management decisions with respect to tourism development should reflect that local context rather than external, often Western-centric values. This is, of course, particularly the case in less-developed countries where the need for socio-economic development and modernization may be considered more important than providing tourists with traditional or authentic experiences; the example of Singapore where, in the 1970s, traditional Chinese shops and markets were replaced with contemporary buildings and experiences (Lea 1988) has since been repeated in many other places.

It is also important to consider the impacts of tourism within a wider national and global context. Tourism destinations do not exist in isolation but are part of an international tourism system within which major multinational corporations frequently play a dominant role, resulting in a condition of dependency in many destinations (see Chapter 1). At the same time, the international tourism system is part of a global political, economic and sociocultural system. Consequently, two points should be noted:

- The impacts of tourism in a particular destination may often be influenced by factors beyond that destination's control (Telfer and Hashimoto 2015). For example, the US's embargo on Americans travelling to Cuba has undoubtedly affected the development of the island's tourism sector, while the more recent improvement in relations between the two countries may well prove a boost to tourism on the island (Sharpley and Knight 2009). Similarly, campaigns by a number of international organizations and pressure groups to boycott tourism to Myanmar in the late 1990s also had some influence on the decline in international tourist arrivals in that country (Henderson 2003), although again, moves towards democracy within Myanmar since 2010 have reversed that trend.
- Impacts that are attributed to tourism are often the result of wider, global influences. For example, and as we shall see shortly, international tourism is often blamed for weakening local

culture, threatening traditional social structures or for introducing Western cultural practices to local communities. However, the alleged globalization of Western culture (Schaeffer 2009), underpinned by dramatic advances in information and communication technologies, is seen as a more powerful influence on sociocultural change in many less-developed countries.

Figure 7.1 *Impacts of tourism: a framework for analysis*

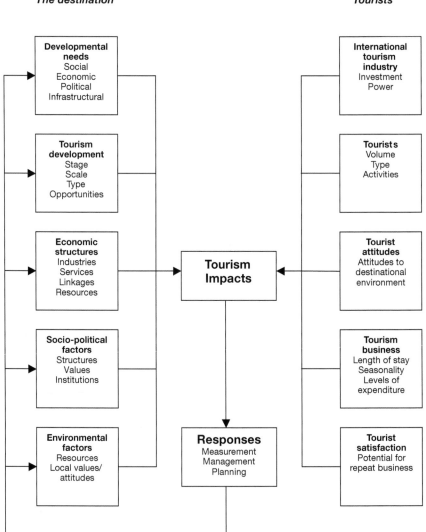

Figure 7.1 provides a simple framework for assessing the impacts of tourism at the level of the destination. This illustrates that the nature of tourism impacts, the manner in which they are perceived and subsequent planning and managerial responses are a function of both destinational and tourism developmental factors. Implicit in this model is the influence, referred to above, of local needs and values with regards to perceptions of and responses to the impacts of tourism, although the extent to which such influence may be exercised will reflect local political and social structures.

However, Figure 7.1 also illustrates the tangible factors that may determine the nature and extent of tourism's impacts in destination areas. These fall under two broad headings, and may be summarized as follows:

Tourists

There is not necessarily a causal effect between the volume of tourists and degree of impact. In some instances, such as culturally or environmentally fragile areas, a very small number of tourists may have a major negative impact, whereas in other areas large numbers of tourists may cause relatively limited negative impacts yet provide significant economic/developmental benefits. Nevertheless, greater negative impacts are usually associated with high volume tourism, particularly when such tourism is associated with supply dominated by international tourism businesses, hence a condition of dependency (see Chapter 1). More certain is the fact that, as discussed in detail in the previous chapter, the type of tourists, their attitudes and their subsequent activities will have a direct influence on the nature and extent of destinational impact. Figure 7.2 summarizes the relationship between tourists' environmental values and their subsequent actions or impacts.

Destination factors and characteristics

There are a variety of destination characteristics that determine both the positive and negative impacts of tourism. These are considered elsewhere (Burns and Holden 1995; Wall and Mathieson 2006) but include the following:

- *Character and sensitivity of the environment*: Some environments or ecosystems are more fragile, less robust or more sensitive to

Figure 7.2 *Tourist experience of destination environment*

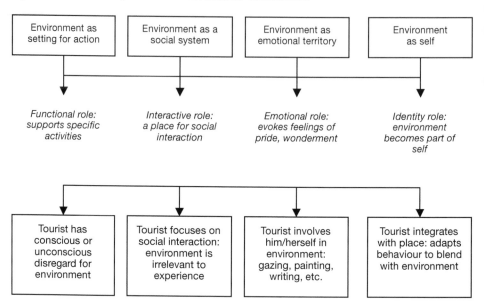

Source: After Holden (2000: 49–50)

change than others, or may take longer to recover from physical damage.

- *Economic structures and stage of development*: The diversity of the local economy, the availability of investment funding, the import-export balance and the ability to meet the tourism sector's requirements, as well as the overall extent of economic development, will determine the degree of economic benefit enjoyed.
- *Political structures and processes*: Tourism development is influenced by local and national political structures, the extent of political engagement or influence in tourism planning and management, and the nature of policies directly and indirectly affecting tourism (Sharpley and Ussi 2014).
- *The nature, scale and rate of growth of tourism development*: An evident link exists between the character and scale of tourism development and the subsequent impacts on the destination. However, if tourism development outstrips appropriate infrastructural development, such as sewage systems, negative impacts may be enhanced.
- *Social structures*: The size and structure of local communities, their cultural practices, moral codes, religious affiliations,

language and so on (particularly in comparison to those of tourists) are fundamental to the degree of 'felt' impact at the destination.

There are, then, numerous influences on the extent to which tourism impacts both positively and negatively on a destination and the ways in which those impacts are perceived and responded to by local communities and visitors alike. The next task is to actually identify the 'typical' impacts of tourism.

The impacts of tourism: an overview

A detailed discussion of the impacts of tourism is well beyond the scope of this chapter, and readers should refer to more detailed reviews of the topic in the literature. Many general tourism textbooks cover the topic more than adequately (for example, Hall and Lew 2009; Holloway and Humphreys 2012; Fletcher *et al.* 2013; Page and Connell 2014), while innumerable books and articles address it specifically (in particular, Wall and Mathieson 2006; Mason 2008). Nevertheless, the relationship between tourism and development cannot be fully understood without an appreciation of the principal impacts of tourism, both positive and negative. Indeed, 'development', as defined at the beginning of this book, is the intended overall outcome or impact of tourism; the challenge is to manage the more specific impacts effectively so that tourism's contribution to development is optimized. The purpose of this section, therefore, is to introduce the main impacts of tourism.

Typically, tourism's impacts are assessed under three broad headings, namely, economic impacts, physical (environmental) impacts and sociocultural impacts. For convenience, this section does likewise.

Economic impacts

The driving force behind the development of tourism is its potential contribution to destination economies. This is particularly so in less-developed countries, where tourism is seen as an effective (and, sometimes, the only) catalyst of economic growth and wider socio-economic development. For many such countries throughout the developing world, tourism represents an economic lifeline, its contribution usually measured in terms of earnings from tourism

(tourism receipts), export earnings (balance of payments), contribution to GNP and employment generation. In some cases, such as Mexico, Thailand, the Seychelles and Fiji, tourism has proved to be an economic success, both as a specific sector and as a driver of development; elsewhere, however, it has failed to achieve its developmental potential, despite providing an important source of foreign exchange earnings and employment. As noted in Chapter 2, Zanzibar has developed little over the last two decades, despite the relative importance of tourism to its economy, while the Gambia is another example of a country that has enjoyed limited success in exploiting its tourism sector as an engine of national development (Sharpley 2009b).

In other words, although tourism undoubtedly makes a measurable contribution to destination economies, neither the magnitude of tourism's economic impacts nor its role in stimulating wider socio-economic development can be taken for granted. Not only are there various economic costs associated with tourism development that, in effect, limit its net economic benefits, but also there are many factors that may reduce its broader developmental contribution (see Figure 7.3). As a result, care should be taken in assessing the economic impacts of tourism (Wall and Mathieson 2006: 79);

Figure 7.3 Factors that influence the economic impacts of tourism

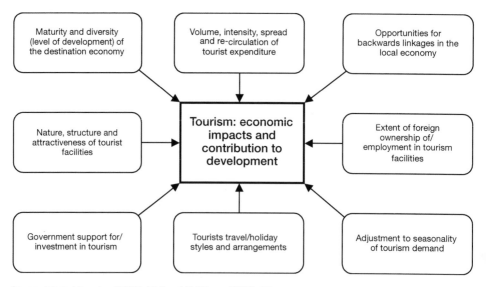

Source: Adapted from Lea (1988); Wall and Mathieson (2006: 90)

short-term economic benefits must be measured against economic (and non-economic) costs and the extent of the longer-term contribution of tourism to development.

The principal economic impacts of tourism are as follows:

Economic benefits

(I) CONTRIBUTION TO BALANCE OF PAYMENTS/FOREIGN EXCHANGE EARNINGS

In 2013, total international tourism receipts amounted to US$1,159 billion, representing 6 per cent of global exports of goods and services (or 8 per cent when international fare receipts are included), and over 29 per cent of global exports of services alone. Therefore, tourism is a valuable source of foreign exchange earnings, particularly earnings of 'hard' currencies. The development of tourism in Cuba since the early 1990s, for example, has been driven almost entirely by the country's need for hard currency. However, three points should be noted. First, developing countries collectively benefit from less than 40 per cent of total international receipts, reflecting the inequitable redistribution of (Western) wealth through tourism. Second, a country's international tourism receipts must be measured against its own expenditure on overseas travel. For many developing countries, the 'travel balance' shows a healthy surplus, although as a country becomes wealthier and its citizens travel overseas more frequently, this surplus may become smaller. Finally, and as explained below, it is net contribution to the balance of payments that must be considered – many countries have significant import costs to meet the needs of tourists.

(II) INCOME GENERATION

Tourism, both international and domestic, is a source of income for businesses and individuals that supply goods and services to tourists. The primary source of such income is direct expenditure by tourists on goods and services, including accommodation, transport, entertainment, food and beverages, and shopping; however, there are also indirect (secondary) and induced (tertiary) effects of tourism spending. Indirect effects relate to the expenditure by tourism businesses on goods and services in the local economy; hotels, for example, purchase food, beverages, equipment, power and water supplies and so on, as well as the services of the construction industry. These suppliers themselves need to purchase goods and services in the local economy, and thus the process of expenditure continues through successive rounds. Eventually, income earned by local people as a result of these rounds of expenditure is spent in the local economy as induced spending. In other words, the original amount spent by tourists is multiplied by a particular amount that reflects the extent of subsequent economic activity, which itself is determined by the characteristics of the local economy. This 'multiplier effect' is an important tool in calculating the overall economic benefit of tourism, and identifies the value of the overall contribution of a country's 'tourism economy' as opposed to its tourism receipts. The process is summarized in Figure 7.4.

Tourism may also be a significant source of revenue for governments. In addition to income tax paid by local tourism workers, for example, sales taxes are often imposed on goods and services sold in tourist establishments or imported goods may be subject to import duties. Development fees may also be levied for tourism development companies on new hotels, for example. Frequently, such revenue is used to fund the further development or

Figure 7.4 *The tourism multiplier process*

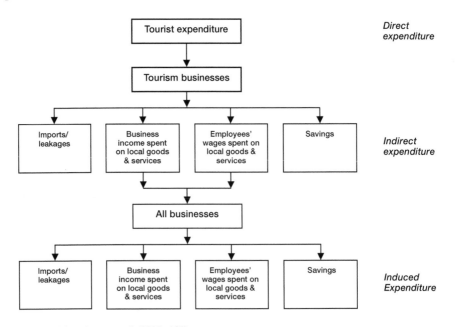

Source: Adapted from Cooper *et al.* (2005: 165)

promotion of tourism, although a key question is to what extent government income from tourism is spent in other areas of the economy (for example, schools, hospitals, etc.) in order to promote wider development in the destination.

Importantly, data identifying the income generated by tourism (often expressed in terms of contribution to GDP) do not reveal the spread of income around an economy, or, in other words, the extent to which income from tourism is shared among the local population. For example, those excluded from the formal tourism economy may be unable to benefit from tourism, hence the potential contribution of pro-poor tourism schemes designed to enhance the earning opportunities of people who are not formally employed within the tourism sector (for example, Bah and Goodwin 2003; Scheyvens 2011, 2015).

(III) EMPLOYMENT GENERATION

Tourism is widely considered a labour-intensive industry, and thus an effective generator of employment opportunities in both the formal and informal sector. Moreover, the full employment generation impact of tourism may be revealed by an 'employment multiplier', similar to that described above. However, care must be taken in assessing the contribution of tourism to employment, for a number of reasons:

- The number of jobs created is dependent on the nature and scale of tourism development; some forms of tourism are more labour-intensive than others.
- Jobs in tourism tend to require lower levels of skills and training, while opportunities for promotion may be limited.
- Tourism employment also tends to be characterized by lower-paid, casual or part-time jobs. In many destinations, it is also highly seasonal. Thus, the contribution to full-time, permanent employment may be more limited than is apparent, particularly as many jobs may be taken by people not belonging to the official workforce (for example, students, retired people or 'informal' workers).
- Similarly, tourism may simply attract workers from other, more traditional sectors of the economy, such as agriculture, not only having little impact on unemployment levels, but also leading to labour shortages in those other sectors.

(IV) ENTREPRENEURIAL ACTIVITIES/BACKWARD LINKAGES

A key contribution of tourism as an agent of development is its potential to stimulate backward linkages or entrepreneurial activity throughout the local economy (see Plate 7.1). That is, the development of tourism requires a variety of goods and services to both establish the sector and also to meet the needs of tourists. Indeed, such linkages are a fundamental element of sustainable tourism development (Telfer and Wall 1996), but are dependent on a number of factors, including:

- the types of goods and services required and the ability of local producers/suppliers to provide them in terms of both quantity and quality;
- the scale and rate of tourism development; rapid, large-scale development tends to outstrip limited local supplies of goods and services; and
- the type of tourism in the destination, hence the type of goods and services required.

Plate 7.1 *Indonesia, Bali: local entrepreneurs are making dyed textiles to be sold as souvenirs to tourists; they are laid out on the riverbank to be dried after the dyeing process*

Source: Photo by D. Telfer

Economic costs

Although attention tends to be focused on the economic benefits of tourism, it is also important to recognize that the development of tourism is associated with a variety of economic costs.

(I) LEAKAGES/PROPENSITY TO IMPORT

Related to tourism's contribution to the balance of payments, an obvious cost is the import of goods and services, or so-called 'leakages', to meet the needs of the tourism sector. Such leakages may be significant in smaller, developing countries with limited local economic sectors, and are greater if tourists travel with overseas tour companies and/or stay in foreign-owned hotels.

(II) OVERDEPENDENCY ON TOURISM

An inherent danger of developing tourism is that the local economy, perhaps once dependent on a single primary product affected by global commodity prices, becomes overdependent upon tourism, and therefore highly susceptible to changes in tourism demand. Such changes in demand reflect the vulnerability of tourism to a variety of influences, such as natural disasters, terrorist activity, global economic downturn or, simply, changes in fashion. Dependence on tourism may be measured by its contribution to GDP, which, as Table 7.1 demonstrates, may be significant, particularly in small island states.

Table 7.1 *Travel and tourism economy as % of total GDP, 2013*

Rank	Country	% GDP
1	Maldives	94.1
2	Macau	86.2
3	Aruba	84.1
4	British Virgin Islands	76.9
5	Vanuatu	64.8
6	Antigua and Barbuda	62.9
7	Anguilla	57.1
8	Seychelles	56.5
9	Former Dutch Antilles	46.7
10	Bahamas	46.0

Source: WTTC (2014: 2)

In the Gambia, for example, a military coup in 1994 resulted in the collapse of its tourism industry during the 1994–1995 season, causing widespread economic problems (Sharpley *et al.* 1996). The coup in Fiji in December 2006 also had an immediate impact on the tourism sector. It was reported, for example, that in the week following the coup, hotel occupancy had fallen to 25 per cent and that the country was losing some $1.3 million a day in tourist expenditure (TVNZ 2006). Similarly, the terrorist bombing in Bali in October 2002 had devastating economic consequences for the island's economy, which, at that time, depended upon tourism for some 50 per cent of its income and 40 per cent of direct employment (Hitchcock and Darma Putra 2005). Elsewhere, visitor numbers in Tunisia in 2014 had almost reached pre-revolution numbers after the 2011 Arab Spring. However, terrorist attacks on visitors at a museum in March 2015 and then in June on tourists sunbathing on a beach, together leaving a total of 55 dead, has led to concern among tourism officials that numbers will drop again, and some cruise ship companies have stated they will not return (Stephen 2015).

(III) INFLATION

Tourism development may lead to inflation, particularly with respect to retail prices during the tourist season and to property/land values in popular tourist areas.

(IV) OPPORTUNITY COSTS

Opportunity costs refer to the economic benefits of tourism compared with the potential benefits foregone of investing in an alternative economic sector (or opportunity). Little research has been undertaken into opportunity costs, although ignoring them may overemphasize the assumed economic benefits of tourism.

(V) EXTERNALITIES

The development of tourism inevitably results in externalities, or incidental costs borne by local communities. These include costs such as additional refuse collections during the tourist season, policing, traffic management and additional health services.

Physical impacts

It is inevitable that tourism brings about physical or environmental impacts. The development of tourism infrastructure, facilities and attractions transforms natural environments, whilst the presence of tourists and their various activities have further and continual impacts on both the natural and built environment. It is not surprising, therefore, that not only have the physical impacts of tourism long been recognized, but also that a significant degree of attention has been paid to them in the literature (for example, Hunter and Green 1995; Mieczkowski 1995; Holden 2007; Wall and Mathieson 2006). Moreover, concern for the physical impacts of tourism lies at the heart of the sustainable tourism development debate – fundamental to the sustainability of tourism is the maintenance and health of its physical resource base.

Typically, the study of tourism's physical impacts focuses upon negative impacts or environmental costs. At the same time, the analysis is frequently structured around particular impacts, such as pollution or erosion, or around the constituent elements of the natural or built environment, often in isolation from the wider political or socio-economic contexts of the destination. However, a number of points deserve mention:

- Tourism development may in fact encourage environmental conservation and improvement; that is, tourism may have positive environmental consequences. Research has demonstrated, for example, that tourism has contributed positively to conservation in the Galapagos Islands (Powell and Ham 2008).
- There is often no baseline for measuring or monitoring environmental change that results from tourism development, particularly against local community perceptions of 'acceptable' change or damage.
- Tourism development may have both immediate and secondary environmental impacts and it is sometimes difficult to isolate one type of impact from another.
- Environmental change in destination areas may not in fact be caused by tourism development but by other human or economic activity.

Rather than simply isolating and describing specific types of impact, therefore, it is more useful to adopt a holistic approach to the study of tourism's physical impacts within the overall context of tourism development at the destination. One framework for doing so was developed by the OECD in the late 1970s, and this remains, perhaps, the most comprehensive and integrated model for assessing tourism's physical impacts (Lea 1988; Pearce 1989). Key to the model is the identification of a number of tourism-generated stressor activities (for example, resort construction, generation of wastes and tourist activities), the nature of the stresses themselves and the primary (environmental) and secondary (human) responses to this environmental stress. Figure 7.5 presents an adapted version of the OECD framework embracing the destinational influences discussed earlier in this chapter.

The principal physical impacts of tourism can be summarized under the four 'stressor activities' highlighted in Figure 7.5.

(i) Permanent environmental restructuring

The development of tourist destinations requires the construction of facilities and attractions (hotels, resorts, restaurants, etc.) and

Figure 7.5 *A model for assessing tourism's physical impacts*

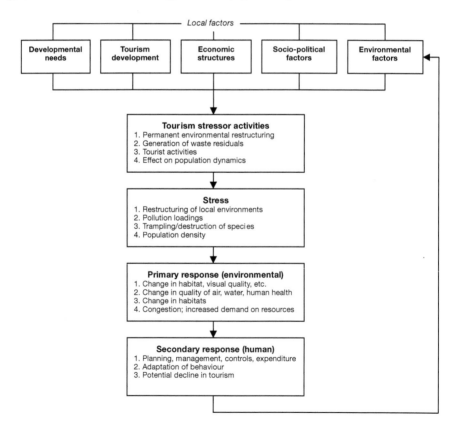

Source: Adapted from OECD (1981)

associated infrastructure, including roads, railways, airport terminals and runways, harbours and marinas and so on. Such development results, of course, in the permanent transformation of the physical fabric of an area as the built environment expands, 'using up' natural ecosytems or taking large areas of land out of primary (agricultural) production. An immediate effect is a change in the visual quality of an area (sometimes referred to as 'architectural pollution') but, in the longer term, significant changes may occur within the natural environment as wildlife habitats are threatened or as ecosystems are damaged by physical construction or by associated impacts, such as pollution. For example, fragile marine environments, including coral reefs, may be seriously damaged by increased sediment loads in coastal waters surrounding tourist developments (Mowl 2002). In response, conservation measures, environmental improvement schemes or visitor management programmes may be needed to prevent further environmental deterioration.

(ii) Generation of wastes and pollution

Tourism is a significant generator of waste materials, manifested primarily in the pollution of land, air and water resources. By definition, tourism involves transport and it is not surprising that different modes of tourist transportation collectively represent a major source of air and noise pollution. In recent years, increasing attention has been paid in particular to the environmental impact of aviation emissions, air travel being seen by many as the world's fastest-growing source of greenhouse gases. Currently, the global commercial jet aircraft fleet generates over 700 million tonnes of carbon dioxide, the major greenhouse gas, annually (ATAG 2014), although the aviation industry is quick to point out that just 5 per cent of global warming is accounted for by commercial air transport. However, it is estimated that continuing growth in air travel will increase this contribution to global warming to 15 per cent by 2050. In comparison, cars currently produce 10 per cent of global greenhouse gas emissions, though half of global car emissions are produced just in the US (Borger 2006). Nevertheless, it is evident that tourism-related transport may make a significant contribution in the longer-term to air pollution and, hence, climate change (see Chapter 8 for more discussion). Water pollution also occurs as a result of tourist transport. In the 1990s, for example it was estimated that the Caribbean region received a total of 63,000 port calls from cruise ships each year generating some 82,000 tonnes of garbage

(Campbell 1999), while a more recent report claims that, overall, 'cruise ships dumped more than 1 billion gallons of sewage in the ocean' in 2013 (FoE 2014).

Resort and other tourism-related built environments also create waste and pollution. For example, marine and freshwater pollution is a common problem resulting from inadequate sewage treatment facilities in resort areas, meaning that raw or untreated sewage is pumped into lakes or the sea. As Lea (1988) notes, this not only impacts negatively on marine ecosystems, but may also represent a health hazard for tourists swimming in dirty water. Tourism also generates significant amounts of solid wastes, the improper disposal of which may lead to land contamination, unattractive or degraded environments, and health hazards for humans and wildlife. For example, Holden (2007) observes that elephants in the Masai Mara game park in Kenya have reportedly been poisoned by eating zinc batteries left on a rubbish tip outside a tourist lodge.

(iii) Tourist activities

Perhaps the most widespread and most frequently documented physical impacts of tourism result from the activities of tourists themselves. Some such impacts are caused either wilfully or through ignorance; for example, tourists walking on or plundering coral reefs is a major problem in many countries. Indeed, it has been reported that of the 109 countries possessing coral reefs, in 90 of them the reefs have been damaged by cruise ship anchors and, in particular, tourists breaking off chunks of coral (UNEP 2002), while, according to Goudie and Viles (1997), 73 per cent of coral reefs off the coast of Egypt have been adversely affected by tourist activities. More recent studies in that region confirm the potential damage to coral reefs by tourism and the need for appropriate management and diver education programmes (Hasler and Ott 2008).

More commonly, the impacts of tourist activities are unintentional, resulting simply from tourists 'being there' in numbers that exceed an area's 'carrying capacity' (see below). This may include, for example, chemicals from sunscreens polluting beaches. Both natural and built environments are adversely affected by the presence of tourists, with the erosion or destruction of fragile areas through walking, trekking, cycling or other activities being a common problem. In some instances, particular activities have a variety of environmental impacts; trekking in remote mountain regions not only results in physical damage to paths and trails, but litter is also a serious problem. For example, some trekking trails in the Peruvian Andes have been nicknamed 'Coca-Cola trails', reflecting the volume of discarded empty drinks bottles along the routes. At the same time, as the case of the Nepalese Himalayas demonstrates (Box 7.1), tourism may also lead to significant deforestation problems that, in the longer term, have wider social and environmental consequences.

There are, of course, numerous other environmental consequences arising from the activities of tourists, from forest fires caused by discarded cigarettes or graffiti on historic structures, to the disruption of breeding or feeding patterns of wildlife in safari parks (Shackley 1996). The challenge for most destinations, however, is how to manage or control environmentally damaging tourist activities without discouraging or preventing tourism in the first place.

(iv) Effect on population dynamics

The final environmental stress identified by the OECD is that associated with population dynamics. More specifically, the geographic flows of tourists, governed by a variety of factors that influence such flows (see Chapter 6) and manifested in significant seasonal increases in the population densities of destinations, may create a number of environmental impacts. The most evident of these is congestion and overcrowding experienced both at key sites/attractions and in resort areas in general, a subsequent effect of which may be greater physical damage, enhanced pollution levels and so on. Of equal, if not greater, importance, however, are the increased demands placed on natural resources. That is, the tourism industry competes with other sectors for scarce natural resources, such as land and water, and during the high season, tourism may place excessive strain on such resources. The demand for water is a particular problem (Gössling *et al.* 2015). Tourists tend to consume a much higher quantity of water while on holiday than at home, both directly through, for example, regular showering/bathing, or indirectly through the expectation that sheets and towels will be laundered daily. Consequently, the tourism sector frequently consumes significantly more water than other local industries. In fact, one early study suggested that in some less-developed countries, 100 luxury hotel guests consume as much water in 55 days as 100 urban families consume in two years (Salem 1994), while a more recent review highlights the need for effective water management, particularly in destinations with a high level of tourism water

Box 7.1

Impacts of trekking in the Nepalese Himalaya

The small Himalayan kingdom of Nepal lies landlocked between India to the south and Tibet to the north. It is one of the poorest countries in the world – per capita income is just US$250 – and it faces many of the challenges of underdevelopment. For example, average life expectancy is 60, infant mortality is 61 per 1,000 births and the literacy rate (adults over 15 years of age) is 44 per cent. With few natural resources, the country relies heavily on international aid, which, in 2004, amounted to US$420 million. However, Nepal boasts a unique and diverse geological formation, from the subtropical jungle of the Terai to the high peaks of the Himalayas – eight of the world's 10 highest mountains, including Mount Everest, are to be found there. Its culture is equally diverse; the dominant religions are Hinduism and Buddhism, and the population is made up of 12 major ethnic groups. It is not surprising, therefore, that since it first opened its borders to outsiders in the 1950s, not only has Nepal attracted visitors keen to experience its environment and culture, but also that tourism has become an increasingly important element of the national economy. In 2004, for example, tourism generated US$230 million, representing 15 per cent of the country's foreign exchange earnings and 3.5 per cent of GDP.

Initially, few tourists were able to visit Nepal – in 1961, just 4,000 tourist arrivals were recorded. The first organized mountain trekking commenced in 1966, and by the mid-1970s arrivals had increased to over 100,000 a year. Since then, tourism has continued to grow, albeit erratically given the ongoing political instability in the country. In 1999, almost half a million tourists visited Nepal, and in 2004 this figure had fallen back to 385,000 arrivals. By 2012, the number of visitors had reached 803,092, with the top five markets visiting being India, China, Sri Lanka, the US and the UK.

One of the principal attractions of Nepal is, of course, the opportunity for mountain trekking – approximately 25 per cent of tourists participate in trekking, either independently or in organized groups. There are four main trekking regions, the most popular of which are the Annapurna Circuit area, which attracts more than half of all trekkers, and the Everest base camp route, which attracts around 20 per cent. The remaining treks take place on the Kanchenjunga route and in the more recently opened Mustang region. All these areas benefit from some degree of protection or control. For example, the Mount Everest (Sagarmatha) National Park was established in 1976, while the designation of the Annapurna Conservation Area in 1984, and its subsequent success as a community-based conservation project, has received widespread acclaim. However, despite the protection afforded by these designations, all trekking areas suffer from a number of tourism-related impacts.

From a positive perspective, tourism has provided an important source of income and employment, particularly for communities alongside the popular routes where trekkers require accommodation and food. Nevertheless, relatively few communities benefit from tourism expenditure (the remote western regions of Nepal, for example, remain untouched by tourism), resulting in severe inequalities in the distribution of wealth. Moreover, employment opportunities in tourism have led to labour shortages in other areas and sectors, in particular in traditional agriculture. Greater attention, however, has been paid to the physical (negative) impacts of tourism in the Nepalese Himalaya, with all the major trekking areas suffering similar problems, including:

- *Deforestation*: The demand for firewood to support tourism-related cooking and heating needs has had a significant impact upon forests and vegetation, while the building of lodges along trekking routes has led to a major increase in the demand for timber (there are, for example, over 700 lodges and tea rooms in the Annapurna area alone). Within the national parks and other controlled areas, tree felling is strictly controlled; this has, however, resulted in extensive deforestation beyond the boundaries of protected areas.
- *Litter*: The accumulation of litter alongside trails and around popular settlements/campsites along the trails is a significant issue. Many studies draw attention to the huge volume of litter, much of it non-burnable or non-biodegradable, including oxygen tanks, left by trekkers and mountaineering expeditions.
- *Trail damage*: The erosion of trails is a common problem throughout the trekking regions. This has been exacerbated by the rapid growth in the number of trekkers over the last two decades, particularly in the Annapurna region.
- *Pollution*: The siting of toilets close to rivers and streams and the use of soaps/detergents for bathing or washing clothes/dishes in streams is a major cause of water pollution. This is exacerbated by the disposal of untreated human waste in rivers and streams.

Throughout the trekking areas in the Nepalese Himalaya, efforts are being made to address these impacts. Numerous schemes exist to control deforestation, to mange the collection of litter, to repair trails and to introduce alternative energy sources for cooking and heating. The Annapurna Conservation Area, in particular, benefits from income derived from the fees that all trekkers are charged – this is channelled directly into conservation and other local projects. However, the need remains to balance the economic benefits of tourism with effective environmental management in order to prevent further damage to the fragile mountain ecology. Specifically, effective policies and controls are required; for example, despite regulations requiring the use of kerosene for cooking, deforestation is still a problem in the Mustang region. At the same time, however, Nepal dramatically reduced the fees for climbing Mount Everest in 2014 to attract more visitors, although the hope is to also better manage the climbing teams. Without better management, the longer-term environmental and social fabric of the region may be irreversibly compromised.

Source: MacLellan *et al.* (2000); Nepal (2000); Government of Nepal (2013); Coldwell (2014)

consumption compared with domestic demand and where rainfall levels are projected to decline in the future (Gössling *et al.* 2012). Similar demands may also be made on power supplies; electricity power cuts are not uncommon in some resort areas with limited generating capacity, necessitating hotels to operate their own backup generators.

As noted above, the physical impacts of tourism are not always negative. In other words, it is important to recognize that the development of tourism may act as a catalyst for environmental protection and improvement. In particular, the designation of nature reserves, national parks, wildlife reserves and other categories of protected areas/landscapes (see Plate 7.2) is often, though not always, directly related to tourism development, while the expenditure on environmental improvements, such as the 'greening' of run-down areas, the cleaning and renovation of the built environment, or water

Plate 7.2 *Argentina, Iguazu Falls: a World Heritage Site that lies on the Argentina and Brazil border; parks protect the jungle ecosystem on both sides of the falls; parks can play an important role in resource protection*

Source: Photo by Tom and Hazel Telfer

quality improvement schemes may all be driven by the need to enhance the attraction of an area to tourists. Similarly, specific historic sites often benefit from tourism. For example, Angkor Wat in Cambodia, which, since 1992, has been a UNESCO World Heritage Site, now attracts over two million visitors a year (see Plate 7.3) and almost one-third of ticket revenues is spent on restoration work and managing the site. Nevertheless, tourism remains a double-edged sword, with rapidly increasing numbers of tourists to the site posing a potential threat to the physical fabric of the temple complex in the longer term (MacKenzie 2006) (see Plate 7.4).

Sociocultural impacts

It has long been recognized that tourism has an impact upon destination societies and cultures; indeed, some earlier tourism impact studies focused specifically on this topic (Smith 1977). It has also long been accepted that such impacts are likely to be more

Plate 7.3 *Cambodia, Angkor Wat: visitors at the UNESCO World Heritage Site of Angkor Wat*

Source: Photo by R. Sharpley

Plate 7.4 *Cambodia, Siem Reap: new hotel under construction*

Source: Photo by R. Sharpley

evident or keenly felt in tourist destinations in developing countries, where the difference in cultural and economic characteristics between local people and, primarily, relatively wealthy Western tourists is likely to be greatest (WTO 1981). At the same time, in some respects it would be considered unfortunate for tourism *not* to have some sociocultural consequences on destinations; as a catalyst of development, tourism is usually promoted with the purpose of economic and social betterment. Moreover, tourism is seen by some as a means of achieving greater international harmony and understanding (WTO 1980), although, perhaps inevitably, it is the negative (and frequently emotive) sociocultural impacts of tourism that attract most attention: 'tourists seem to be the incarnation of the materialism, philistinism and cultural homogenisation that is sweeping all before it in a converging world' (Macnaught 1982).

Although it is likely, if not inevitable that destinations experience social and cultural change as a result of tourism development, the extent of that change is dependent on a number of factors. The socio-economic 'gulf' between tourists and local communities has already been mentioned but, in addition, the degree of sociocultural change may be influenced by:

- *The types and numbers of tourists/tourist behaviour*: It is usually thought that higher numbers of mass tourists will impact more on host societies than smaller numbers of independent/responsible travellers, although smaller numbers of tourists in places relatively untouched by tourism may have a significant impact.
- *The size and structure of the tourism industry*: The larger the tourism industry relative to the local community, the greater its sociocultural impact is likely to be.
- *The relative importance of the tourism industry*: The consequences of tourism will be more keenly felt in destinations that are highly dependent on tourism, although established resorts may have a variety of controls in place to limit such consequences.
- *The pace of tourism development*: Research has shown that sociocultural impacts are more likely to be experienced when the development of tourism is rapid and uncontrolled.

It is also important to note, of course, that all societies and cultures are dynamic; they are all in a constant state of change and no society is immune from external influences. Tourism is undoubtedly one such influence, but frequently tourism contributes towards, but does not cause, sociocultural impacts and change. Nevertheless, tourism development is often blamed for what are seen as undesirable changes in destination societies, and therefore care must be taken to determine the precise components of sociocultural change.

Tourist–host encounters

The basic context for the sociocultural impacts of tourism is the so-called 'tourist–host encounter'. In other words, tourists inevitably come into contact with local people (hosts) in destination areas, and despite the usual brevity of such encounters, a variety of social processes are at work that determine the nature of that encounter (see Plate 7.5). This, in turn, goes some way to determining or explaining the potential sociocultural impacts of tourism because, generally speaking, the more unbalanced or unequal the encounter or relationship, the more likely it is that negative impacts will occur.

As Lea (1988) notes, there are potentially as many types of encounter as there are tourists and hosts. Nevertheless, a number of common characteristics can be identified with tourist–host encounters, foremost of which is the wealth of tourists relative to that of local people, particularly in developing countries. This may lead to feelings of resentment or inferiority on the part of local people. Four other principal characteristics are usually highlighted (Wall and Mathieson 2006: 223):

● Most encounters are transitory or fleeting, the resultant relationships being shallow and superficial.
● Most encounters are constrained temporally (the two weeks of the holiday or the tourist season) and spatially (the location/separateness of tourist facilities).
● Most encounters are pre-planned or lack spontaneity as hospitality becomes commercialized.
● Tourist–host encounters tend to be unbalanced, local people perhaps feeling inferior or subservient to tourists.

To this list a further characteristic may be added, namely that many tourists travel with an apparent lack of knowledge, understanding or sensitivity to local culture or customs in destination areas. This represents another potential barrier to balanced or meaningful tourist–host encounters, and supports the argument for more responsible behaviour on the part of tourists (Chapter 6).

Typically, the above characteristics refer to encounters that occur between tourists and local people in organized tourist settings, such as in hotels or restaurants, in shops or on the beach. However, the context of tourist–host encounters may vary significantly, from structured, commercial exchange-based encounters to spontaneous meetings or even relations that involve no contact or communication at all – that is, simply sharing the same space (Sharpley 2014). Hence, it is possible to conceptualize tourist–host encounters as lying on a continuum, with different forms of encounter implying varying experiences for the tourist and varying degrees of impact on the host (Figure 7.6).

It is also important to point out that different categories of tourists come into contact with local communities in contexts other than the holiday. In other words, there are various ways in which mobile populations (including tourists) interact with 'static' communities (Hannam *et al.* 2006) and, as a result, increasing attention is being paid to the issue of resident tourists, second-home owners and migrant workers who serve the needs of new tourist populations in

Plate 7.5 *Argentina, Estancia Santa Susana near Buenos Aires: cultural performance at historic ranch*

Source: Photo by Tom and Hazel Telfer

Figure 7.6 *A continuum of tourist–host encounters*

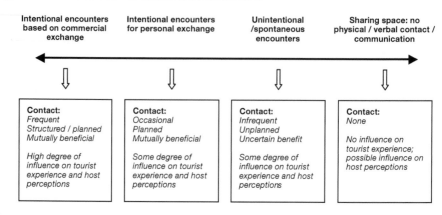

Source: Sharpley (2014)

destination areas. For example, it has been observed that conflicts exist between British migrants and local communities in southern Spain (O'Reilly 2003), while, more generally, complex relationships exist between local people and different types of tourist-migrants (Williams and Hall 2000). However, here we are concerned with the more 'traditional' sociocultural consequences of tourism.

The way in which local communities respond to the characteristics outlined above and, indeed, to the impacts of tourism in general is discussed shortly, but first it is important to review the principal socio-impacts of tourism development. Typically, social and cultural impacts are referred to collectively, though it is useful to distinguish, albeit somewhat artificially, between social and cultural impacts. Thus, sociocultural impacts can be thought of as the more immediate effects of tourism on local people and their lifestyles, whereas cultural impacts refer to longer-term changes that occur in the context of social values, attitudes and behaviour, as well as changes in the production and meaning of cultural art forms and practices.

Social impacts

From a positive perspective, tourism may have a variety of beneficial consequences for destination societies. These include infrastructural developments, the improvement of the physical environment and the provision of facilities that benefit tourists and local communities alike. More specifically, tourism provides employment opportunities that, in some countries, have brought a new freedom and independence, and improved social condition, to many women. There are also other potential benefits. In the Gambia, for example, many schools benefit from charitable donations from tourists. In fact, many visitors, on their return home, establish small charities to raise funds for school buildings and materials, providing educational opportunities to many Gambian children (www.friendsofgovi.org.uk).

Conversely, tourism also has a number of less beneficial impacts on host societies. In general, for example, traditional community structures may be transformed as younger people are drawn from inland rural areas to the coast or cities to work in the tourism industry. Such migration patterns often result in a population imbalance in rural areas and the polarization of societies between younger, more affluent groups and the older, more traditional generations. More specifically, the presence of tourists and their activities impact on local people in a number of ways:

- *The demonstration effect*: Tourism introduces new or alien values or lifestyles in destination areas. Local people may attempt to emulate behaviour or styles of dress, or strive to achieve levels of wealth demonstrated by tourists.
- *Crime*: Although there is little evidence to directly link increases in crime with the development of tourism, there is little doubt that where there are significant numbers of tourists, criminal activity is also evident. This may result in increased expenditure on law enforcement, the growth in activities such as gambling and black-market operations, increased crime against residents and, potentially, a reduction in tourism.
- *Religion*: In many tourism destinations, religious buildings, shrines and practices have become commoditized. That is, they have become an attraction and a part of the tourism product, gazed upon and 'collected' by tourists. As a result, there is frequently conflict between local communities, devout visitors and tourists as religious rituals or places of worship are disrupted by tourists.
- *Prostitution/sex tourism*: Although care must be taken in apportioning the blame on tourism, there is no doubt that in many destinations, such as Thailand, Cuba, Sri Lanka and the Philippines, tourism has led to an increase in prostitution. The social impact of this can be devastating, particularly in the case of child prostitution and the spread of sexually transmitted diseases (Ryan and Hall 2001; Bauer and McKercher 2003).

Cultural impacts

Over time, the culture of host societies may change and adapt either directly or indirectly as a result of tourism. Much of the literature is concerned, in particular, with the way in which cultural forms, such as arts and crafts or carnivals, festivals and religious events, become adapted, trivialized, packaged and commoditized for consumption by tourists, and there is certainly no doubt that this occurs on a wide scale. Many art forms become mass-produced as souvenirs (that is, 'airport art'), while, frequently, cultural rituals are transformed and staged for tourists, becoming devoid of all meaning to the participants. For example, the style, production and cultural significance of the raksa dance masks in Sri Lanka have undergone significant change as a result of their production as souvenirs (see Box 7.2).

It is also true, however, that tourism also encourages the revitalization or resurgence of interest in traditional cultural practices, and there are many examples of how tourism supports the redevelopment of traditional art forms and production techniques. For example, the traditional craft of greenstone carving in the southern Indian town of Mahabalipuram has been revitalized by the demand for souvenirs. The products, though intended for sale to tourists, are no less authentic than those produced by similar techniques hundreds of years ago.

Less visibly, tourism also contributes to broader, deeper cultural transformations in destination societies. These are changes that occur in a society's values, moral codes, behavioural modes and identifying characteristics, such as dress and language. As mentioned earlier, it is difficult to separate the influence of tourism from other factors that induce cultural change, but, nevertheless, it is generally accepted that tourism can accelerate this process, largely through what is known as acculturation.

Acculturation is the process whereby, when two cultures come into contact (for example, through tourist–host encounters), over time they will become more like each other through a process of borrowing. By implication, if one culture is stronger or more dominant than the other, then it is more likely that this borrowing will be a one-way process. The extent to which tourism contributes to the acculturation process will vary depending on a variety of factors, such as the cultural gulf between the tourist and the host and the influence of other forces. In some cases, cultures are more

Box 7.2

The commoditization of dance masks, Sri Lanka

One of the most popular and widely available tourist souvenirs in Sri Lanka are colourful 'raksa' or 'devil dance' dance masks; at the same time, they are also a striking example of the way in which cultural artefacts may be adapted and commoditized by tourism. Dance masks fulfil an important role in many forms of ritual and ceremony in Sri Lanka. Indeed, representing the images of a variety of deities and demons, they have long been used in a number of different contexts, such as folk dramas, festivals, and rituals of exorcism and healing. They have a fundamental meaning and significance to the performance of such rituals and, at the same time, the production of the masks is also a recognized and socially important activity within the local community. However, the rapid growth of tourism to Sri Lanka during the late 1970s and early 1980s resulted in the masks being appropriated as an appealing and commercially attractive representation of Sinhalese culture for the tourist market. As they became mass-produced, they were removed from the traditional system of production and performance, and, as a consequence, not only did they lose their cultural authenticity, but also the manufacturers of the masks suffered a decline in their social status. More specifically, dance masks produced for tourist consumption began to be produced in a variety of different sizes, decoration and colours to suit the taste of tourists rather than in a more traditional or authentic style. At the same time, the manufacturers of masks (normally small, family enterprises) sacrificed their traditional social role in the complex production–performance ritual as they began to mass-produce inauthentic dance masks for commercial gain.

Source: Simpson (1993)

resilient (see Plate 7.6), yet, in many instances, cultural change can be directly attributable to tourism.

Impacts on tourists

While most attention is focused upon the impacts that tourists have upon local communities, it is important to note that there exists a number of 'reverse' social impacts; that is, tourists themselves may experience impacts as a result of visiting tourist destinations. At a basic level, such impacts occur when tourists become the victims of crime or health problems; indeed, both have always been regarded as risks associated with travel and tourism. More recently, however, the opening up of more distant or exotic locations to mass tourism has

Plate 7.6 *Indonesia, Bali: traditional cultural ceremonies remain very important in Bali though it is a major tourism destination*

Source: Photo by D. Telfer

resulted in increasing incidence of tourists contracting serious diseases, such as malaria. Similarly, an increase in the number of cases of 'old' diseases, such as diphtheria, which have been virtually eradicated in Western societies, and of sexually transmitted diseases, has been directly associated with the growth in tourism to previously less-accessible areas, particularly the transitional economies of eastern Europe. However, a variety of other 'reverse' sociocultural impacts can be identified, including:

- the internationalization of fashions and tastes in clothing, music, art forms, cuisine and so on;
- a reduction in national xenophobia, resulting in a greater awareness and acceptance of different cultures;
- the verification or changing of tourists' perceptions of different places, peoples and cultures; studies have also shown that tourists often return home with a more positive attitude towards their own society and culture;
- the adoption by tourists, on either a temporary or permanent basis, of new cultural practices.

The impacts of tourism: local community responses

Despite the usual focus on the impacts of tourism and tourists on destination environments and societies, local communities do not, of course, passively accept these impacts as an inevitable consequence of tourism development. Equally, nor do local communities always respond to the impacts of tourism in ways that, from a Western-centric perspective, might be expected. For example, although the development of tourism in Cyprus has resulted in significant environmental degradation, this is not considered a problem by many Cypriots for whom entrepreneurialism and wealth creation for their families is a strong cultural value that outweighs environmental concerns (Sharpley 2001).

In other words, local communities respond to the development of tourism and its associated impacts in a variety of ways; these responses, in turn, may reflect local culture and values as well as more specific factors, including:

- the nature and scale of tourism development;
- the structure/ownership of the local tourism industry;
- the stage of development/maturity of the tourism sector; and
- at the individual level, the degree of involvement in/benefit from tourism.

The perceptions and responses of local residents to tourism development and to tourists themselves have long attracted academic attention, to the extent that it has been described as 'one of the most systematic and well-studied areas of tourism' (McGehee and Anderek 2004: 132). Driving this interest in the subject is the belief that understanding resident perceptions and responses is fundamental to the successful and sustainable development of tourism. Within the now substantial literature, a number of models have been developed that explore local community responses and attitudes to tourism development that provide a useful framework for exploring ways of managing the impacts of tourism. These are considered in some detail in Wall and Mathieson (2006: 227–35), while recently, research into resident perceptions has been subject to detailed criticism (Deery et al. 2012; Nunkoo et al. 2013; Sharpley 2014).

With respect to the models referred to above, there are common elements that may be summarized as follows. Typically, the models suggest a continuum of attitudes or behavioural responses to tourism development, from negative to positive. For example, in Doxey's (1975) widely cited framework, it is suggested that local residents in tourist destinations suffer varying degrees of irritation dependent upon the extent that their lives are disrupted by tourism. These levels of irritation progress from euphoria, through apathy and irritation, to antagonism and, finally, a resigned acceptance of tourism impacts. Doxey also suggests that the level of irritation will grow as the destination develops, in a manner similar to Butler's (1980) resort life cycle. In other words, there is a unidirectional increase in irritation reflecting the stage of tourism development. This, however, may not necessarily be the case.

For example, studies have shown that communities' responses to tourism's impacts may follow a continuum from welcome to rejection, though in different behavioural contexts (Figure 7.7). Research by Dogan (1989) and Ap and Crompton (1993) identifies a variety of responses from favourable through to unfavourable responses, largely dependent upon the extent to which communities feel that the inconveniences or impacts of tourism are balanced by the benefits it brings. Conversely, Carmichael's (2000) study of resident attitudes and responses to a large casino development revealed a distinction between attitudes and responses. That is, the majority of those who supported the casino accepted it 'silently', while those who expressed negative attitudes demonstrated 'resigned acceptance', calling into question the extent to which local residents have manifested positive or negative attitudes in the behaviours described by the models. Investigating resident attitudes of imperialism in tourism in Jamaica, Sinclair-Maragh and Gursoy (2015) found that

Figure 7.7 *Community responses to tourism impacts*

Dogan (1989)	Ap and Crompton (1993)
Adoption: replacement of traditional host social structures with the adoption of tourist culture	*Embracement: enthusiastic welcoming of tourists*
Revitalisation: tourism is used to preserve, promote or revitalise local culture for display to tourists	*Tolerance: acceptance of the consequences of tourism in recognition of its benefits*
Boundary maintenance: physical/social boundaries are established between tourists and local communities	*Adjustment: alteration of behaviour to avoid the inconveniences resulting from tourism*
Retreatism: local community avoids contact with tourists/develops an increased cultural consciousness	*Withdrawal: physical or psychological distancing from tourism and tourists*
Resistance: hostility and aggression against tourists/tourism industry.	

resident perceptions of political and cultural imperialism have positive impacts on how residents view tourism, while economic imperialism negatively influenced how residents view tourism.

Similarly, other studies have looked at the differing attitudes towards tourism within communities, the purpose being to identify factors that may explain why different members of destination communities may respond in different ways at any point in time (or stage of tourism development). Typically, and perhaps not surprisingly, these demonstrate that those members of local communities who are more involved in or economically dependent on tourism will view it more favourably or embrace tourism development more positively; conversely, those less dependent on tourism, but nevertheless affected by its impacts, will respond negatively. Research in Cyprus confirms this, although even among those involved in the tourism sector, resentment against tourists emerges when local tolerance levels are exceeded (Akis *et al.* 1996).

What these studies, and indeed the analysis of tourism impacts in general, demonstrate is the fact that the consequences of tourism development cannot be viewed from a simplistic, descriptive and, implicitly, Western-centric perspective. For tourism to contribute to sustainable development, there is a need to manage local resources in a manner that reflects local values, the local political economy and local development needs. This is the focus of the final section of this chapter.

The impacts of tourism and sustainable development

Tourism must, of course, be effectively planned and managed in order to optimize its developmental potential. Tourism planning is concerned, as Wall and Mathieson (2006: 293) note, with 'the process of making decisions about future desired states and how to attain them'; its focus tends to be longer term and, in the case of many developing countries, it is embodied in overall tourism master plans that provide a broad vision for the development of the destination as a whole. Many books address the tourism planning process, albeit from different perspectives (for example, Murphy 1985; Inskeep 1991; Hall 2007; Kastarlak and Barber 2011). Tourism management, in contrast, has a much shorter time scale and is concerned with processes and techniques designed to accommodate tourists and tourism development, but, at the same time, to minimize their negative consequences. Such processes and techniques typically fall under two headings, namely: managing physical resources (Newsome *et al.* 2000), including land designation, spatial planning strategies and site management; and managing visitors (Leung *et al.* 2014). In both cases, a variety of management and regulatory measures are usually proposed.

An overview of tourism planning and management is beyond the scope of this chapter. The important point, however, is that, if progress is to be made towards achieving sustainable development through tourism, effective means are required to assess or measure whether such progress is being made. In other words, while appropriate planning and management is, of course, a vital ingredient of the tourism development process, there is a need to measure and monitor the extent to which tourism development is achieving local sustainability needs and objectives. Consequently, attention is now increasingly focused on a systems approach to tourism planning (for example, Tribe *et al.* 2000), a dynamic process that designs, implements, monitors and adapts tourism policy according to local developmental needs and resource limitations. Inherent in the systems approach is the requirement for establishing indicators of sustainability.

Sustainability indicators

Sustainability indicators provide a basis for monitoring and measuring the extent to which the key sustainable developmental issues of a tourism destination, such as economic and social benefits, particular resource use or tourism-specific goals, such as tourist satisfaction or seasonality, are being met. Not only do they help to clarify goals, but they provide a focus for identifying and assessing stresses on the environment, for measuring the impacts of tourism and for measuring the effects of management actions. More broadly, sustainability indicators may fulfil the following purposes within the planning process (UNEP/WTO 2005: 73):

- They provide a baseline for measuring changes in the condition of resources and for assessing progress in satisfying local community needs.
- They represent a set of targets that form the basis of tourism development policies and actions.
- They provide a framework for assessing the effectiveness of actions.

- They enable the evaluation, review and modification of tourism development plans and policies.

Key to this process, of course, is the identification and selection of appropriate indicators, a task that should ideally be built into the process of local consultation and participation. The WTO (2004a) identifies numerous such indicators; however, indicators should be relevant to the local context, be easily measurable and provide clear and credible information. In *Making Tourism More Sustainable: A Guide for Policy Makers* (UNEP/WTO 2005), 12 baseline issues with corresponding baseline indicators are suggested. These are summarized in Figure 7.8.

In the context of measuring and managing tourism impacts, a fundamental requirement is the establishment of limits to environmental or sociocultural change. The question is: How might those limits be established? Two concepts deserve consideration.

(I) CARRYING CAPACITY

The concept of carrying capacity refers, quite simply, to the number of tourists that a destination or site can accommodate (or 'carry') without negative impacts on the local environment/society or a reduction in the quality of the tourist experience. A number of different carrying capacities may be measured:

- *physical capacity*: the actual number of tourists a place can physically accommodate;
- *ecological capacity*: the extent to which the local ecology can withstand the impacts of tourism;
- *sociocultural capacity*: the limit of sociocultural impact and change a local community will accept;
- *psychological capacity*: the amount of congestion that will be tolerated by tourists before they feel that their experience is being impaired.

Carrying capacity remains the subject of intense debate in the tourism literature (McCool and Lime 2001; Coccossis and Mexa 2004; Salerno *et al.* 2013). In particular, questions remain over how particular capacities are established, how they are measured and monitored, how they influence the tourist experience, what limits of damage or change are acceptable, and on what basis capacities are

Figure 7.8 *Baseline issues and indicators for sustainable tourism development*

Baseline issue	Baseline indicators
Local satisfaction with tourism	• Local population satisfaction levels
Effects of tourism on communities	• Ratio of tourists to local people at different periods
	• Recognition of tourism benefits for local communities (services/infrastructure)
Tourist satisfaction	• Levels of tourist satisfaction
	• Number/proportion of repeat visitors
Seasonality	• Arrivals by period
	• Occupancy levels by period
	• Proportion of tourism employment that is permanent/full-time
Economic benefits of tourism	• Numbers/proportion employed in tourism
	• Net economic benefits of tourism (income)
Energy consumption	• Per capita energy consumption
	• Proportion of energy from renewable sources
Water usage	• Water consumption per tourist/establishment
	• Water saving/recycling
Quality of drinking water	• Proportion of establishments providing drinkable water
	• Number of water-related illnesses amongst tourists
Sewage treatment	• Number of tourist establishments treating sewage
	• Proportion of sewage per establishment being treated
Solid waste management	• Volume of waste produced
	• Volume of waste recycled
Development control	• Existence of land-use /development policy
	• Proportion of land subject to development controls
Visitor management	• Total number of tourist arrivals
	• Density of tourist numbers at specific locations

Source: Adapted from UNEP/WTO (2005)

set. Nevertheless, carrying capacity does provide an appropriate measurement/indicator tool in particular circumstances. For example, in the late 1990s, the island of Malta revised its tourism development policy on a carrying capacity assessment linked to the supply of accommodation. It was decided that the development of tourism should be based upon the then existing bedstock supply rather than an expansion of the accommodation sector.

(II) LIMITS OF ACCEPTABLE CHANGE

An alternative approach to establishing limits for environmental and sociocultural change is the concept of limits of acceptable change (LAC). Rather than focusing on the numbers of tourists or scale of activity, it recognizes that it is the impact or degree of change that is the problem. Therefore, LAC establishes limits to the impacts of tourism on destination environments and societies, limits that should be decided upon by local consultation and measured by indicators. In this sense, LAC precedes or is inherent within the sustainability indicator process.

To summarize, then, the development of tourism is inevitably accompanied by impacts on destination environments and societies. The nature and extent of those impacts is determined by a variety of factors, but, for tourism to contribute to the sustainable development of the destination area, negative impacts should be managed and controlled within local levels of environmental and social tolerance, while positive impacts should be optimized. Contemporary approaches to this challenge favour the sustainability indicators approach, although, as suggested throughout this book, there is no single solution to the tourism development dilemma. The following chapter examines a range of challenges to tourism development, further illustrating the complexity of the tourism development dilemma.

Discussion questions

1   Is mass tourism the best development option to maximize the economic impact of tourism?
2   What are the challenges in using various measures to minimize negative impacts such as carrying capacity or sustainability indicators?
3   If a proposed tourism development in a remote location can generate a great many new jobs, should it proceed if there will be negative social and environmental impacts?
4   If elements of a local culture can be preserved by putting them on display for tourists, should they be appropriated for the tourism market?

Further reading

Holden, A. (2007) *Environment and Tourism*, 2nd edn, Abingdon: Routledge.

Focusing specifically on the environmental consequences of tourism, this book provides an in-depth analysis of the nature and extent of, and potential solutions to, the impacts of tourism on the physical environment. Numerous relevant case studies are a particular feature of the book.

Wall, G. and Mathieson, A. (2006) *Tourism: Change, Impacts and Opportunities*, Harlow: Pearson Education.

This book is an updated and revised edition of Mathieson and Wall's 1982 book *Tourism: Economic, Physical and Social Impacts*. Combining the original detailed analysis of tourism's impacts with additional sections addressing contemporary issues and challenges, this remains the most comprehensive book on the subject of the impacts of tourism.

Websites

http://statistics.unwto.org/en

The World Tourism Organization website contains information relating to the Tourism Satellite Account.

www.unep.org/climatechange/

This United Nations Environment Programme website focuses on climate change. There are a variety of useful links on the page, including one to a programme related to development and climate. The programme acknowledges the pressing issues for many developing countries, such as poverty and food security, to name a few; however, the project aims at identifying development paths linked to positive
climate outcomes, as well as facilitating dialogue and decision-making with key stakeholders both nationally and internationally.

8 Challenges to tourism and development

Learning objectives

When you have finished reading this chapter, you should be able to:

- recognize the diversity of global threats facing tourism in developing countries
- understand the challenges of climate change for tourism in developing countries
- identify key resource constraints facing the tourism industry
- be familiar with the implications of economic and political instability and demographic change for tourism

The principal focus of this book is on the relationship between tourism and development. More precisely, this book is concerned primarily with what is referred to in Chapter 1 as the tourism development dilemma. That is, over the last half century or so, tourism has not only become an increasingly significant global economic sector but has also come to be considered an effective vehicle of economic and social development in destination areas. Indeed, in the developing world in particular, tourism both occupies an important position in the economy and is an integral element of the development policies of many countries. Moreover, it has been long seen as a means to 'eliminate the widening economic gap between developed and developing countries and the steady acceleration of economic and social development and progress, in particular in developing countries' (WTO 1980: 1). Yet, as observed in Chapter 1, the establishment of a tourism sector does not inevitably result in wider economic and social development in the destination; frequently, the assumed developmental benefits of tourism fail to materialize, benefit only local elites, or are achieved at significant economic, social or environmental cost to local communities. Hence, the dilemma lies in the challenge of managing these negative consequences for the potential longer-term benefits offered by tourism development, a challenge that the chapters in this book thus far have explored in some detail.

Importantly, however, tourism does not exist in isolation from the world that surrounds it. In other words, tourism has been described as a discrete 'system' (Leiper 1979, 1990) involving the tourist-generating region (where tourists come from), the destination region (where tourists go to) and the transit region, and the study of tourism in general, focuses on the relationship between different elements within this system. Equally, many of the issues considered within this book in relation to tourism and development have been explored within the context of the tourism system. But that tourism system may, of course, be influenced by a variety of external factors. Putting it more simply, it has long been recognized that tourism is highly susceptible to external forces and events, such as natural disasters, health scares, political turmoil, terrorism, economic downturns and so on. Typically, such events are reflected in declines in tourist arrivals either nationally or regionally. For example, the AH1N1 ('swine flu') influenza outbreak in Mexico in 2009 had an immediate though short-lived downward impact on international tourist arrivals to that country (Speakman and Sharpley 2012); similarly, the island of Bali suffered a decrease in arrivals for a number of years following the nightclub bombing in 2002 (Darma Putra and Hitchcock 2006). Conversely, the severe acute respiratory syndrome (SARS) outbreak in China in 2002–2003 had a wider impact on tourism on countries in the region (McKercher and Chon 2004). On occasion, the decline in international tourist arrivals might be experienced worldwide, as was the case in 2009 following the global financial crisis of 2007–2008, although this is a relatively rare occurrence. Whether national, regional or global, however, such declines – and their impact on development more generally – tend to be temporary.

In contrast, there exists a number of issues and trends in tourism's external environment that represent longer-term challenges to tourism and development, challenges that may either restrict the scale, direction and nature of tourism itself or the extent to which destinations might achieve developmental benefits from tourism. In other words, as Smil (2008) suggests more generally, predicting the future can only be undertaken on the basis of understanding change and drawing on knowledge and history to identify the probability and outcome of factors and events that will shape the global future. On the one hand, such events might be instantaneous with immediate impact, such as the external forces on tourism referred to above, the prediction and management of which is widely addressed in the literature (for example, Faulkner 2001; Glaesser 2006; Ritchie 2009). On the other hand, there may be gradual change or emerging influences over time, trends that slowly but increasingly come to represent risks to the established order of things or, in the context of this book, risks that represent challenges to the future of tourism and development.

These global risks are explored in the context of tourism in some detail by Telfer and Hashimoto (2015). Drawing on the annual *Global Risks Report* produced by the World Economic Forum, they list five themes under which major global risks are categorized: economic, environment, geopolitical, societal and technological. Interestingly, in the latest report (World Economic Forum 2015), cited by Telfer and Hashimoto (2015: 404), the five risks most likely to emerge over the next decade are: the economic risks of (i) increasing income inequality; and (ii), excessive government debt; the environmental risk of (iii) rising greenhouse gas emissions; and the societal risks of (iv) water supply crises; and (v) mismanagement of population ageing.

It is beyond the scope of this chapter to consider the potential impact on tourism and development of every risk identified by the World Economic Forum. Nevertheless, a number of key issues that have emerged since the publication of the first edition of this book, and which challenge tourism and its contribution to development in the longer term, demand attention. A number of these are widely recognized and discussed at length in the literature, not least climate change, which, as discussed in the following section, is perhaps one of the most immediate and certainly the most publicized global challenge that tourism both contributes to and is impacted by. The tourism industry relies on key resources such as oil and water, both of which are under threat. While the number of those regarded as food-insecure is declining, there are still significant numbers of people in developing countries malnourished, and the tourism industry not only operates in this environment; it also uses a significant amount of food and generates food waste. Economic and political instability continue to cause uncertainty for tourism, and it is often developing countries that are most vulnerable. Other trends, such as demographic change and, in particular, population ageing are, arguably, considered to be less pressing but nevertheless may impact significantly on tourism over time. Either way, the challenges introduced in this chapter, though not intended as an exhaustive list, collectively represent an agenda for the future of tourism and development.

Climate change

In late 2015, a UN Climate Change Conference will take place in Paris with the aim of producing a universal, legally binding agreement to take effect in 2020 with the goal of limiting global warming to below 2 degrees Celsius (Briggs 2015). To meet this goal, greenhouse gas emissions in 2050 will have to be 40–70 per cent lower than they were in 2010. An additional objective of the conference is the mobilization of $100 billion annually from developed countries to help developing countries combat climate change while promoting fair and sustainable development (UN Climate Change Conference 2015).

The Paris meeting is part of the efforts within the United Nations Framework Convention on Climate Change (UNFCCC) originally signed in 1992 at the Rio Earth Summit and operationalized by the Kyoto Protocol of 1997. The Kyoto Protocol set binding emission reductions from 2008 to 2012, while the Doha Amendment extended the commitment period to 2020 (UN Framework Convention on Climate Change 2014). Not all member countries have committed to the emission reduction targets. The UNFCCC defines climate change as 'a change of climate which is attributed directly or indirectly to human activity that alters the composition of the global atmosphere and which is in addition to natural climate variability observed over comparable time periods' (UNFCCC 1992: 7).

The Intergovernmental Panel on Climate Change (IPCC) is an important source of information for research on climate change and issues reports supporting the UN Framework Convention on Climate Change. Their most recent report, the 5th Assessment Report published in 2014, consists of a synthesis report as well as three working group reports on (1) physical science, (2) impacts, adaptation and vulnerability and (3) mitigation of climate change. While it is beyond the scope of this chapter to examine these reports in detail, Table 8.1 contains a list of observed impacts, vulnerability and exposure from the 2014 IPCC report on *Climate Change, Impacts, Adaptation and Vulnerability*. These impacts are very complex and highly interrelated, and impact tourism in multiple ways, as will be explored below. A key finding from the report in the context of this chapter is that as a result of uneven development processes, marginalized people are more vulnerable to the impacts of climate change. Developed countries have greater capacity to adapt to climate change compared to developing countries due to their wealth (Hall *et al.* 2015). Citing a report for the Global Humanitarian Forum, Hall *et al.* (2015) report that four billion people are vulnerable to climate change, 500 million people are at extreme risk and approximately half a million lives can be expected to be lost per year by 2029.

Along with the work at the UN, there are a wide variety of governments, NGOs and corporations studying climate change and implementing mitigation strategies. The Earth League, for example, is an alliance of climate researchers from 17 institutions that released a statement with the following eight calls for action (Briggs 2015):

- limiting global warming to below 2 degrees Celsius;
- keeping future CO_2 emissions below 1,000 gigatonnes (billion tonnes);
- creating a zero-carbon society by 2050;
- equity of approach – with richer countries helping poorer ones;
- technological research and innovation;
- a global strategy to address loss and damage from climate change;
- safeguarding ecosystems such as forests and oceans that absorb CO_2; and
- providing climate finance for developing countries.

The challenge is to develop strategies to be put in place before it is too late because, as commentators such as Dwyer (2008) claim, wars will be fought as a result of climate change because of the impacts it will have on resources. Mitigation, adaptation and geo-engineering are seen as the main ways to address climate change, as noted below:

Table 8.1 *Observed impacts, vulnerability and exposure to climate change*

- In recent decades, changes in climate have caused impacts on natural and human systems on all continents and across the oceans.
- In many regions, changing precipitation or melting snow and ice are altering hydrological systems affecting water resources (quality and quantity):
  - Glaciers shrinking affecting run-off.
  - Permafrost warming and thawing in high latitude and altitude regions.
- Many terrestrial, freshwater and marine species have shifted geographic ranges, seasonal activities, migration patterns, abundances and species interactions.
- Negative impacts on crop yields are more common than positive impacts:
  - Production aspects of food security a concern.
  - Crop prices sensitive to climate extremes.
- Worldwide impact of climate change on human ill health is relatively small compared with other stressors and is not well evidenced. However:
  - Increased heat-related mortality.
  - Local changes in temperature and rainfall have changed distribution of some water-borne illnesses and disease vectors.
- Differences in vulnerability and exposure to climate change arise from non-climatic factors and from multidimensional inequalities often produced by uneven development processes:
  - People who are socially, economically, culturally, politically, institutionally or otherwise marginalized are especially vulnerable to climate change and also to some adaptation and mitigation responses.
- Impacts from recent climate-related extremes (for example, heatwaves, drought, floods, cyclones, wildfires) reveal significant vulnerability and exposure of some ecosystems and many human systems to current climate vulnerability:
  - Impacts include alteration of ecosystems, disruption of food production and water supply, damage to infrastructure and settlements, morbidity and mortality, and consequences for mental health and well-being.
- Climate-related hazards exacerbate other stressors, often with negative outcomes for livelihoods, especially for people living in poverty:
  - Impacts on livelihoods, crop yields, destruction of homes, increased food prices, food insecurity.

Note: the report rates the above impacts on confidence levels

Climate-related drivers of impacts include:

• warming trends	• snow cover
• extreme temperature	• damaging cyclones
• drying trend	• rising sea level
• extreme precipitation	• ocean acidification
• precipitation levels	• carbon dioxode fertilization

Source: IPCC (2014)

- *mitigation*: ongoing reduction of greenhouse gas emissions from all sectors;
- *adaptation*: adapt and protect critical assets (for example, power stations, transportation routes, built environment) from flooding and sea level rise and it could mean planned abandonment of settlements and infrastructure;
- *geo-engineering*: use technology to slow global temperature rise by removing carbon dioxide from the atmosphere or reflecting solar radiation back to space (Institution of Mechanical Engineers 2015).

The UNWTO's (2009a) proposed strategies to address greenhouse emissions include reducing energy use, improving energy efficiency, increasing the use of renewable energy and sequestering carbon (for example, carbon sinks such as forestry schemes, some of which are linked to carbon offsetting).

Climate change and tourism in developing countries

The dilemma for many developing nations where tourism is an agent of development is that many rely on tourists arriving by airlines or cruise ships, modes of transport that contribute to global warming and climate change. Global warming brings increased intense weather events, such as hurricanes or typhoons, as well as rising seawater levels. Some developing destinations that rely on tourism, such as Kiribati, are slowly going underwater (see Box 8.1). Other countries that are landlocked may rely on meltwater from annual snowfall and glaciers in mountainous regions. With global warming, these glaciers are disappearing. So, in effect, what is bringing an economic revenue stream (tourism) is also bringing with it the potential for catastrophic environmental change. The changes not only impact the tourism industry, but also the destination. As noted in earlier chapters, tourism can be viewed as a system with generating regions, transport routes and destination regions. There is greater recognition that the impacts of tourism occur at all points in the tourism system. If tourism operates under a 'business as usual' scenario, carbon dioxide emissions from global tourism are projected to increase by 130 per cent, with most of the increase attributed to air travel (UNWTO 2009a). In Chapter 1, the growth in visitor numbers was discussed, and if domestic travel is factored into this, the contribution of emissions over the long term will be significant. In addition, many developing nations are also launching low-cost air carriers (see Chapter 3); this is democratizing travel but also adding to global emissions. Table 8.2 examines the effects of climate change and variability on tourism destinations and operators. For those in the tourism industry and those that indirectly rely on tourism in developing countries, any negative climate change event, gradual or sudden, will reduce visitor demand resulting in lost income and, eventually, a reduction in GDP.

Developing nations, including the Caribbean, small island developing states (SIDS), South East Asia and Africa are seen as most-at-risk nations as they have: high exposure to multiple climate change impacts (see Table 8.2); distance to major markets (long-haul travel over five hours) and are exposed to more stringent aviation emission policies; lower adaptive capacity; limited domestic markets; and high dependency on international tourism (UNWTO 2009a). More generally, many forms of tourism are under threat due to climate change. Beach tourism is threatened by sea level rise and extreme storms. Niche tourism products, such as ecotourism in natural environments or scuba diving on coral reefs, are threatened by the loss of biodiversity as a result of climate change. Climate change can then have a negative impact on the role tourism can have on poverty reduction (Holden 2013). Examining the cases of Barbados in the context of sustainable tourism, climate change and sea level rise, Mycoo (2014) calls for policy reforms in building construction, coastal management, water resources management, waste management, physical planning and sustainable land management for climate change adaptations for most SIDS. Hall et al. (2015) argue that climate change vulnerability must be included in decision-making regarding any future strategy for tourism to contribute to poverty alleviation and the UN Millennium Development Goals.

It is clear human activity, including tourism, is causing climate change and many developing nations relying on tourism are the least able to deal with the impending consequences. Initial steps are being taken by various sectors of the tourism industry to cut emissions but more needs to be done. The potential for tourism numbers to drop for some destinations may pale in comparison to the potential environmental, social and economic consequences for the host nation as a result of climate change. Hence, Hall et al. (2015) raise a vital question in regards to tourism and climate change: If we travel now, will we pay later?

Box 8.1

Kiribati and climate change

Kiribati is an independent republic with a population of 110,000 in the central Pacific Ocean. It has become the centre of attention in discussions and reports on climate change. The country is situated on 33 low-lying coral islands (21 inhabited), all of which are atolls or ring-shaped islands with a central lagoon, except for the island of Banaba. The atolls are divided into three main island groups over 2 million km^2 of ocean and many of the atolls are only 6 metres above sea level. In addition to offering beach tourism, other tourism activities advertised on the government web page include fishing, birdwatching, cultural experiences, surfing, outdoor adventures, scuba diving and visiting Second World War sites. Other markets include visiting yachts and cruise tourism. The vision statement in the Kiribati National Tourism Action Plan is 'For tourism to become the largest and most sustainable economic sector driving employment, growth and the Kiribati economy'. The focus of tourism development in the islands is to be on low numbers and high yield. Tourism in the country accounts for approximately 20 per cent of the country's GDP, although in 2013 the island only had 6,000 visitors. The key targets from the National Tourism Action Plan include:

- tourism generating 40–50 per cent of GDP of the country;
- 1,200 employees (full-time, part-time, casual) in tourism and hospitality;
- 40 new micro, small and medium-sized businesses;
- six enterprises of 'national significance';
- cruise ship visitation with a minimum of 32,500 tourists;
- visitation on average of 10,000 land-based travellers;
- increased seat capacity of airline routes to Kiribati; and
- 15 per cent increase in visitor expenditure.

While the plans are in place to increasingly rely on tourism, climate change is having wide ranging impacts on the island. The government is reporting since 1950 increases in: annual and seasonal maximum temperatures; annual rainfall in the wet season but also drought; sea levels; coastal erosion; wave heights; number of storms; and ocean acidification. Future predictions place sea level rise by 2030, under a high-emission scenario, to be in the range of 5–14 cm, which will increase the impacts of storm surges and coastal flooding. The coastal erosion is displacing people and placing stress on agriculture, while storm surges bring saltwater intrusion into the freshwater supply. The government website states, 'In Kiribati, the entire nation faces real danger – our own survival is at stake as a people, as a unique and vibrant culture and as a sovereign nation.' The government has in place a number of actions and steps with the focus on mitigation, adaptation and relocation. The Kiribati Adaptation Program is centred on a number of fronts, including working with local communities to develop locally managed adaptation plans, improving the management of water resources and strengthening coastal resilience. In anticipation of what may become the inevitable, the government recently purchased 20 km² on Vanua Levu, one of the islands of Fiji, which is approximately 2,000 km away, for possible relocation of residents. In the immediate future, the land will be used for agriculture and fish-farming projects to guarantee the nation's food security. With rising tides, seawater is increasingly contaminating the atoll's groundwater, and in addition there is coral bleaching, which is all having an impact on food supply. The relocation strategy has two parts. The first is to help those wanting to migrate now who may set up expatriate communities helping future migrants and sending back remittances, and the second is to raise the level of qualifications of those in Kiribati so they can apply for jobs in places such as New Zealand and Australia. The aim of the relocation strategy is to allow for 'migration with dignity' and to avoid humanitarian evacuation in the future.

Source: Caramel (2014); Kiribati Climate Change (2015); Kiribati Tourism (2015); Kiribati National Tourism Office Action Plan (n.d.)

Table 8.2 *Effect of climate variability on tourism destinations and operators*

Climate change impacts	Examples
Length and quality of tourism season	• Climate is principal resource tourism is based on • Small tropical island developing states require stable weather (few storms) • Ski resorts (for example, India, Bolivia) require enough snow and season length
Various facets of tourism operations	• The following can impact profitability: water supply and quality, heating-cooling costs, snowmaking requirements, irrigation needs, pest management, evacuations and temporary closures
Environmental resources	• Environmental resources are attractions and are sensitive to climate change, including wildlife, biodiversity, water levels and quality, snow conditions, glacier extent • Heritage attractions at risk • Coastal erosion due to sea level rise or extreme storms • Expansions of deserts
Environmental conditions	• Conditions can deter tourists, including infectious diseases, wildfires, algal blooms, insects or water-borne pests (for example, jellyfish), extreme events (for example, hurricanes, floods, heatwaves)
Tourist decision-making	• Seasonal fluctuations at destinations and departures locations can impact travel demand decisions, travel experience, tourist spending, holiday satisfaction
Destination security	• Conflicts over resource shortages (for example, water) may lead to political instability

Source: UNWTO (2009a)

Sustainability of resource supply

Oil

In late 2014, global oil prices fell to their lowest level for four years; having reached a high of US$150 per barrel in 2008 and subsequently stabilized at around US$110 per barrel since 2010, they more than halved to under US$50 per barrel. Although the result of lower demand and excessive production, this fall in the price of oil might be seen as an indication that growing concerns over the sustainability of oil supplies have been prematurely alarmist. That is, almost 10 years ago, it was suggested by some that the world had already reached the position of so-called 'peak oil', a theory initially proposed by the geophysicist Dr King Hubbert (Schindler and Zittel 2008; Leigh 2011). According to Hubbert, peak oil occurs when the maximum rate of oil production is arrived at; beyond peak oil, the rate of oil production (as opposed to total oil reserves) inexorably declines, with inevitable rises in price as demand exceeds supply. Hubbert's prediction of US peak oil in the 1970s initially proved correct. However, with the introduction of 'fracking' to produce shale oil, US oil production has since returned to the level of the 1970s, although production costs are significantly higher than conventional drilling methods. Nevertheless, although global production levels continue to increase (see Table 8.3), many now claim that peak oil has in effect been reached, as production levels at many major oilfields are in decline while new oilfields are being discovered more slowly.

Table 8.3 *Global oil production ('000 barrels per day)*

Year	Production	Year	Production
2000	77,725	2008	86,570
2001	77,672	2009	85,739
2002	77,101	2010	88,158
2003	79,606	2011	88,583
2004	83,402	2012	90,485
2005	85,101	2013	90,923
2006	85,153	2014	93,003
2007	85,167		

Source: International Energy Statistics (2015)

In practice, predicting or identifying global peak oil accurately is complex, not least because it is linked to trends in global demand for oil. Recent declines in economic growth in China, for example, may dampen demand, which, in turn, should lead to cuts in production to maintain price levels; however, it may also lead to less investment in oil exploration and extraction technologies. Nevertheless, if peak oil has not yet been reached, it is inevitable that it will be in the relatively near future, with significant long-term implications for tourism and development. Indeed, declining oil production/supply is arguably the single greatest challenge facing tourism, and therefore it is surprising that, with some notable exceptions, relatively little academic attention has been paid to it (see, for example, Becken 2008, 2011, 2015; Leigh 2011; Logar and van der Bergh 2013).

Ironically, declining oil production has led some oil-producing countries, most notably Dubai but also Qatar and other Gulf States, to develop tourism as an alternative economic sector (Henderson 2014). However, tourism has long been recognized as an energy (oil) intensive activity (Gössling et al. 2005); the most popular forms of transport (cars/buses, aeroplanes and ships) are oil dependent, while accommodation and other sectors are also significant consumers of oil and oil-based products. Hence, as the production and supply of oil declines and prices rise and with a lack of viable alternative energy sources, travel in particular will become much more expensive. Consequently, there is likely to be an overall decline in the demand for tourism, initially to more distant destinations but, as oil becomes more scarce/expensive, to regional destinations also. At the same time, there is also likely to be increased use of more energy efficient forms of public transport, growth in local domestic tourism, and a shift towards 'simpler' greener, less energy-intensive forms of travel and tourism, such as cycling and camping holidays.

The implications for tourism and development are clear, particularly for many developing nations that are dependent on long-haul tourism markets. Not only is it inevitable that international tourism will begin to decline, eventually becoming what it once was, the preserve of a wealthy elite. It will also, quite simply, no longer be possible to think of tourism as a driver of economic and social development unless in the future a viable alternative energy source to oil is found.

Water

According to the United Nations (UN 2015: 2):

> Water is at the core of sustainable development. Water resources, and the range of services they provide, underpin poverty reduction, economic growth and environmental sustainability. From food and energy security to human and environmental health, water contributes to improvements in social well-being and inclusive growth, affecting the livelihoods of billions.

Putting it more starkly, water is fundamental to human existence. Not only is 'access to safe drinking water and sanitation . . . a human right' (UN 2015: 3), a right that remains denied to millions in the developing world, but it is also essential to the production of most goods and services, not least agriculture and food production. Hence, as the global population and associated demand for products grows, so too does the demand for water. Over the last 50 years, for example, global freshwater consumption has more than tripled (Becken *et al.* 2013), while the UN (2015) predicts a further 55 per cent growth in demand by 2050. As a consequence, not only will there be an increasing incidence of what is referred to as 'water stress' in many parts of the world, stress that may be exacerbated by the effects of climate change (Vörösmarty *et al.* 2000), but there is also likely to be increasing competition for limited water supplies.

One economic sector that competes for water is, of course, tourism. As McKercher (1993) notes, not only does tourism generally compete for scarce resources, but it also has the potential to overconsume such resources. Water is one such resource. Lehmann (2009) observes that fresh water is one of the most critical and scarce resources for tourism, noting that the development of tourism may put additional strain on local water supplies, particularly where there is a relatively high density of accommodation facilities (Hof and Schmidt 2011). Indeed, the accommodation sector accounts for the highest use of water in destinations and there is 'a tendency for higher-standard accommodation to consume significantly higher water volumes' (Gössling *et al.* 2012). That is, hotels with spa facilities, multiple swimming pools, landscaped grounds requiring irrigation, higher volumes of laundry and so on tend to consume the most water, while overall, tourists typically use significantly more water per day than local residents, particularly in warmer or tropical climates where they are likely to shower frequently. Even in more temperate climates, such as the Mediterranean, tourists' use of water may significantly exceed that of local residents. One study, for example, suggests that tourists in Spain daily consume more than double the amount of water than the average Spanish citizen (De Stefano 2004, cited in Lehmann 2009), while, ironically, the development of lower volume, higher-quality tourism in the Spanish island of Mallorca has actually increased water consumption (Hof and Schmidt 2011).

Nor is water consumption by the tourism sector limited to its use accommodation (see Plate 8.1). There are a variety of forms of what Lehmann (2009) refers to as consumptive water use, while water also plays a 'non-consumptive' role in tourism. In other words, water is essential to certain tourist experiences, such as water sports, but its use does not reduce the overall supply of water. Figure 8.1 summarizes the consumptive and non-consumptive roles of water in tourism.

Although it is recognized that tourism is a major consumer of water, care must be taken not to overstate the demand of tourism on water resources. In their detailed review, for example, Gössling *et al.* (2012) point out that, globally, 70 per cent of freshwater

Plate 8.1 *UAE, Dubai: aquarium in Dubai Mall*

Source: Photo by R. Sharpley

Figure 8.1 *Consumptive and non-consumptive water use*

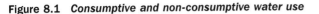

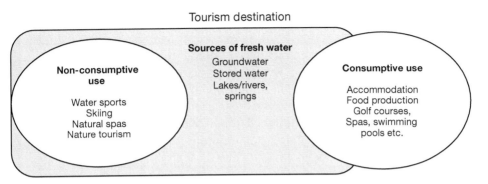

Source: Adapted from Lehmann (2009: 3)

consumption is accounted for by agriculture, 20 per cent by industrial uses and 10 per cent by domestic household use. Moreover, on average, international tourism accounts for just 1 per cent of national water use, suggesting that efforts to reduce water consumption by the tourism sector may have little impact on overall water supplies. Nevertheless, generalizations cannot be made. The availability of fresh water varies considerably, of course, from one country to another and depends on a variety of factors, including rainfall patterns, ground supplies, water storage and supply infrastructure and so on, though islands (which often support high levels of tourism) tend to face greater water scarcity than other landforms (Black and King 2009). Consequently, many islands, including those with significant tourism sectors, now rely on imports of fresh water and/or desalination plants. Equally, the proportion of water consumed by the tourism sector varies according to the nature, scope and scale of tourism, with islands again being particularly vulnerable.

However, there is no doubt that the sustainability of water resources represents a major challenge for tourism and development, particularly in the developing world. Many developing nations, specifically small island developing states, already suffer significant water stress; that is, the supply of fresh water has become a major consideration in the tourism development dilemma. And, given the fact that much of the growth in international tourism is occurring in developing countries, many of which face the twin challenges of population growth and relative freshwater scarcity, such water stress is likely to become more acute, with certain regions being particularly at risk. For example, in recent years, international tourism has increased most rapidly in the Asia-Pacific region, and continuing growth in arrivals combined with the construction of tourist infrastructure, such as spa resorts and golf courses, is likely to put increasing strain on already limited water resources to the detriment of development more generally (Gössling *et al.* 2012; Becken *et al.* 2013). Thus, in many destinations, not only will there be a requirement for effective water planning and management for the tourism sector, from better monitoring and measurement of water consumption and processes to improve water efficiency in hotels to investment in technologies to sustain water supplies, but also for fundamental decisions regarding the nature and extent of tourism development in the destination. However, as Gössling *et al.* (2012: 13) note, this will require effective policies and regulation 'because the tourism industry is not likely to make water use a key priority by itself, given the low cost of water in comparison to other operational costs'.

Food

The world faces significant challenges in food production; we will need to produce at least 50 per cent more food to feed the nine billion people by 2050, yet climate change could reduce crop yields by more than 25 per cent (World Bank 2015d). Land, biodiversity, forests and oceans are all being depleted, food prices continue to rise, and those most food-insecure in developing countries, the world's poorest, are at greatest risk (World Bank 2015d). Food insecurity is a situation when 'people lack secure access to sufficient amounts of safe and nutritious food for normal growth and development and an active and healthy life' (FAO 2015: 52). It can be chronic, seasonal or transitory, and causes of food insecurity include the unavailability of food, inappropriate distribution system, insufficient purchasing power, or inadequate use of food at the household level (FAO 2015). Humans have a right to food. Indeed, the Universal Declaration of Human Rights (UN 1948) Article 25 states that 'Everyone has the right to a standard of living adequate for the health and well-being of himself and his family, including food'. In recent decades, tremendous progress has been made in reducing hunger and the Food and Agricultural Organization of the United Nations (FAO 2015) reports that 805 million people were estimated to be chronically undernourished in 2012–2014, some 100 million less than a decade earlier. However, there are marked differences in regions across the globe (see Table 8.4). Sub-Saharan Africa has the highest prevalence of undernourishment (nearly one in four people), while developing nations as a whole were not on track to meet the UN Millennium Development goal of halving the number of undernourished by 2015. Asia, with the largest proportion of the world's population, has the highest number of undernourished people and, in particular, Southern Asia has been making slow progress in reducing hunger (FAO 2015).

Hunger is caused by a diversity of interconnected reasons, including the poverty trap, lack of investment in agriculture, climate and weather, war and displacement, unstable markets and food wastage (WFP 2015). To this list, Patel (2012) adds the structure of the global food system, the cost of oil, commodity markets, the shift to biofuels, population growth and the cost of land. Traditionally, food aid has been the response to hunger and food insecurity; however, humanitarian aid is coming under scrutiny from food policymakers and practitioners, focusing on how to make positive food security outcomes through globalization of relief, improved food aid governance and development assistance, as well as economic progress (Van Uffelen 2013). Global food prices have fluctuated, and sharp increases in early 2008 caused a range of 'humanitarian, human rights, socio-economic, environmental, developmental, political and security-related consequences' (UN 2011). For example, the doubling of prices of staple foods such as rice and wheat in some markets resulted in food riots (Scheyvens 2011). In Haiti, these food riots resulted in travel warnings just as Haiti was looking to invest in tourism (ETN 2008). Food prices rose again in 2011–2012, leading to more riots.

In the destination, then, what is the relationship between tourism and food? In the context of tourism, food is produced, packaged, transported and prepared for consumption by tourists, tourism employees (formal and informal) and those in the destination. Food represents an attraction as well as a potential source of income for those able to sell to the tourism industry or open a restaurant or food stall. However, food is also wasted. In developing countries, resort developments may represent islands of affluence surrounded by those going hungry. Kenya, for example, is well known for safari tours, ecolodges and the Maasai people. In 2012, the country received 1.619 million international visitors and US$935 million in international tourists receipts (UNWTO 2014a), yet it also received 91,272 metric tons of food aid along with US$113.7 million through the Food for Peace programme through USAID. As of April 2014,

Table 8.4 *Undernourishment around the world, 2008–2010 and 2012–2014*

Number of undernourished (millions) and prevalence (%) of undernourished

	2008–2010		2012–2014*	
	No.	*%*	*No.*	*%*
World	840.5	12.1	805.3	11.3
Developed region	15.7	<5	14.6	<5
Developing region	824.9	14.5	790.7	13.5
Africa	216.8	20.9	226.7	20.5
Northern Africa	5.6	<5	12.6	6.0
Sub-Saharan Africa	211.2	24.5	214.1	23.8
Asia	565.3	14.1	525.6	12.7
Caucasus and Central Asia	7.4	9.5	6.0	7.4
Eastern Asia	185.8	12.7	161.2	10.8
South-Eastern Asia	79.3	13.4	63.5	10.3
Southern Asia	274.5	16.3	276.4	15.8
Western Asia	18.3	9.1	18.5	8.7
Latin America and the Caribbean	41.5	7.0	37.0	6.1
Caribbean	7.6	20.7	7.5	20.1
Latin America	33.9	6.1	29.5	5.1
Oceania	1.3	13.5	1.4	14.0

* Estimate

Source: FAO, IFAD and WFP (2014)

40 per cent of Kenyans lived below the poverty line; 75 per cent of food was produced by small-scale farmers of which 70 per cent are women with little or no access to farm inputs; there was below average rainfall; and the country was hosting 600,000 refugees from surrounding countries in conflict (USAID 2014). In March 2015, Kenya was listed along with 36 other countries as requiring external food assistance under the category of 'severe localised food insecurity'. It is estimated that 1.5 million people are severely food-insecure, mainly in the north-eastern and central counties (GIEWS 2015). Kenya also hosts slum tours in Kibera, the largest slum in East Africa next to Nairobi. Tourists can now visit those who are food-insecure. Kibera is similar to other slums where urban migration has led to the development of slum communities and food deserts where there is a lack of access to arable land, low incomes and increasing food insecurity (Pires 2014).

One only has to look at the advertising for cruise ships docking in developing countries to see that food is abundant on these ships. Many of these ships travel with all the food they require for the journey. Moreover, they continue to get larger; Royal Caribbean's *The Allure of the Seas*, for example, can hold up to 6,360 passengers and offers multiple restaurants. Large tourism resorts have the ability to import significant quantities of food, generating a tremendous number of food miles contributing to global warming, and yet these may be next to communities with high rates of poverty and limited access to food. The presence of tourism resorts can lead to inflation in local food prices and the best-quality food may be sold to hotels and restaurants.

The demand for food may also lead to a change in what is produced as farmers cater to the demands of the tourism industry. Tourism also draws migrants from agriculture, leaving fewer to work the farms. In Hainan, China, the central government established plans for a high-end resort, leading to an investment boom in five-star hotels, golf courses and marinas. The island's economy grew 35 per cent faster than the rest of the country; however, it has generated inflation and family farms are being sold off for luxury homes (Wang and Edwards 2012). If conditions are right, tourism can potentially stimulate the local agricultural sector. Telfer and Wall (1996) examined the purchasing patterns of a large resort on the island of Lombok, Indonesia, and found the hotel purchasing local food and fish. In 2008, the World Responsible Tourism Award for 'Best for Poverty Reduction' went to the 'Gambia is Good Project', which had 1,000 farmers (90 per cent of them women) producing 20 tonnes of fruit and vegetables in the tourism season to supply hotels and restaurants, making a contribution to poverty reduction (World Responsible Tourism Awards 2015). The challenge facing any such project, however, relates to food quality and quantity and how the project can be institutionalized so that it can endure.

Increases in food insecurity and rising food prices can have significant implications and costs for the tourism industry. Yeoman (2012) looked to the future in examining food tourism in 2050 and explored scenarios where food was abundant and where food was in short supply. Exploring a scenario where food was limited and only primarily accessible to the wealthy, he listed 10 possible drivers of this type of scenario, which included:

- the corporation controlling the food supply chain;
- the role of food in society;

- climate change;
- diet;
- food inflation – a world problem;
- wealth distribution – haves and have-nots;
- urbanization;
- scramble for food;
- role of food supplements and a healthy lifestyle; and
- advancement of science (Yeoman 2012).

Food is a resource used by the tourism industry in developing countries, many of which are food-insecure. The industry typically has the economic power to source food even where there is limited food. However, it is also a resource subject to price fluctuations and control by large food corporations. What happens if supply is reduced and prices rise? Will the cost of travel then increase, causing a drop in visitor numbers and residents in the destination may be even more food insecure? The dilemma for host governments in developing countries in pursuing tourism is that tourism can often be located in destinations where local people struggle to find enough to eat.

In concluding this discussion on resource supply, the section turns to the work of Chossudovsky (2008), who argues that fuel, water and food are three of the necessities in a civilized society. Many of the world's poor have limited access, and he argues that the prices of these three resources are set through the market at the global level and not by national governments. Generally, then, the inability to have much influence on the prices of these highly valuable commodities represents significant challenges for the tourism development process.

Global economic instability

The 2008 global economic crisis had a significant impact on global trade and financial markets, affecting all sectors of the global economy, including tourism, and exposed the vulnerability of overreliance on the market. Facilitated by advances in electronic communication, the global economy is becoming increasingly interconnected and, as a consequence, global, national, regional, local and individual economies have become intertwined. Economic shocks originating in a bank, country or region can quickly be felt globally as international markets immediately react. Individuals, corporations, banks, investment companies and governments can rapidly trade or move funds through a variety of financial markets, including:

- capital markets (stocks and bonds);
- money markets;
- derivatives markets;
- foreign exchange markets;
- insurance markets; and
- commodity markets (Economy Watch 2010).

The resulting downturns can have significant impacts worldwide on trade volumes, currency values and debt levels, and can lead to bank failures and other institutional problems (see Chapter 3 for a further discussion on the economic aspects of globalization). According to the IMF, the most basic lesson learned from the 2008 crisis was that 'flawed incentives and interconnections in modern financial systems can have huge macroeconomic consequences' (IMF 2009). Economic crises, however, are not new, and have been a recurring feature of boom and bust economies dominated by the forces of the market. Recent past examples of economic crises include the 1994 economic crisis in Mexico, the 1997 Asian financial crisis, the 1999–2002 Argentine economic crisis and, of course, the 2008 global economic crisis, which fed the eurozone crisis from the end of 2009. Within the eurozone, some nations have been receiving significant financial bailouts from the IMF and the European Union, a surprising turn of events given that, previously, it was typically developing countries that received these massive bailouts. The 2008 global economic crisis has arguably been the most significant of recent crises, having had a ripple effect causing other economic shocks that many countries are still dealing with today.

It is beyond the scope of this chapter to examine all of the causes and implications of the 2008 economic crisis; however, the lessons learned are of value and are addressed briefly here. The IMF (2009) argues that the root cause of the market failure was optimism that was fuelled by a long period of high growth, volatility, low real interest rates and a number of policy failures, including: (a) financial regulation not seeing risk concentration and offering flawed incentives; (b) macroeconomic policies that did not see the systematic risks in the financial system and housing market; and (c) the global architecture of the financial system where surveillance was highly fragmented (IMF 2009: 1). With the long period of 'rapid credit growth, low risk premiums, abundant availability of liquidity, strong leveraging, soaring asset prices and the development of bubbles in the real estate sector' (European Commission 2009: 1), financial institutions with overstretched leveraged positions were extremely vulnerable to market corrections. A correction came in the US sub-prime market, especially the housing market, which was 'sufficient to topple the system' (European Commission 2009: 1), leading to bank failures, a drop in the stock market and credit restrictions, all causing further shocks through the global economy. When housing mortgages were renewed at high rates, large numbers of people defaulted on their loans, starting a chain reaction. Responses included greater regulation in global finance, government bailouts and financial assistance from a variety of organizations, such as the World Bank, the International Monetary Fund and the European Union. Moving forward, the IMF developed a range of recommendations based on lessons learned that centred on broader financial regulations, macroeconomic policies of central banks taking a macro-prudential view, and a re-examination of the global architecture of the world economy that enhances regulation, collaboration and promotes stability (IMF 2009). The IMF has also restructured many of its lending programmes in order to better deal with future crises (see Chapter 3 on structural adjustment).

In the context of this chapter, the global economic crisis of 2008–2009 had a major impact on international tourism, resulting in a 4 per cent decline in international tourism arrivals and a 6 per cent decrease in international tourism revenues (UNWTO and ILO 2013). Tourism has become an important industry for developing countries, and any decline can have major consequences. As an indication of this, in 2009, tourism accounted for 45 per cent of service exports in least-developed countries and, in the same year, emerging economies collectively received 410 million international visitors (47 per cent of the world total) and US$306 billion in international tourism receipts (36 per cent of the world total) (UNWTO and ILO 2013). Given the nature of tourism-related employment and its potential to contribute to poverty alleviation, it is important to understand and learn from an economic crisis. Reduced demand can bring empty hotel rooms, cancelled flights and cancellations of tourism development projects, thereby having a negative impact not only on the GDP of a country dependent on tourism, but also on the livelihoods of communities and individuals who rely on a steady flow of visitors. The economic crisis hit the advanced economies, many of which are major tourism source markets, and then spread rapidly to emerging markets (UNWTO 2009b). The crisis had a particularly negative effect on poor and vulnerable groups in developing countries (UNWTO and ILO 2013). Various governments took direct action to support their tourism industries, although, at the time, they were aimed at marketing, public–private partnerships along with monetary and fiscal measures as opposed to specific policies aimed at poor and vulnerable groups (UNWTO and ILO 2013). The government of Barbados, for example, established a US$12.5 million Tourism Industry Relief Fund in 2009 to assist the tourism industry, while in Grenada and Jamaica the governments lowered the General Consumption Tax by 50 per cent for the tourism sector and also provided loans to protect tourism employment (UNWTO and ILO 2013). In late 2009, the UNWTO issued a 'Roadmap for Recovery' in response to the economic crisis centred on three action areas, as summarized in Table 8.5.

Table 8.5 *UNWTO's road map for recovery*

Resilience

1. Focus on job retention and sector support
2. Understand the market and respond rapidly
3. Boost partnerships and 'coopetition' [sic]
4. Advance innovation and technology
5. Strengthen regional and interregional support

Stimulus

6. Create new jobs – particularly in small and medium enterprises (SMEs)
7. Mainstream tourism in stimulus and infrastructure
8. Review tax and visa barriers to growth
9. Improve tourism promotion and capitalize on major events
10. Include tourism in Aid for Trade & Development Support

Green economy

11. Develop green jobs and skills training
12. Respond effectively to climate change
13. Profile tourism in all green economy strategies
14. Encourage green tourism infrastructure investment
15. Promote a green tourism culture in suppliers, consumers and communities

Source: UNWTO (2009b)

A report by the UNWTO and ILO (2013) examined how the economic crisis impacted employment of poor and vulnerable groups. The report presented a macro-analysis as well as case studies on Costa Rica, the Maldives and Tanzania, and in each location the poor and vulnerable were hit hardest. In Costa Rica, for example, international visitors in 2009 dropped by 9 per cent, large chain hotels and domestic tour operators instituted lay-offs, reduced working hours and used forced unpaid leave for up to three months. Over 5,000 people working in the tourism sector lost their jobs, and those who continued to work saw their purchasing power decline by 12 per cent. Women and unskilled workers were more exposed to the crisis than males and qualified personnel. Low-skilled people resorted to savings, support from neighbours or setting up small businesses as they had difficulties finding a new job (UNWTO and ILO 2013). The UNWTO and ILO (2013) report provides a number of recommendations in the areas of public policy and business measures to mitigate the impacts of an economic crisis, and in particular they include recommendations on the poor and vulnerable groups, as noted below:

- Develop more advanced tourism and employment data indicators so governments can make timely policy decisions.
- Improve ways to increase the resilience and recovery of the industry through strategies of marketing, diversification of market segments, easing visa restrictions and collaboration between public, private sector and local communities.
- Improve ways to mitigate impact on poor and vulnerable groups in the tourism sector, including: evaluation of tourism-related taxes; granting tourism companies access to additional liquidity (loans) tied to employee retention; training in particular for youth, women and low-skilled as they are at high risk of unemployment; minimize shocks on low-income households by promoting equal distribution of tourism revenue; implement social insurance schemes; promotion of pro-poor tourism products.
- Improve communication between public and private sector in the area of crisis monitoring and crisis reaction plans.

Since the 2008–2009 global economic crisis, international tourism arrivals and receipts have rebounded, showing the long-term resilience at the global level. However, tourism relies heavily on the free market, which typically follows a boom and bust cycle;

inevitably, therefore, there will be future economic downturns. The poor, vulnerable and low skilled in developing countries are particularly vulnerable to economic crises, and developing countries often do not have the resources to weather these financial storms. Referring to a series of recurrent crises after the turn of the millennium, including the 2008 economic crisis, Cohen (2012: 103) states that 'These unforeseen developments raise a fundamental question about the long-term stability and ultimate viability of the current global politico-economic system, based as it is on a neo-liberal economic philosophy, credit based consumerism, and growing sovereign debt'. This raises questions as to which directions global finance should take in the future and whether there will be stronger government intervention or greater reliance on the part of global financial institutions to avoid similar economic crises of the scale of 2008.

Political instability/ineffective governance

The successful development of tourism (and, indeed, successful development through tourism) is dependent on both political stability and effective governance, the former being a prerequisite for the latter. That is, for any government or regime to be able to govern effectively, to do the things that governments should do, including implementing appropriate policies and processes to support and promote the development of tourism (see Chapter 4), it must exist in a stable political environment. In other words, according to one widely cited definition, governance is:

> the exercise of political and administrative authority to manage a country's affairs at all levels. It comprises the mechanisms, processes and institutions through which citizens and groups articulate their interests, exercise their legal rights, meet their obligations and mediate their differences.
>
> (UNDP, cited in World Bank 2015e)

Hence, for 'political and administrative authority' to be exercised effectively, the relevant processes, mechanisms and institutions must be stable. Therefore, it is not surprising that, typically, a stable political environment is commensurate with 'good' governance, although improvements in governance result not only from stability, but also the interaction of a variety of other factors, including the rule of law, regulatory quality, the control of corruption, and voice and accountability (IBRD/World Bank 2007). In other words, a

Table 8.6 *The world's most politically unstable countries, 2009–2010*

Rank	Country	Rank	Country
1	Zimbabwe	7=	Afghanistan
2	Chad	7=	Central African Republic
3	Congo Kinshasa	7=	Côte d'Ivoire
4=	Cambodia	7=	Haiti
4=	Sudan	7=	Pakistan
6	Iraq	7=	Zambia

Source: *The Economist* (2009)

stable government may not necessarily govern effectively or fulfil its obligations to its citizens. It is not surprising, therefore, that those countries considered to be most politically unstable based on scores derived by combining measures of economic distress and underlying vulnerability to unrest also arguably suffer from poor governance (see Table 8.6).

The relationship between tourism, politics and governance embraces a broad range of issues well beyond the scope of this chapter (see, for example, Burns and Novelli 2006; Hall 2008; Matthews and Richter 1991). Nevertheless, with regards to longer-term challenges to tourism and development, two key issues demand attention, namely the need to maintain political stability in the destination and the need for the appropriate governance in order to optimize the developmental benefits accruing from tourism.

Political instability

It has long been recognized that political instability, or what Sönmez (1998: 417) terms 'political turmoil', may have a detrimental impact on tourism development. Political instability refers, in essence, to the propensity for a government to collapse, although typically it is not the actual collapse of a government, but the events that cause or surround it that may negatively impact on tourism. For example, since it reconstituted itself as a republic in 1946, Italy has had more than 60 governments, each surviving on average little more than a year, yet the country retains the aura of political stability as well as a successful tourism sector. Conversely, in other countries, more usually in the developing world, political instability is manifested in a variety of ways that either result from or cause the collapse of or change in government, or are evidence of either attempts to destabilize/challenge an existing government or the use of force by the government to counter such challenges.

Whatever the case, there is significant evidence that when political instability is manifested in political violence, a decline in tourism ensues (Neumayer 2004). Putting it another way, tourists tend to be risk-averse, and when a destination is known or perceived to suffer potential political instability/violence, they are likely to seek alternative, substitute destinations that are considered to be less risky (Sönmez and Graefe 1998; Neumayer 2004). Moreover, even when there is no evidence of violence in politically unstable destinations, such as following the bloodless military coup in the Gambia in 1994, travel advisories in major generating markets may lead to a decline in tourism. In the Gambia, for example, the almost complete collapse of tourism over the winter of 1994–1995 led to the temporary loss of 2,000 jobs and the closure of eight hotels, with significant implications for the local economy (Sharpley *et al.* 1996). Equally, non-politically related violence, where tourists are potential victims of criminally motivated violence or where destinations gain a reputation for such violence occurring, is also like to deter tourists, with inevitable consequences for the local tourism sector (Pizam and Mansfeld 1996). In Mexico, for example, the drug wars have moved into tourist areas and, while tourism remains strong in the country, the potential exists for the drug wars to negatively impact on tourism development in the longer term. However, such violence results not from political instability (though it may be a contributing factor), but from the ineffective rule of law.

Of particular relevance to this chapter, not only is there a long history of tourism-related terrorist incidents (Ryan 1991b), but evidence suggests that the potential exists for tourists to be increasingly targeted for terrorist purposes (see Plate 8.2). Certainly, since the hijacking of the cruise ship Achille Lauro and subsequent murder of one American passenger by members of the Palestine Liberation Front in 1985, there have been numerous examples of the intentional targeting of tourists by terrorist groups, most recently when 17 tourists from a visiting cruise ship were killed while visiting the Bardo Museum in Tunis. As in previous incidents elsewhere, there was an almost immediate decline in tourist arrivals in and bookings to Tunisia, compounding the fall in tourism to the country following the so-called Arab Spring revolution there in 2011. However, a recent report identifies that since 2000, there has been a fivefold increase in the annual number of deaths worldwide resulting from terrorism, from 3,361 in 2000 to 17,958 in 2013 (IEP 2014: 2). Three points should be noted. First, the total number of deaths from terrorism is relatively small compared to the global total of homicide victims (437,000 in 2012); second, over 80 per cent of deaths through terrorist attacks occurred in just five countries (Iraq, Afghanistan, Pakistan, Nigeria and Syria), none of which have major tourism sectors; and third, the number of tourists killed in terrorist attacks is only a small proportion of the total. Nevertheless, as the IEP (2014: 2) report notes, 'The threat of terrorist activity is a major if not the major national security risk for many countries', and, for some, a potential threat to their tourism industries.

Failing states

A number of factors limit the extent to which developing countries may exploit tourism as a catalyst of economic growth and development, including the limited scope and maturity of the local economy, a lack of infrastructure and so on (see Figure 1.1). In particular, inappropriate policies for tourism compounded by ineffective institutional structures and poor governance are significant challenges facing tourism and development (UNCTAD 2001), reflecting the more general recognition that the nature of governance or the effectiveness of state intervention is a significant factor in the development process in developing countries. In other words, 'the structure and functioning of the state . . . primarily explains both the failures and the relatively few success stories' in development (Mouzelis 1994: 126), with a slow or underdevelopment increasingly being associated with the concept of the failing or failed state (Helman and Ratner 1993; Ghani and Lockhart 2008).

Although definitions abound, states are considered to be 'failing' if they fail to fulfil their two fundamental obligations: (i) to provide effective security for their citizens; and (ii) to deliver positive public goods, such as education, healthcare, basic infrastructure, economic opportunities and so on (Chauvet *et al.* 2011). More topically, Torres and Anderson (2004: 12) suggest that a state is failing if it 'is unable or unwilling to harness domestic and international resources for poverty reduction'. Thus, in the context of tourism, a state may be considered to be failing if, for whatever reason, it does not provide appropriate policies, resources and other support mechanisms to fully exploit tourism for its developmental potential.

Plate 8.2 *Indonesia, Bali: memorial for victims of the 2002 Bali nightclub bombing*

Source: Photo by R. Sharpley

Both the causes and outcomes of state failure are diverse (Di John 2010). However, Torres and Anderson (2004) usefully summarize state failure as a function of the capacity (resources, policies, regulations, institutions, control, experience and so on) and willingness of governments. In other words, although some states may be committed to providing security and public goods to their citizens, they may simply lack the capacity to do so. Conversely, others might possess the capacity to fulfil their obligations to their citizens but lack the desire to do so, as manifested in what Reno (2000) refers to as the 'shadow state', or personal rule behind the facade of legitimate government to the political and/or financial benefit of those in power (see Funke and Solomon 2002). Bayart (1993) describes this phenomenon as the 'politics of the belly' when discussing post-independence government in many African states. More technically, the World Bank lists Low-Income Countries Under Stress (LICUS) as states 'with particularly weak policies, institutions, and governance', states that also score a low average for 'public sector management and institutions' on the World Bank's annual Resource Allocation Index (see www.worldbank.org/ida/IRAI-2011.html).

It is in this latter context that state failure represents a threat to tourism-related development. That is, although a country may have a relatively thriving tourism sector in terms of arrivals and receipts, poor governance may limit the extent to which this is translated into wider socio-economic development. Putting it another way, through state failure, citizens are denied the potential benefits of tourism. Thus, as tourism becomes increasingly important to countries such as Cambodia, which has witnessed significant growth in arrivals but remains a 'core' LICTUS state, there is the risk that tourism's developmental potential will remain unrealized (see Box 8.2).

Demographic change

The global population is demographically dynamic; that is, its structure and characteristics are constantly transforming in both absolute and relative terms, not least with respect to size. Not only is the world's population growing, but the rate of growth is itself dynamic. For example, at the beginning of the nineteenth century, the world's population was approximately one billion. However, it took just 130 years to grow to two billion (1930), 30 years to reach three billion (1959), 15 years to reach four billion (1974), with the five

Box 8.2

Tourism and governance in Cambodia

Despite being home to the ancient Khmer city of Angkor, a vast complex that includes Angkor Wat, the world's largest religious building, it is only since the early 1990s that Cambodia has built a successful tourism sector. This is not to say that the country had not previously been an international tourist destination; indeed, by the end of the 1960s, Cambodia was perhaps the most popular destination in South East Asia, attracting more tourists to its beaches, jungles and, of course, Angkor Wat than neighbouring Thailand. However, 20 years of political upheaval, through civil war (1970–1975), the Khmer Rouge period (1975–1979) and the subsequent Vietnamese-backed government (1979–1989) meant that the country became off-limits to tourists, and it was only after the Paris Peace Accords in 1991 and the re-establishment of the democratic Kingdom of Cambodia in 1993 that it opened its doors to tourism once again. Since then, the tourism sector has flourished. In 1993, just over 118,000 international tourist arrivals were recorded; by 2000, the figure had risen to almost half a million, and throughout the last 15 years remarkable growth has been achieved, establishing Cambodia as one of the world's fastest-growing tourism destinations.

Table 8.7 *Cambodia Tourist Arrivals, 1993–2013*

Year	Arrivals	Receipts (US$mn)	Year	Arrivals	Receipts (US$mn)
1993	118,183	n/a	2004	1,055,202	578
1994	176,617	n/a	2006	1,700,041	1,049
1996	260,489	118	2008	2,125,465	1,595
1998	289,524	166	2010	2,508,269	1,786
2000	466,365	228	2012	3,584,307	2,210
2002	786,524	379	2013	4,210,165	2,547

Source: Adapted from MoT (2014)

The rapid growth in tourist arrivals and receipts highlighted in the table above has led to tourism becoming a significant sector of Cambodia's economy. After the garment industry, tourism is the second largest economic sector, directly accounting for 11.5 per cent of GDP, with the wider tourism economy contributing more than 25 per cent of GDP. It also accounts directly for 782,500 jobs, or 9.7 per cent of total employment. However, these headline figures mask a lack of progress in development, which, arguably, reflects poor management and governance of the tourism sector. Tourism itself is spatially defined; although the country offers a wealth of natural and built attractions, Ankor Wat and the Siem Reap region in which it is located remains the principal destination for tourists, with Siem Reap airport receiving almost 60 per cent of all international air arrivals. The country's other international airport is at the capital, Phnom Penh, which also draws tourists to visit the Royal Palace, the Tuol Sleng Genocide Museum and the 'Killing Fields' at Choeung Ek, but despite the official policy manifested in the 2012–2020 Tourism Development Plan to diversify tourism into four areas (Siem Reap, Phonm Penh, the coast and the north east) – and to increase annual arrivals to seven million – tourism in Cambodia remains firmly entrenched in Siem Reap. It is also notable that average length of stay is just 6.8 days; the average tourist does not stay long enough to tour the country more widely, while a lack of investment in infrastructure means that travel outside Siem Reap and Phnom Penh remains difficult. In short, there has been a failure to exploit the wider tourism potential of the country, while Siem Reap/Angkor Wat has become a victim of its own success (Winter 2008).

In terms of wider socio-economic development, tourism has had relatively little impact. Overall, 17.7 percent of Cambodia's 15 million population lives below the poverty threshold and, with a per capita GDP of just US$1,006 (that is, an average income of $2.80/day), 11 million Cambodians, mostly rural dwellers, live in poverty or near-poverty. Even Siem Reap Province, 'despite being a major tourist destination, is still one of the poorest provinces in the country' (Mao *et al.* 2013). Although tourism provides significant employment opportunities in the accommodation, transport and crafts/souvenirs sectors – for example, Siem Reap city boasts 175 hotels and 275 guest houses offering a total of 15,805 rooms – wages remain low. Indeed, a lack of planning and control has resulted in the oversupply of craft shops and hotels, further depressing earning opportunities. It is also recognized that there has been a failure to support or promote the development of economic linkages (Beresford *et al.* 2004) and, as a consequence, the tourism sector suffers approximately 40 per cent leakages. Moreover, controversy surrounds the operation and management of the country's main tourism asset, Angkor. Although officially managed by an independent public body, ASPARA (Authorité pour la Protection du Site et l'Aménagement de la Région d'Angkor), entrance fees to Angkor (amounting to US$200 million annually) are collected by a major private company, Sokha, which has significant leisure and hospitality interests in Cambodia. Though no official figures are available, it is suggested that just 10 per cent of entrance fee income goes towards restoration at Angkor and 30 per cent to ASPARA. Similar arrangements exist at other tourist sites, while, certainly in Siem Reap, many of the properties leased by bars and restaurant are thought to be government-owned, reflecting perhaps the claim that 'the patronage system and corruption are the main barriers to fair economic development and business activities, adversely impacting the investment environment and poverty reduction [in Cambodia]' (Chheang 2008).

Source: Chheang (2008); Grihault (2011); Mao *et al.* (2013); R. Sharpley personal communications

billion mark being passed just 13 years later, in 1987 (UN 1999). At the time of writing (mid-2015), the world's total population is just over 7.3 billion, roughly double the figure in 1970, although current projections indicate a slower growth rate throughout the twenty-first century (UN 2013).

Population growth is not, of course, uniform across the world. As can be seen from Table 8.8, different rates of population growth have been experienced in different regions. Developing countries tend to have the highest growth rates, whereas developed countries have stable, if not declining, population levels (some European countries, for example, are experiencing a decline in indigenous populations because of low birth rates but overall population growth as a result of immigration). Hence, both Europe and North America have, since 1950, had a declining share of the (growing) world population, whereas Asia in particular has had a growing share, with Asia and Africa accounting for 75.5 per cent of the world's population in 2013. This is projected to increase to 79.2 per cent by 2050 (and 91.4 per cent by 2100), but with Africa's population growing more rapidly than that in Asia.

In the context of this chapter, however, it is not population size that is of relevance, but the demographic characteristics of populations, specifically those in tourism-generating countries and regions. In other words, when considering future challenges facing tourism development, it is necessary to examine demographic trends in order to identify where tourists will be coming from and, indeed, what kind

Table 8.8 Actual and projected population growth/share by region

	Population (millions) and percentage share of total							
	1950	%	1980	%	2013	%	2050*	%
World	2,526	100	4,449	100	7,162	100	9,551	100
Africa	229	9.1	478	10.8	1,111	15.5	2,393	25.1
Asia	1,396	55.3	2,634	59.2	4,299	60.0	5,164	54.1
Europe	549	21.7	695	15.6	742	10.4	709	7.4
Latin America	168	6.6	364	8.2	617	8.6	782	8.2
North America	172	6.8	255	5.7	355	5.0	446	4.7
Oceania	13	0.5	23	0.5	38	0.5	57	0.6

* Based on medium fertility levels

Source: UN (2013)

of tourism experiences they might demand. Two trends are of particular importance, namely the relative wealth of populations (disposable income being a principal driver of participation in international tourism) and transformations in the age structure of populations, specifically the ageing of populations in tourism-generating countries.

Wealth and emerging markets

As noted in Chapter 1, according to the UNWTO (2014b: 13), 'by 2030, the majority of all international tourist arrivals (57 per cent) will be in emerging economy destinations'. Moreover, much of this growth will be generated regionally; that is, the growth in arrivals to developing countries/emerging economies will, to a great extent, be accounted for by arrivals from neighbouring or nearby countries. For example, in 2013, China was the fourth most popular international tourism destination in the world, attracting almost 56 million international visitors. However, more than half of all international visitors were from Hong Kong, Macao and Taiwan, while another 16 million were from the Asian region. Similarly, international tourism to Cambodia (see Box 8.2) is dominated by regional markets, with Vietnam accounting for 20 per cent of arrivals and almost three-quarters of international tourists originating from North and South East Asia (MoT 2014).

This represents a fundamental shift in the balance of world tourism as the rapidly growing economies of the Asia-Pacific region, India and, to a lesser extent, Russia and Brazil become major sources of international tourism. In other words, as the populations of these countries become wealthier, they will have a greater propensity to travel overseas. In particular, China is considered to be the market with greatest growth potential; already the world's largest source of international tourists, the number of Chinese travelling overseas is predicted to double by 2020, reaching around 200 million. Moreover, spending by Chinese international tourists is expected to reach $264 billion annually by 2019. Thus, China will be at the forefront of what has been referred to as the 'Asian wave', with tourists from China, Hong Kong, Japan, South Korea, Taiwan, Thailand and other countries in the region becoming increasingly important markets.

Also, within these rapidly growing economies, it is possible to find the 'super-rich', business people who are willing and able to spend significant sums of money on tourism. For example, it was recently reported that a Chinese billionaire treated 6,400 employees from his business to a four-day holiday in France. This necessitated booking 140 hotels in Paris and 4,700 rooms on the Cote d'Azur, and, overall, the group was expected to spend more than €13 million (£9.5 million) on hotels, food and excursions (Davison 2015). Hence, significant opportunities exist for countries around the world to exploit the new-found wealth in emerging economies. However, with much travel being regional, developing countries elsewhere will need to compete effectively and provide products and experiences that meet the specific needs of these markets. Moreover, long-term growth in tourism from these markets is not assured, and the challenge for many developing countries will be to maintain and build more traditional tourism markets in Europe and North America. However, as the next section suggests, the demographics of these markets is also changing, with implications for the future of tourism development.

Population ageing

Perhaps the most distinctive global demographic trend is the ageing of the world's population. That is, as a consequence of declining fertility rates and increased life expectancy in both developed and developing countries, the age structure of the global population is undergoing a significant transformation manifested primarily in a relative increase in the number of older people. Specifically, 'the number of people over the age of 60 is projected to reach 1 billion by 2020 and almost 2 billion by 2050, representing 22 per cent of the world's population' (Bloom *et al.* 2010: 583). Over the same period, the number of people aged 80 or over is expected to increase from 1 to 4 per cent of the global population.

This shift in population age structures will not, of course, be the same in all countries. The combination of decreasing birth rates and life expectancy varies significantly from country to country and is influenced by a wide variety of factors, though it is not surprising that, in the shorter term at least, population ageing will be most evident in developed countries (Table 8.9).

Nevertheless, less-developed countries will also tend to have increasingly ageing populations. Arguably, this will represent a significant challenge given the current lack of established welfare and pension schemes in many of those countries.

The implications of population ageing are displayed in Figure 8.2. Although based on projections for European countries, it is likely

Table 8.9 *Population ageing by 2050*

Countries with highest share of 60+ population	% of population over 60	Countries with highest growth in 60+ population and share	Increase 2011– 2050 (%)	% of population over 60
Japan	42	United Arab Emirates	35	36
Portugal	40	Bahrain	29	32
Bosnia and Herzegovina	40	Iran	26	33
Cuba	39	Oman	25	29
Republic of Korea	30	Singapore	23	38
Italy	38	Republic of Korea	23	39
Spain	38	Vietnam	22	31
Singapore	38	Cuba	22	39
Germany	38	China	21	34
Switzerland	37	Trinidad and Tobago	21	32

Source: Adapted from Bloom *et al.* (2011: 2–3)

Figure 8.2 *Age group share of population in the EU, 1950–2050*

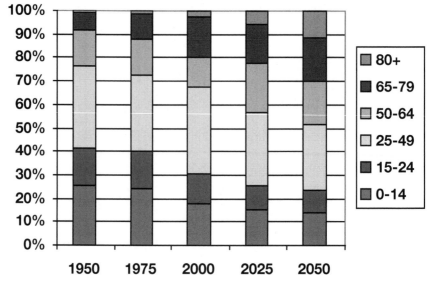

Source: Adapted from Zaidi (2008: 3)

that similar patterns will be evident in many other countries, including China, where, as is evident from Table 8.9, the proportion of the population over 60 years old in 2050 will not be dissimilar to that in many European countries. Specifically, the shift to an older population will be compensated for by a decline in the size of all younger age groups, perhaps most significantly the 25–49 group, upon which financial and social burdens, such as caring for elderly relatives, is most likely to fall.

The implications of population ageing are also of relevance for tourism, particularly with regard to the ageing of populations in key tourism markets. As observed in Chapter 1, with the exception of China, Brazil and Russia, the world's top tourism generators remain, at present, Western nations that are most likely to experience significant population ageing, though emerging markets will similarly witness the trend towards older populations. The implications can be seen from two perspectives. On the one hand, there exists the 'alarmist' school; that is, population ageing is seen in some quarters as a threat to economic stability and social cohesion (Gee 2002). It is argued that the greater costs of social and healthcare and state pensions (where provided) needed to support the expanding older population will not only put an unsustainable strain on public finances, but will also require younger people to work longer and to pay higher taxes. At the same time, many consider that the generous pensions and benefits of the economic growth years enjoyed by the 'baby boomer' generation will not last, and, as a consequence, both younger and older generations in many Western countries will, by the mid-twenty-first century, have much lower disposable incomes, and hence spend less on tourism, with evident implications for more expensive/distant destinations or forms of tourism. In other words, the value of the 'grey market' to tourism has long been recognized (McGuire *et al.* 1988), yet there is the threat that this market, as well as younger markets, might decline significantly. In short, population ageing may threaten the growth of tourism and its consequential role in development.

On the other hand, others argue that such concerns are misplaced (Gee 2002; Bloom *et al.* 2010), suggesting that older generations will not only enjoy greater life expectancy, but also better health. Consequently, they will remain economically active, contributing to public finances through tax on earnings but also maintaining levels of disposable income to fund increasingly leisure-focused lives. From this perspective, the 'grey market' will become increasingly important to tourism destinations; it is a flexible market, less constrained by children, school holidays and work commitments but, at the same time, a market with specific needs with regard to products and services. It may also, however, be a market less inclined towards longer-distance travel. Hence, destinations may need to adapt to the needs and demands of an ageing clientele.

Of course, projections of population ageing are just that: projections. Nevertheless, all the evidence suggests that it will increasingly become both a reality and a challenge for all countries. Moreover, one way or another, it will impact on tourism destinations, and thus it is a trend that all destinations in both the developed and developing worlds must be aware of.

Conclusion

Tourism does not exist in isolation from global trends. If tourism is to be used effectively as a development tool, greater recognition is needed of how global challenges and risks may interact with the tourism development dilemma. The World Economic Forum publishes an annual report of Global Risks, and it is important not only to identify the nature of these potential risks but also to understand how interconnected these risks are and the potential cascading effects that can result when they occur (World Economic Forum 2015). Climate change, for example, is having wide-ranging environmental, social and political consequences and developing countries will be among the hardest hit. Climate change is also closely linked to the resources discussed in this chapter, namely oil, water and food. The tourism industry relies heavily on oil in the transportation sector, which subsequently contributes to global warming. Climate change has impacted water supply, with many locations experiencing drought, which, in turn, has reduced agricultural productivity, thereby driving up food prices. Economic shocks, such as the 2008 global economic crisis, have led to recession with political and social consequences and, while the global economy is recovering, another economic crisis is likely given the boom and bust nature of market economies. These economic crises often hit the poor and vulnerable in developing countries the most, which raises the challenge as to how tourism can help in poverty reduction. Political instability, ineffective governance, terrorism and failing states are all challenges that prevent the establishment of the safety and security needed to allow tourism to flourish. In some destinations, all of these politically related risks are highly interconnected and conflicts from one country can quickly spread into neighbouring countries. A final challenge examined in the chapter is demographic change. The global population continues to rise and there is new wealth in emerging economies such as in Asia. Populations are also ageing in tourism-generating regions. Tourism destinations must be aware of changing demographics in order to be competitive.

It has not been possible to present every global risk and challenge facing the tourism industry in developing countries in this chapter. Nevertheless, it is important to understand both the vulnerability and resilience of tourism in developing destinations if tourism is to be used as an agent of development. The list of issues addressed here will surely evolve and change as new challenges come to the forefront. Smil (2008: 253, original emphasis) reminds us of the continuity of history, observing that 'terms such as *demise* or *collapse* or *end* are often merely categories of our making, and that catastrophes and endings are also opportunities and beginnings'. The purpose of this chapter has been to set out an agenda for the challenges facing tourism and development, and it is clear planners must take a broader view. The final chapter of this book now considers that there is no single solution to the tourism development dilemma.

Discussion questions

1   How will climate change impact tourism in developing countries?
2   What are the implications of political instability for tourism?
3   How will changing demographics impact tourism in the future?
4   What will happen in the future as key resources such as oil and water come under threat?

Further reading

Yeoman, I. (2012) *2050: Tomorrow's Tourism*, Bristol: Channel View Publications.

This book explores possible futures for tourism through a variety of scenarios in different destinations, and includes topics such as human trafficking, transport and technological innovation.

Becken, S. (2015) *Tourism and Oil*, Bristol: Channel View Publications.

The book examines oil constraints and tourism, including how much oil the tourism industry uses along with information on oil reserves. An analysis is provided on the economic implications of rising oil prices and the implications for tourism in a post-peak oil world.

Gössling, S., Hall, C. and Scott, D. (2015) *Tourism and Water*, Bristol: Channel View Publications.

The book provides a comprehensive review of the knowledge on tourism and water, including an assessment of the impact of global tourism and water management approaches.

Websites

www.weforum.org/reports/global-risks-report-2015
World Economic Forum Global Risk Report 2015 explores 28 global risks divided into the categories of economic, societal, geopolitical and technological, as well as the drivers of these risks.

http://touristkilled.com
Tourist Killed is a website that documents tourist deaths, attacks, robberies, rip-offs and arrests.

9 Conclusion: the tourism development dilemma

Learning objectives

When you finish reading this chapter, you should be able to:

- recognize the pros and cons of pursuing tourism as a development option
- understand the debates surrounding the various paradigms of tourism development
- appreciate the range of controlling external forces that may hinder or help development through tourism
- identify the potential conflicts and ways that multinational corporations, local elites and local residents can work together in tourism development

Since the publication of the first edition of this book, tourism has continued to be an increasingly favoured development tool in many developing countries. With the relative ease of entry into the tourism market and its purported ability to generate foreign exchange, create employment and reduce the poverty gap, it is no wonder that it is being pursued. However, like any development option or avenue of economic endeavour, it comes with a cost. This, then, is at the heart of the tourism development dilemma. Tourism represents an attractive, and perhaps the only, means of stimulating economic and social development for some developing nations. However, in many instances, that development fails to materialize, the benefits only going to multinational corporations or the local elite, or it is achieved with a very high environmental, economic or social cost. In the developing world, tourism is usually implemented through a top-down planning approach and decision-making is 'predominately based on the interventions of government agencies and large tourism firms, resulting in the dominance of external, often foreign capital and the marginalisation of local people' (Liu and Wall 2006: 159). Developing countries opting into the tourism industry will encounter both the positive and negative consequences of this globally competitive industry, and the challenge lies in accepting or managing the negative consequences in the hopes of obtaining the potential long-term benefits of tourism. The complexities of using tourism as a development tool and the dilemma that many countries face in coping with the uncertainty that tourism brings have been the focus of this book. Understanding the vulnerability and resilience of developing destinations to the tourism industry is important. The tourism development process intersects with the economic, political, environmental and social conditions in the destination, and is also framed by the global political economy. The first section of this chapter will focus on the development imperative and its relationship to tourism, and will examine the scope and realities of the developing world where tourism takes place. The second section will then focus on the imperative of sustainability, which has come to the forefront of the development debate. The chapter will then examine the tourism development dilemma through a proposed framework that examines current influences on tourism, the form of tourism that is developed in the destination, and the responses to tourism. The framework then focuses on the resulting trade-offs that are made in the tourism development process, which are at times both competing and conflicting. These trade-offs illustrate the differences between tourism planning and management in developing countries and the idealism of sustainable development.

The development imperative and tourism

The UN is in the process of charting a development agenda for the post-2015 UN Millennium Development Goal era. Box 1.1 contains the UN Millennium Development Goals and Box 2.1 outlines some of the steps taken so far to creating the new development agenda in which the Millennium Development Goals continue to play a central role. If achieved, the UN Millennium Development Goals have the potential to lift half a billion people out of poverty and a further 250 million will no longer suffer from hunger. The goals are ambitious, a challenge to meet and yet necessary targets. Table 9.1 displays selected achievements made towards the UN Millennium Development Goals and areas where more work is need. It is clear that while significant achievements have been made, there is still a tremendous amount of work to be done. It is also important to consider that while there may be improvements at the global level, there are still extreme differences at national, regional, community,

Table 9.1 *Selected progress on the UN Millennium Development Goals*

UN Millennium Development Goals	Progress	More work needed
1. Eradicate extreme poverty and hunger	Extreme poverty rates have been cut in half since 1990.	Worldwide, one in nine people remain hungry.
2. Achieve universal primary education	Enrolment in primary education in developing regions has reached 90 per cent.	58 million remain out of school.
3. Promote gender equality and empower women	The world has achieved equality in primary education between girls and boys.	In many countries, women still face discrimination in access to education, work and participation in decision-making.
4. Reduce child mortality	17,000 fewer children die each day than in 1990.	Over 6 million children still die before their fifth birthday each year.
5. Improve maternal health	Maternal mortality fell by 45 per cent since 1990.	Only half of women in developing regions receive recommended healthcare during pregnancy.
6. Combat HIV/AIDS, malaria and other diseases	9.7 million people were receiving life-saving medicines for HIV in 2012.	Every hour, so many young women are newly infected with HIV.
	3.3 million malaria deaths were prevented in the span of 12 years.	In 2012, malaria killed an estimated 627,000 people.
7. Ensure environmental sustainability	2.3 billion people gained access to clean drinking water since 1990.	2.5 billion do not have basic sanitation such as toilets or latrines.
8. Develop a global partnership for development	Debt service has declined for developing countries. Trade continues to improve.	Aid money hit a record high of $134.8 billion in 2013, but shifted away from the poorest countries.

Source: UN (2015)

local or individual levels of development. The complexity of the task ahead is inherent in the new evolving development agenda outlined in Box 2.1 with the new 17 goals to take us to 2030.

In describing the status of the world's population, Sachs (2005) places people on the 'ladder of economic development', with the higher rungs representing steps up the path to economic well-being. Approximately one billion people around the world or one-sixth of humanity are too ill, hungry or destitute to even place a foot on the first rung of the development ladder. They are the 'extreme poor' of the world and are fighting for survival. If they fall victim to a natural disaster (drought or flood), serious illness or collapse of the world market price for their agricultural cash crop, the result will be extreme suffering and perhaps death (Sachs 2005). A few steps up the ladder are roughly 1.5 billion people who are at the upper end of the low-income world. Sachs (2005) describes these people as 'the poor', living above mere subsistence and, while daily survival is virtually assured, they struggle to make ends meet in cities and rural areas. They face chronic financial hardship and lack basic amenities, including safe drinking water and functioning latrines. Together, 'the poor' and 'the extreme poor' make up approximately 40 per cent of humanity. Further up the ladder are another 2.5 billion people in the middle-income world earning a few thousand dollars per year; however, they are not to be confused with the middle class in rich countries. Most live in cities, have adequate clothing and their children go to school. They may have some amenities in their homes such as indoor plumbing, and they may be able to purchase some means of transportation, such as a scooter and later possibly an automobile (Sachs 2005). They have adequate food, although some are following the trend in developed countries of eating fast food. The remaining one billion people, or one-sixth of humanity, are higher up the development ladder and are in the high-income world. The people in the high-income households include the approximately one billion in the rich countries but also a growing number of affluent people living in middle-income countries. There are tens of millions of high-income people in cities such as Shanghai, São Paolo and Mexico City (Sachs 2005). On the positive side, Sachs (2005) states that more than half the world is climbing the development ladder, as is evident through measures of economic well-being such as increases in income, life expectancy, education, access to water and sanitation, and falling infant mortality rates. However, one-sixth of humanity is not even on the development ladder and remains caught in a poverty trap. From a basic needs perspective, the Food and Agriculture Organization of the United Nations (FAO) releases an annual report entitled *The State of Food Insecurity in the World*. The 2014 report indicates that while tremendous progress had been made, 805 million people remain chronically undernourished, especially in regions such as sub-Saharan Africa and southern Asia (FAO, IFAD and WFP 2014). In exploring pro-poor tourism, Chok *et al.* (2007) refer to the UNHSP (2003) report on slums, which indicates that the urban population in developing countries is expected to double to four billion in 30 years' time and currently almost 80 per cent of the urban population in the world's 30 least-developed nations live in slums. It is with the people of the developing world that the well-off, domestic and foreign tourists interact. It is then easy to see how the simple haggling over the price of a souvenir in a local market in a developing country can have profound repercussions for someone struggling to make ends meet.

It is important to also recognize that those who are disadvantaged in the labour market must not be thought of as passive acceptors of their fate (Potter *et al.* 1999). Low-income households display a wide range of coping mechanisms (Rakodi 1995, cited in Potter *et al.* 1999). Rakodi (1995) identifies three main categories for urban households' strategies for coping with worsening poverty. Under changing household composition, the strategies include migration, increasing household size to maximize earning opportunities or not increasing household size by fertility controls. Under consumption controls, Rakodi (1995) lists reducing consumption, purchasing cheaper items, removing children from school, putting off medical treatment, delaying repairs to property, and limiting social contacts, including rural visits. The final category of strategies relates to increasing assets. These include having more household members in the workforce, starting up enterprises where possible, increasing subsistence activity such as growing food or gathering fuel, and increasing scavenging and subletting rooms and/or shacks. Rakodi (1995) states that while not all of these strategies are available to all households, they should be linked to policy responses. In the context of these strategies, how would a member of a disadvantaged household respond to tourism development in an urban area? Selling souvenirs, acting as an informal tour guide or moving to a new tourist area are just a few of the options related to tourism. In a study of tourism employment in Bali, Indonesia, Cukier (2002) found that most informal sector workers were not marginalized, but rather were earning above minimum wages and often more than formal front-desk employees. These strategies listed by Rakodi (1995), along with

Cukier's findings, reflect the importance of small enterprises, the informal sector and local entrepreneurial activity.

The contribution of tourism also needs to be put within the wider development issues that any specific country may be facing. Writing on the topic of nature tourism and climate change in southern Africa, Preston-Whyte and Watson (2005: 141) made the following comment revealing the complexity of development:

> At present the major challenge facing southern African governments is to deal with the colonial land legacy, HIV/AIDS and associated poverty so that the region's inhabitants will find themselves in a position to actually benefit from the potentially favourable effects of climate change on nature tourism.

Evolution of development theory and tourism

The challenges of the developing world are complex, and the search continues for strategies, programmes and initiatives from within and outside the developing world to help improve the situation in developing countries and overcome barriers to move up the development ladder. As illustrated in Chapter 1, the concept of development has changed over time and has been a source of controversy. There has been a shift from top-down economic models to more broad-based approaches that are more bottom-up and focus on satisfying basic needs, sustainability, poverty reduction and human development. Recent trends place global development as an emerging development paradigm with global climate change at the centre of attention. How development is measured has also moved beyond only economic indicators, such as GNP/capita to more broad-based indicators such as the UNDP Human Development Index. Box 9.1 includes a discussion of the relationship between tourism and human rights, which has been brought into the broader scope of development. Telfer (2015a) summarizes these changes in development thinking over time and their influence on tourism development in terms of modernization, dependency, economic neo-liberalism, alternative development, the impasse and post-development, human development, and global development (see Chapter 1 and Table 1.2). Initially in the 1960s, the focus of tourism under modernization was primarily economic, with the belief that tourism generated increases in foreign exchange, employment and engendered a large multiplier effect, stimulating the local economy. In time, however, the benefits of tourism were questioned, due to high rates of leakages and lower than expected multipliers. The negative impacts of tourism in developing countries were documented, paralleling the dependency critique of modernization. In the 1980s and 1990s, the economic neo-liberal paradigm gained prominence with a focus on international markets and globalization. Multinational tourism corporations extended their operations around the world, looking for locations with attractive destinations and lower production costs. Within the alternative development paradigm, sustainable tourism development came to the forefront as concerns continued to grow over existing patterns of economic growth, which could not be sustained by the environment (Telfer 2015a). The 1980s also saw an impasse in development thinking and the rise of post-development, as previous development paradigms could not explain all of the difficulties that developing countries were encountering (Schuurman 1996). Sachs (1996: 1) argues, 'development stands like a ruin in the intellectual landscape. Delusion and disappointment, failures and crimes have been the steady companions of development and they all tell a common story: it did not work'. Thomas (2000) quotes Robert Chambers (1997: 9), who, in a response to Sachs' comments above, suggests, 'That is no grounds for pessimism. Much can grow on and out of ruin. Past errors as well as achievements contribute to current learning'.

The continued presence of poverty and inequality also raised concerns over the development process and globalization (Sachs 1996; Saul 2005). As a critic of globalization, Saul (2005: 3) argues:

> that very clear idea of globalisation is now slipping away. Much of it is already gone. Parts of it will probably remain. The field is crowded with other competing ideas, ideologies and influences ranging from positive to catastrophic. In this atmosphere of confusion, we can't be sure what is coming next, although we could most certainly influence the outcome.

The 1990s saw the emergence of human development, with a focus on human rights, civil society, security, social capital, poverty reduction, pro-poor growth and good governance. Pro-poor tourism emerged as a pro-poor growth strategy advocating that all forms of tourism need to be made more pro-poor. The most recent and still evolving development paradigm is global development. The main argument within global development is that new forms of global governance are required to deal with complex global issues. Global climate change is perhaps the best example of how a global response is required to deal with the emerging threat. Climate change is increasingly being recognized as a challenge facing tourism as the industry not only contributes towards global emissions, but destinations are also facing a variety of threats because of climate change (see Chapter 8).

The question for this book is: What can tourism do to help move people, groups, communities, regions or countries climb up the ladder of economic development (see Plate 9.1)? Development is defined in Chapter 1 as 'the continuous and positive change in the economic, social, political and cultural dimensions of the human condition guided by the principle of freedom of choice and limited by the capacity of the environment to sustain such change'. Tourism is only one industry, and while it is the focus of this book, it cannot possibly be the sole solution to the challenges of developing countries. If adopted, tourism needs to be viewed as part of a broader-based approach to development in conjunction with other economic activities. Along with the global UN Millennium Goals, there are also national development goals and goals at the levels of regions, cities, towns, villages, communities and eventually individual goals. What opportunities does an individual have to take part if he/she so desires, and are they able to participate directly or indirectly in the economy of tourism? Can he/she take up a position in the formal sector such as a front-desk position at a beach resort in Sri Lanka, or can he/she enter the informal sector and perhaps sell souvenirs in the streets of Rio de Janerio? At the community level, individuals may come together perhaps with the help of an NGO or local government and launch a village tourism initiative such as the Kasongan pottery village in Central Java, Indonesia. At a regional or national level, the goal may be to establish tourism to attract visitor numbers, development charges and taxes, which can be used for broader development goals so the nation can climb further up the development ladder.

Tourism is a complex, evolving, dynamic and sometimes volatile industry, and judging its role in the development process is not an easy task, as highlighted in this further example. An initial visit to the north-west coast of the island of Lombok, Indonesia, in 1994 saw an open beach with no tourism development. A year later, a return visit to the area saw the same beachfront property behind a large fence with a sign on the wall advertising a new hotel under construction. Access was now restricted. Over 20 years have now passed and in place of the fence stands a new hotel and others have followed. How has the community changed? Do the locals benefit

Box 9.1

Tourism development and human rights

The concept of human rights intersects with tourism and development issues on a number of different fronts, and it is useful to examine tourism within the context of the *Universal Declaration of Human Rights* (Hashimoto 2004; Nkyi and Hashimoto 2015). Reported human rights issues often occurring in tourism include inhuman treatment of people, labour rights violations, restrictions on people's movements and freedoms of settlement, unfair business competition between multinationals and local tourism businesses, unwanted security checks of tourists and lack of adequate security for tourists in destinations (Nkyi and Hashimoto 2015). The United Nations (UN 2007) adopted the Universal Declaration of Human Rights in December 1948. Other more specialized human rights conventions have also been developed, such as *The International Covenant on Economic, Social and Cultural Rights* in 1966 and *The Convention on the Rights of the Child* in 1989 (Freeman 2005). With the notion of development broadening out from an economic focus to a more 'human development' approach, human rights and development can be seen as conceptually overlapping (Freeman 2005). As an example of this, the United Nations Development Programme has a section on the protection of human rights in its policies (Freeman 2005). Sen (1999) argues from the perspective of 'development as freedom', and notes five specific types of rights and opportunities that can help to advance the general capability of a person. The five rights and opportunities include political freedoms, economic facilities, social opportunities, transparency guarantees and protective security. Tourism Concern advocates for human rights, and has had a number of campaigns addressing human rights abuses, including a call for a boycott on travel to Burma (recently modified in alignment with the wishes of Aung San Suu Kyi and her party, the National League for Democracy, for those travelling to travel independently and in solidarity with locals and to boycott package tours, cruise ships and so on, as this type of travel may only benefit the ruling junta (Tourism Concern 2014b)). The remainder of this box will now focus on selected issues that arise when considering tourism in the context of the Universal Declaration of Human Rights, as suggested by Atsuko Hashimoto (2004). The Declaration has 30 articles, and all can be examined in the context of tourism. A full discussion of all 30 articles is beyond the scope of this book; however, a few are included below for discussion purposes. The selection of the articles below is not meant to imply that they are of higher importance than any of the other articles in the Declaration.

Article 1

All human beings are born free and equal in dignity and rights. They are endowed with reason and conscience and should act towards one another in a spirit of brotherhood.

- Tourism presents the opportunities for the understanding of different cultures.

Article 4

No one shall be held in slavery or servitude; slavery and the slave trade shall be prohibited in all their forms.

- The sex tourism industry can result in a form of slavery and slave trade.
- The potential remains for forced labour, including child labour, in tourism.
- Some tourists travel to purchase the freedom of slaves.

Article 13, Parts 1 and 2

Everyone has the right to freedom of movement and residence within the borders of each state; and everyone has the right to leave any country, including his own, and to return to his country.

- In some countries, the movement of nationals and tourists is restricted.
- In some areas, locals are not permitted to go to certain locations, such as beachfronts, where resorts have been built.
- Some communities have been relocated to make way for tourism developments and not allowed to return.

Article 17, Parts 1 and 2

Everyone has the right to own property alone as well as in association with others; no one shall be arbitrarily deprived of his property.

- Tourism development may displace people to make way for hotels, golf courses and tourism-related infrastructure.

Article 22

Everyone, as a member of society, has the right to social security and is entitled to realization, through national effort and international co-operation and in accordance with the organization and resources of each State, of the economic, social and cultural rights indispensable for his dignity and the free development of his personality.

- Tourism can commodify indigenous cultures to be sold as a tourism product.
- The promotion of cultural tourism may enhance and strengthen local cultures.

Article 23, Part 1

Everyone has the right to work, to free choice of employment, to just and favourable conditions of work and to protection against unemployment.

- Tourism generates employment, but the quality of the work and the payment received needs to be considered as well.
- Fair trade tourism can help ensure small tourism business is treated more equitably.

The few selected examples above illustrate how tourism can be discussed within the context of the Universal Declaration of Human Rights. The edited work by Andreassen and Marks (2006) further explores the concepts of the 'human rights based approach to development' and the 'human right to development'.

Source: Sen (1999); Hashimoto (2004); UN (2007); Tourism Concern (2014b); Nkyi and Hashimoto (2015)

Plate 9.1 *Indonesia, Lombok: local village very close to main tourist resort area of Sengiggi Beach*

Source: Photo by D. Telfer

from the hotel? Is this particular hotel contributing to the development of the island? Is the hotel in a small way helping Indonesia to climb up the development ladder? Is this hotel one piece of a much larger puzzle helping to contribute to the United Nations Millennium Development Goals? In 2011, a new international airport opened on the island of Lombok, accommodating larger aircraft, and so how will the new airport change tourism development and local communities? All of these questions are difficult to answer, yet considering them opens the door to exploring the complex relationship that exists between tourism and development.

Development goals need to be framed within some hard questions – development for whom and by whom? We need to consider the priorities of governments, NGOs, private businesses (both large and small), communities and individuals. How do their goals relate to broader development goals? How do their priorities fit in with the global market and external political forces as well as the conditions in the destination? As noted in Chapter 5, *fakalakalaka* is the Tonga notion of how to be modern and how to do development (Horan 2002), and it does not necessarily coincide with the macroeconomic indices used by Western international lending agencies. Based on the work by Iliau (1997), Horan (2002: 216) states that *fakalakalaka*

> was a more encompassing and in-depth concept than merely the personal gain of material possession, let alone just making money by selling tourists textiles. This notion of development is about total development of the individual: physical, spiritual and mental. Development then extends to all areas of life including personal relationships, family dynamics and the development of the community.

Beyond issues of poverty, indigenous communities such as those found in Tonga may, in fact, have differing perspectives of a development ladder identified by Sachs (2005) (see also Box 2.2 on Bhutan). Clearly, tourism developed under different development paradigms or schools of thought will have a different emphasis and, as noted in Chapters 3 and 4, it is important to understand the values, ideologies and strategies of these paradigms, as well as who has the power and control to implement and/or enforce them.

The sustainability imperative and tourism

The above section has stressed the development imperative and how tourism is used as a development tool in the context of development paradigms. Within the broader development imperative and the overall need for improvements in developing countries, there is also the sustainability imperative. With the recognition that resources are limited and need to be protected for future use, sustainability has become a guiding framework for development. The sole focus on economic growth has been challenged by the more holistic approach theorized under sustainable development encompassing environmental, social and economic concerns. There is increased recognition, in terms of tourism, that local communities need to be brought into the planning process of tourism. Traditionally, sustainability focused on environmental sustainability and only recently has the focus shifted towards poverty alleviation as outlined in the UN Millennium Development Goals (Rogerson 2006). The evolution of sustainability and the challenges associated with the term have been explored in Chapter 2. Liu and Wall (2006: 160) state that:

while tourism has gained prominence in the social and economic agendas of all levels of government and academics continually espouse the need to involve local people, sufficient attention is rarely accorded to the means by which the capabilities of local people to respond to tourism opportunities can be enhanced.

Redclift (2000) outlines five spheres of environmental activity in which to analyse sustainability. These are presented in Table 9.2 along with a series of questions to consider in terms of tourism development. The spheres are all highly interrelated, but the questions posed ask about the interrelationships between tourism and sustainable development.

Proops and Wilkinson (2000) explore the relationship between sustainability, knowledge, ethics and the law. They outline four areas of knowledge and understanding that are necessary to establish well-formulated policies for sustainability. These four areas for sustainability are adapted here to tourism, with cautionary notes with regards to the tourism development dilemma. The first is an understanding of the natural world and how the activities of production and consumption impact on it. In the context of the developing world, it is important for destinations to understand the power structures of the international tourism industry. The invitation to foreign-controlled companies to operate in the destination can leave the host country in a very dependent position. The creation of beach resorts or community-based village tourism will take different forms and attract different types of consumers. The understanding of the interaction of the built form and consumers in the destination and the resulting impact on the environment may be difficult to manage over longer periods of time.

The second is an understanding of human perceptions and motivations so that we can know why people indulge in behaviour that is destructive to nature. Tourists themselves, as consumers, bring with them not only their home culture, but they enter the realm of tourist culture. Different tourists will have different values, behaviours and disposable income. Some tourists may be willing to adopt more environmentally friendly forms of behaviour; however, for many, travelling is a holiday and they may be more interested in relaxing than following more restrictive codes of conduct. Will tourists be willing to travel less if climate change continues to accelerate?

The third area of knowledge and understanding necessary to establish well-formulated policies for sustainability is an understanding of

Table 9.2 *Spheres of environmental activity and questions for sustainable tourism development*

Sphere of production	• What are the environmental impacts of the production of tourism?
	• What are the effects of employment in tourism on health, welfare and community life?
	• How does tourism contribute to income levels?
	• What are the associated risks with the tourism industry?
	• What are the indirect consequences of tourism production activities such as waste, toxins, and pollution and links to climate change?
Sphere of consumption	• How do different tourists consume tourism differently?
	• What risks (that is, health, food) are there in the consumption of tourism?
	• What are the indirect consequences of tourism consumption such as food miles, ghost acres, ecological footprints, demonstration effect and commodification of culture?
	• What energy is generated to meet tourists' demands and the disposal of waste packaging to meet these demands?
	• What is the impact of tourism consumption on climate change?
Sphere of social capital/infrastructure	• What is the built environment for tourism?
	• How do tourists and locals interact?
	• What utilities are required for tourism (energy, water and waste disposal)?
	• How are the activities linked by transportation?
	• What public services (public parks, open areas, recreational services) are used by tourists?
Sphere of nature	• How are the countryside, forests and landscape used by tourists?
	• Do tourists and locals have access to natural areas (beach, wilderness and so on)?
	• Are animal rights and welfare protected in tourism?
Sphere of physical sustainability	• What is the environmental quality for both tourists and residents?
	• What are the background processes that govern environmental quality for tourists and residents such as climate change, air pollution, ozone depletion, destruction of forests, and watershed basins due to anthropogenic causes?

Source: After Redclift (2000)

ethical systems so we can establish if human motivations that are destructive to nature might be morally constrained. There have been attempts by different organizations to develop codes of conduct for tourists such as the one presented in Figure 6.3. However, as Pearce (2005) argues, it is not clear from an evidence-based perspective that all suggested behaviours in codes of conduct will result in sustainable outcomes. An example of this (Pearce 2005) is the appropriateness of encouraging tourists to boycott locations with poor human rights records. He suggests alternatively that tourists and the governments who represent them may be forces for social change in these destination communities. The second problem Pearce (2005) identifies with codes of conduct is that with some prescribed roles, it is difficult to make the necessary judgements. One example given in this situation is the proposed boycott of a business. It may appear that a company is exploiting its workers with low pay and so the code of conduct will call for a boycott. However, the amount of money being paid may seem trivial in a developed country context, yet it may be 'substantial in Indonesia, and jobs that look menial and almost pointless in populated destinations such as India or China may provide dignity and some income of value to poverty stricken regions' (Pearce 2005: 23).

Finally, Proops and Wilkinson (2000) state that we need an understanding of the effectiveness of various systems of incentives and restrictions on human actions so appropriate measures can be put into law. Governments may enact policies and laws to regulate the tourism industry. Alternatively, governments may proactively offer investment incentives to attract developers. In a globalized economy, a developer may chose to go where the incentives are higher and the environmental and social regulations are lower.

Having presented these four statements, Proops and Wilkinson (2000) then argue that the statements are very problematic and impose significant difficulties in trying to establish a sustainable world. The difficulty with all of these statements is related to the complexity of the interactions of humans with the environment, as well as the difficulty in understanding human behaviour towards the environment. The tourism examples provided above, in the context of the four statements, clearly illustrate the challenges associated with sustainable tourism development. While it may not be possible for tourism to be completely sustainable, making all forms of tourism more sustainable becomes the challenge. It is important to also keep in mind that sustainability itself has come under a variety of criticisms, including it being considered a Western notion of development. There are also calls to move beyond sustainability and adopt a more pragmatic approach (Sharpley 2009c). Sharpley (2009c), for example, proposes a destination capitals model whereby tourism planning and environmental sustainability is focused on the destination rather than an overarching blueprint for sustainable tourism. It would allow destinations to exploit their capitals (environment, human, sociocultural, economic and political) according to local needs while protecting their capital base for the future. It would allow, for example, the development of traditional mass tourism that might not meet sustainable tourism criteria, yet might provide a viable approach to tourism development (Sharpley 2009/2010). The chapter now turns to explore a framework for understanding the tourism development dilemma.

The tourism development dilemma framework

In order to better assess the tourism development dilemma, a framework is presented in Figure 9.1. The framework is based on the work of Potter (1995) in studying urbanization in small island Caribbean states. The framework is briefly outlined before the various components are described in more detail. In the centre of the framework is the tourism destination. It is presented based on the shape of an island as the original framework was based on a hypothetical Caribbean island; however, the main concepts are transferable to other locations and scales. Some of the main influences on the tourism development process include modernization, globalization, production, dependency, postmodernism, consumption, sustainable development and climate change, and these issues and others have been explored throughout this book. These influences can have a major impact on how tourism develops in a destination. They may come from outside the destination, from within the destination, or both. It is important to recognize that this list of influences is not comprehensive and may differ from destination to destination. While the influences are discussed individually below, they are not isolated but very much interrelated as they interact and overlap in the destination. Within the context of these influences, local and external tourism development agents take advantage of opportunities to develop tourism, resulting in built form and reactions by individuals, groups, communities, regions and governments. The built form and the type of tourism

Figure 9.1 *The tourism development dilemma framework*

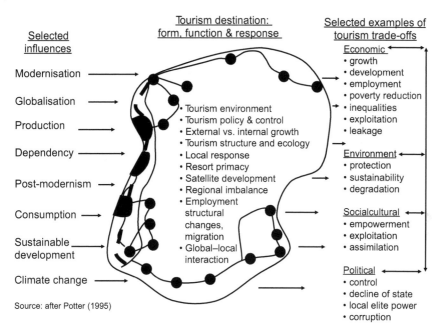

Source: after Potter (1995)

developed, such as mass beach tourism or remote village tourism, will have an impact on the contribution of tourism to development.

Tourism not only has the potential to contribute to broader based development goals, but, as illustrated in Figure 9.1, the resulting interactions also have a geographic dynamic. In the context of the Caribbean islands, such as Barbados, St Lucia and Grenada, tourism has obviously focused development and change in coastal areas, and further development of the industry has been along the pre-existing coastal urban zone (Potter 1995: 334). While the main tourism zone develops and expands, satellite resorts or attractions such as village tourism or ecotourism will open up if there are adequate transport routes, as indicated in Figure 9.1. In the case of ecotourism resorts, the periphery may become more significant than the initial main tourism resort area. The geographic location of the built form will also influence the extent to which tourism can contribute towards development. Tourism concentration will result in migration.

The final section of Figure 9.1 is the trade-offs that are at the heart of the tourism development dilemma. The outcomes are listed under the four headings of economic, environment, sociocultural and political; however, the trade-offs can occur within and between these categories. Of course, under sustainable development, the goal is to minimize negative impacts. In an ideal world, however, trade-offs may still occur. It is important to consider who makes the decisions as to what the trade-offs will be. Figure 9.1 not only explores the tourism development dilemma, but also draws together the material presented in the previous chapters of the book.

When examining Figure 9.1, it is important to keep in mind the issues of power and control. In Chapter 4, the tourism development process was examined and the first items to be explored were the values, ideologies, goals, priorities, strategies and resources of tourism development agents. There is a range of competing forces in tourism and the values of those in power will have a strong influence on the overall development of tourism, and therefore will influence the potential to contribute to broader notions of development of the destination. Multinationals, international funding agencies, governments, NGOs, communities and the domestic private sector are a few of the many influencing tourism development. As Bianchi (2015: 290, original emphasis) argues:

> the attempt to conceive of market behaviour in isolation from the ideologies and values of the different actors and interest groups involved, as reflected in the free market notion of *comparative advantage*, underplays both the unequal distribution of incomes and power which may result from 'open' competition in the tourism market, as well as the political nature of markets whereby the state has historically conditioned the activities of economic classes, and furthermore, ignores the uneven consequences of unlimited market competition.

Bianchi (2015: 291) focuses on the political economy of tourism, which is summarized as 'the examination of the systemic sources of power which both reflect and constitute the competition for resources and the manipulation of scarcity, in the context of converting people, places and histories into objects of tourism'. When considering Figure 9.1, it is essential to consider the influence of power and control by various groups and individuals in the tourism development process.

Influences

Some of the main influences on tourism development are listed in Figure 9.1. The aim here is to examine these influences in terms of the tourism development dilemma and how they may have an impact on tourism being a force for overall development. Developing countries face some incredible challenges and strong forces from both beyond and within their borders. Although the influences below are considered individually for the purposes of analysis, many are highly connected and either directly or indirectly have influence on each other. They are presented here for discussion purposes to explore the range of influences interacting with tourism and destinations in a very dynamic global environment. The list is not meant to be comprehensive, but rather illustrative of the complexity of the tourism development dilemma.

Modernization

Modernization is a socio-economic development that follows an evolutionary path from a traditional society to one that is modern such as in Western Europe or North America (Schmidt 1989). Rostow (1967), for example, argued that for development to occur, it had to pass through a series of stages from traditional society to the age of high mass consumption. Some would argue that the goal of development then is to become a modern state and perhaps the next breakout nation (Sharma 2013). There is pressure for developing countries to climb the ladder of economic development (Sachs 2005). Pursuing tourism as an agent of economic growth is, however, not a simple task. Sharma (2013: ix) compares the history of economic development to a game of snakes and ladders:

> There is no straight path to the top and there are fewer ladders than snakes which means that it's much easier to fall than climb. A nation can climb the ladders for a decade, two decades, three decades, only to hit a snake and fall back to the bottom again while rivals pass it by.

Tourism is one tool to help generate the resources to move up the economic ladder and in the direction of a modern state. If maximum financial gain is the target, large-scale mass tourism may be one strategy used to achieve this goal. As indicated in Chapter 1, there are links between modernization and regional economic development theory, and one such theory emphasizes growth poles. Large-scale resort complexes such as Cancún, Mexico, and Nusa Dua in Bali, Indonesia, could be argued to act as growth poles for the economy. Tourism Development Corporations may be established by governments to identify areas for large-scale development, and then provide services to attract developers. Dubai is an example of a destination shifting towards tourism; in January 2015, Dubai

International Airport overtook London's Heathrow Airport to become the world's top hub for international travel. There also is an interesting dilemma within modernization as it relates to tourism. Tourism brings with it conflicting pressures. One end of the spectrum stresses the importance of moving ahead to provide all the luxuries tourists demand, while, at the other end of the spectrum, there are demands that some communities need to stay as they are so that tourists can see the ways of the past or experience the undeveloped. Can tourism dictate to some communities that they need to remain in a non-modern state to attract tourists?

Globalization

As outlined in Chapter 3, globalization has facilitated the movement of goods, people, information, values and finances across political boundaries. In this era of free trade, global markets and economic neo-liberalism, there is a great deal of pressure on destinations to open their borders to trade, including tourism. At times, this has resulted in the reduction of the power of the state, and so, to a large extent, multinational corporations hold a great deal of power in the tourism decision-making process. Globalization has changed the nature of sovereignty by creating overlapping natures of governance. For example, states must not only look after their own internal affairs as they open up their borders, but must also deal with international trade organizations, international funding agencies or, perhaps, regional trading blocs. Multinational tourism corporations are looking for good locations with favourable trade policies and low production costs to open their businesses. Developing countries often then compete to attract developers for tourism. Rodrik (2005) identifies the 'Political Trilemma of the Global Economy'. He suggests that (a) the nation state system, (b) deep economic integration and (c) democracy (mass politics) are mutually incompatible. At most, one could obtain two out of the three; if a country wanted to push for global economic integration, a choice would have to be made to give up on the power of the nation state or democracy (mass politics). Democratic politics would be reduced as, once countries join the global economy, they must follow the financial rules set by the global market, leaving very little room for democratic debate over national economic policymaking. The 2008 global economic crisis and its ongoing repercussions clearly illustrate the dangers of global economic integration and reliance on the free market. In the case of tourism joining the global economy, once borders are open and restrictions are lifted, there are criticisms that multinational corporations such as hotels, airlines or tour operators will operate with high levels of leakage. As noted in Chapter 8, efforts are being made to increase regulation following the 2008 global economic crisis. However, given the volatility of the market, future economic crises will likely occur.

Another aspect of globalization is the international sources of funding. If developing countries want to attract tourism but do not have the necessary funds (say for infrastructure), where can they turn? Easterly (2005) examined the complex process that a developing country has to go through to obtain aid money, such as preparing a 'Poverty Reduction Strategy Paper' that would go forward to the World Bank for evaluation. As is evident in the following statement, Easterly (2005: 187–8) is highly critical of the bureaucracy related to obtaining and receiving foreign aid:

> Frontline staff in aid agencies can barely keep up with the flood of mandates, political pressures, procedural requirements and authorizations required to keep the aid money flowing. In the recipient governments, the relative burden on the management is even greater given the scarcity of skills, the extreme political pressures in factionalized societies and the prevalence of red tape and corruption.

He goes on to argue that foreign aid is continuing to lose support in rich countries just as liberal political and economic ideals are losing support in poor countries. If a developing country includes tourism as part of their strategy for development in an application for foreign aid, then the above comment by Easterly (2005) highlights the challenges ahead. International loans have also been criticized in the past for the accompanied required changes to domestic policy, which have had harsh impacts on recipient countries. Given these restrictions, China has become a significant new source of international development assistance for developing countries, providing funds for tourism-related developments and infrastructure. The loans from China often do not have the same kinds of conditions as those imposed by organizations such as the IMF or the World Bank.

Globalization has a number of positive aspects for developing countries. Multinationals, for example, bring capital, knowledge transfer, business operating systems and technology, as well as stimulating entrepreneurial development. A second generation of multinationals is emerging, based in developing countries and expanding beyond their borders. A rapidly emerging aspect of globalization is technological change and information technology. The Internet has allowed small operators in developing countries to advertise their products online opening the door to new markets for those able to access the Internet in those countries. Online crowdsourcing to generate micro-loans is another new avenue for small-scale tourism entrepreneurs to access credit. While technology has facilitated online bookings for companies, it has also empowered individuals through things such as mobile phones and handheld portable translators. The experience of sitting in a restaurant in Indonesia with an informal sector tour guide and watching him use a handheld translator to communicate in English, Indonesian and Japanese with the people he was leading illustrates how technology can open up new opportunities if one has the funds to access the technology. This also illustrates the globalization process on culture. Continued efforts are needed to bridge the digital divide allowing more people in developing countries access to the Internet.

Production

There is a range of considerations with respect to the factors of production for the tourism development process. Factors of production are typically divided into land, labour, capital and entrepreneurship. However, in the context of Chapter 8, the potential future scarcity of resources such as oil, water and food may raise significant challenges not only for the tourism industry, but also for destinations. The factors of production in tourism will also depend on the type of tourism developed and who is in control of the tourism development process. In community-based tourism, such as in the agritourism complex in the village of Bangunkerto (central Java), Indonesia, the main inputs into the project were community participation and some governmental assistance. In order to pave the roads through the local village so tour buses could get through, the entire village was out working. The men were collecting rocks from the site of a past volcanic eruption to use as part of the new road base while the women were arranging the actual rocks and creating the roadbed in the village (Telfer 2000). In the context of large-scale tourism developments, criticisms focus on the issue that the factors of production may be imported, profits repatriated and decision-making is often external to the destination. However, there is evidence that large hotels can adopt the policy of using local products to increase local multiplier effect (Telfer and Wall 2000).

Increasingly, tourism value chain analyses are being completed (see Figure 4.2) to find ways so the poor can participate in the industry and make tourism more effective in poverty alleviation.

Potter (1995: 334) argues that 'globalisation is not bringing about a uniform world, rather leading to new and highly differentiated localities and regions'. He refers to this as part of the concept of divergence. Looking at the history of urbanization in the Caribbean, Potter (1995) states that development tends to be focused on the pre-existing core urban areas where infrastructure was already in place. Second, development has continued in a top-down, centre-out manner. Making comparisons to the tourism industry in the Caribbean, Potter (1995) indicates that tourism has continued to focus on coastal areas along with further development of pre-existing coastal urban zones. Over time, new extensions to coastal development have occurred in newer places such as Montego Bay and Ocho Rios. This concentration has geographic implications, as reflected in Figure 9.1. The larger resorts continue to evolve along the coast and newer resort areas open up over time. The clustering of resorts will draw not only tourists, but also migrants looking for work resulting in regional imbalances in the country. Potter (1995) also comments that the flow of workers in the tourism sector dampens the availability of labour for agriculture and may lead directly to the creation of idle land.

A final comment that relates to production is in the area of sustainability. There are increasing pressures for companies to operate in a more sustainable manner. More and more companies are publishing statements of corporate social responsibility or developing environmental programmes. Some tour operators have identified this as part of their marketing strategy and offer more responsible travel. Programmes such as Green Globe and the European Blue Flag campaign for beaches all reflect this trend. If these policies and programmes are followed, they represent opportunities to enhance the contribution that tourism can make to development as well as protecting the environment.

Dependency

Problems of dependency relate to external control of the industry, as well as having an overreliance on tourism. Developing countries may have little alternative but to rely heavily on the tourism sector, and national priorities may take a back seat to the priorities of multinationals. Tour operators, for example, hold a great deal of power and can quite quickly decide to change destinations for the year, leaving hotel operators trying to fill their hotel rooms. In some developing countries, national-based hotel chains have opened up, thereby reducing the level of external dependency. The problem of being overreliant on one industry applies not only to tourism, but to any industry. If there is a downturn in the market or increased competition, the industry will suffer. For example, if a tourist from Canada wants to go to the Caribbean or a tourist from the UK wants to go to the Mediterranean to escape the cold winter, he/she has a wide range of destinations from which to choose. As seen in Chapter 8, political stability and safety concerns are all vital to the image of the destination; any negative event will have a large impact on tourism in the destination area. Tourism Concern, a UK-based NGO, ran a campaign entitled 'Sun, Sand, Sea and Sweatshops', which revealed that many hotel workers in developing countries are too often trapped in poverty through poor working conditions. Working without a contract, going for months without pay or having to show up at the gate of the hotel in the morning to see if they are needed for the day are a few of the examples Tourism Concern found that illustrate the perilous situation that people dependent on tourism may face (Tourism Concern n.d.).

Postmodernism

The concept of postmodernism impacts on a variety of dimensions of tourism. Seen in part as a rejection of the more traditional and dominant ways of conceiving or operating, postmodernism has opened up new products and destinations, and tourists are demanding new and exciting experiences. Some have linked this to the shift from Fordism to Post-Fordism, whereby there is a movement away from mass standardized packages. In order for developing countries to remain competitive, they have to offer new products and open up more areas of their country to tourism. Uriely (1997) traces the evolution of the theories of modern and postmodern tourism and, in a review of the literature, states that postmodern tourism has been discussed in terms of (1) the 'simulational' focused on the hyperreal experiences such as theme parks, and (2) the 'other', which is the search for the 'real', reflected in a growing interest in the 'natural' and the countryside. Developing destinations have created products, including theme parks, casinos and iconic shaped buildings, to attract tourists. Places such as Dubai and Abu Dhabi have created new coastal landscapes centred on iconic hotels and buildings (for example, Dubai Palm Islands), resorts and cultural facilities (see Plate 9.2). Macau has reclaimed land in the development of large casino projects and the Sunrise Kempinski Hotel in China near Beijing was constructed in the shape of a sun (a disc). These iconic structures are being used in place-based marketing in the highly competitive tourism industry. Demands for experiential adventure tourism or ecotourism are also taking people to areas previously not open to tourism. As indicated in Figure 9.1, there are resort areas opening up towards the interior of the island. Demand for heritage tourism is increasing as well, with tourists consuming culture, searching for authenticity and wanting to be educated. As noted in Chapter 4, heritage and cultural tourism raise issues of commodification and issues of access and control of heritage monuments. From a development perspective, these types of products may well enhance the opportunities for locals to participate in the tourism industry, as long as they have access. The simulated environments such as the casinos and theme parks are typically large-scale, and therefore require significant investments, potentially limiting the ability of local participation. Again, the development dilemma appears as opportunities may also come with associated costs (economic, environmental, social and political).

With the shift towards changing markets, pressures are put upon existing resort complexes to alter their factors of production. Hotels, for example, need to provide more differentiated activities, which means more equipment. Tour operators who are bringing in the tourists are competing in a very price-sensitive environment. They may place demands on resorts to offer more activities/products, but they do not want to pass these costs on to the consumer. Therefore, the destination resort may have to absorb the new costs in meeting the higher demands of tour operators and tourists.

Consumption

The consumption of tourism, as explored in Chapter 6, has a number of important implications for the extent to which tourism can play a meaningful role in the development process. Tourists were identified in Figure 4.1 as one of the agents of tourism development. How many tourists come, what they demand, how they behave and, specifically, how much money they spend in the destination all are important factors in the development process. A significant milestone

Plate 9.2 *UAE, Dubai: Atlantis hotel; note, this is a copy of the Atlantis hotel in The Bahamas*

Source: Photo by R. Sharpley

was reached in 2012 with international arrivals reaching one billion. It is important to look at the markets that are attracted to developing countries. Often in a discussion of tourism, the focus is on tourists from the developed countries visiting developing countries. What should not be forgotten is the role that domestic tourists can play in the development process (Singh 2009b) along with role of the growing number of wealthy international travellers from developing countries. Domestic tourists may, for example, be more willing to use local forms of transportation, eat in local restaurants and stay in locally run accommodation. If one considers a country such as China, the potential in terms of numbers of future domestic tourists may well be a more important market than the international market. As reported in Chapter 1, there were an estimated 2.61 billion domestic trips in China in 2011, a number double that of worldwide international arrivals (National Bureau of Statistics 2013). The Chinese also spent a record US$129 billion in international tourism expenditure in 2013 as a result of rising disposable incomes, fewer travel restrictions and an appreciating currency (UNWTO 2014c). A report from the UN World Tourism Organization in 2013 focuses on the key outbound tourism markets in South East Asia, including Indonesia, Malaysia, Singapore, Thailand and Vietnam. The study reveals that intra-regional travel of these five South East Asian markets had continued to expand over land, water and, most importantly, air with the expansion of low-cost carriers (UNWTO 2013c; see also WTO 2006). These trends reveal the importance for developing countries of not only targeting their own domestic tourism markets, but also the more affluent sectors of the population in nearby developing countries.

The consumption of tourism is also linked to the discussion of postmodernism in the previous section. New demands for new products will not only create opportunities, but also associated costs. Health and wellness tourism and medical tourism are growing in countries such as Thailand, which received 1.8 million international patients/tourists in 2013, generating $4.7 billion for the Thai economy (Mellor 2014). Perhaps the medical tourism market will grow as many developed countries have ageing populations. This raises a number of questions in terms of development trade-offs. Are resources being diverted from domestic medical care to the care of international tourists?

As tourists become more experienced, their demand profiles can change. There is an argument that is being debated in the literature that tourists are becoming more aware of the impacts of travel and are becoming more environmentally friendly. Figure 6.3 offers a code of ethics; however, as indicated in this chapter, there are debates about the effectiveness of such codes. There has also been a rise in volunteer tourism, with tourists contributing to and participating in development projects such as building schools. While new forms of travel tend to get the most notice in the press and are frequently studied by academics, it is, however, important not to forget the traditional mass tourism product that has been the mainstay of tourism and is likely to remain so. There may be different labels applied to it through marketing, or there may be a perceived increase in the range of choice associated with these trips; however, mass tourism still represents the 'workhorse' of the tourism industry. It is also important to note that not all newer and smaller-scale forms of tourism are sustainable, and in fact it is some of the larger tourism corporations that are taking the lead in developing environmental initiatives. The challenge becomes making all tourism more sustainable so that more benefits will remain in the host destination.

The final issue to explore in terms of consumption is what Potter (1995) terms 'convergence'. He argues there is an increasing convergence on Western models and patterns of consumption. Economic dependency dovetails with cultural and psychological dependency (Potter 1995). In tourism, a great deal has been written about the demonstration effect (see Chapter 6) and the impact that has on local patterns of behaviour and consumption. Festivals and souvenirs have been changed to accommodate tourists. Consumption of some products will increase and other traditional products may fall away. These types of trade-offs are identified in Figure 9.1 and are discussed later in this chapter.

Sustainable development

Sustainable development was explored in detail in Chapter 2, and in this chapter a call was put forward in terms of the 'imperative of sustainable development'. Although the term is highly debated, it is arguably a Western-based concept being imposed on developing countries. While there are conflicts over how it should or can be implemented and measured, the concept continues to receive attention as a driving framework for tourism. According to UNEP (2011), tourism has the potential to contribute to sustainability and poverty reduction. The definition of development used in this book highlights the fact that improvements need to proceed within environmental limits. It is worth noting here that sustainability requires a holistic approach, covering not only environmental concerns, but also social, cultural, political and economic concerns. Addressing the question of what it will take to move our civilization in the direction of long-term sustainable development, Wright (2008) focuses on two sets of themes. Strategic themes include sustainability, stewardship (the ethic that guides actions for the benefit of the natural world and other people) and science, and these are identified as concepts or ideas that can move societies towards a sustainable future. Integrative themes include ecosystem capital (goods and services provided by ecosystems), policy and politics, and globalization, and these three categories deal with the current status of interactions between natural systems and human societies. Tourism as a development tool interacts with all of these themes, and steps are being taken to encourage businesses, tourists and destinations to be more sustainable. There are calls for more local participation in tourism planning and the operation of the industry (Chapters 4 and 5), but, as Mowforth and Munt (1998) stress, sustainability needs to be understood within the context of power relations.

In an interesting study by Jamal *et al.* (2006), the authors state that while sustainable tourism and ecotourism are rooted in the notions of individual/societal and environmental well-being, they encountered conflicting values. They found that the actors and programmes related to ecotourism have institutionalized a modernistic, commodifed paradigm. A question considered later in this chapter focuses on the ability to actually implement sustainability. If sustainability is adopted as a framework, what impacts will it have on the contribution towards overall development of the destination? All forms of tourism need to be made more sustainable, and Figure 9.1 explores the trade-offs that may occur in a destination.

Climate change

Inherently linked to sustainability is climate change. As indicated in Chapter 8, climate change is rapidly becoming one of the key emerging influences on tourism. The industry not only contributes to climate change, but many destinations are experiencing its impacts. A report by the UNWTO (2014c) on tourism in small island developing states (SIDS) suggests that climate change is one of the main challenges for these countries as they face storm surges, rising sea levels, beach erosion and coral bleaching, all of which directly and indirectly affect tourism. In the Kashmir Valley in the Himalayas, for example, climate change is causing rising temperatures and erratic snowfall that will have severe impact on winter tourism (Dar and Rashid 2014). However, while climate change policies are viewed as increasingly urgent, many developing countries have not enacted national climate change policies (Hambira and Saarinen 2015). Hambira and Saarinen (2015) document other challenges developing countries have with climate change and climate change policy. Many developing countries do not have the financial or technological capacity to address climate change and are, therefore, reactive rather than proactive in dealing with it. As a result, the cost of inaction will be higher in the future. Moreover, where climate change policies are being developed, they tend to be more general rather than sector-specific, such as policies for the tourism sector. Greater information is needed for developing countries, as most of the research over the last two decades covers the Global North contexts and views (Hambira and Saarinen 2015). In a study of Botswana policymakers' perceptions of climate change, it was found that the importance of urgently establishing policy is recognized; however, they foresee inadequate information and uncertainties surrounding the impacts of climate change on natural capital (Hambira and Saarinen 2015). As a result of the study, a series of questions were posed directly relating to information needs on climate change, including:

- Which ecosystems or natural capital are crucial to Botswana's tourism sectors and how are they likely to be affected (negatively or positively)?
- How can the probable negative and positive impacts be avoided or enhanced?
- What kind of mitigation and adaptation policies will be most suitable for a developing country such as Botswana?
- What are the cost implications?
- Does Botswana's tourism sector have the required adaptive capacity?
- How will tourism flows in the country most likely be affected?
- How will the tourism sector's contribution to the country's gross national product be affected?
- How can scientific information related to climate change and tourism be made simple for communities and tourism business operators (Hambira and Saarinen 2015: 359)?

All of the above questions are transferable to other developing countries that face similar challenges in developing climate change policy in the context of limited information and financial resources. The challenge for many developing countries is they are dealing with more immediate needs in terms of poverty as well as having to respond to the longer-term impacts of climate change (Hambira and Saarinen 2015). Wright *et al.* (2015) argue that climate change will significantly challenge least-developed countries in their ability to achieve the proposed sustainable development goals in the post-2015 development agenda (see Box 2.1).

Having outlined some of the major influences on tourism in developing countries, the chapter now turns to an examination of the destination.

Tourism destination: form, function and response

The influences described in the previous sections interact with the tourism development process outlined in Chapter 4, resulting in some form of tourism. The island in the centre of Figure 9.1 represents that end product and is illustrated here as having a highly developed resort area that not only has beach development, but also some form of satellite tourism development, such as heritage tourism, ecotourism or nature-based tourism. While the diagram is of an island, many of the principles apply to other forms and scales of tourism. Markets are identified and individuals, communities, companies, NGOs and governments respond to provide the product or provide the finances for the product. These development agents may come from within the destination or they may be external to the destination, which highlights the issues of control and power. Once a developing country decides to adopt tourism as a development option, governments need to decide whether to create and implement a tourism master plan or to allow the industry to evolve with little interference from the government. If tourism plans are created, it is important to consider to what extent local people are involved in the planning process and whether they would actually be able to participate based on the constraints they face (for example, economic, social, political, lack of knowledge).

In the centre of Figure 9.1, a series of elements are considered that reflect on how tourism is developed. All of the agents of development encounter a policy, planning and regulatory environment (see Figure 4.1). Decisions will be taken by a number of different actors on what to build, who should finance it, who should operate it and what, if any, incentives should be provided to developers. For all key stakeholders, there is a need to evaluate what tourism development options are the best and how the industry will compete in a global environment. The decisions made will vary from country to country over what forms of tourism to develop. There will also be a local response to tourism and, given the complex nature of communities, some groups will respond to the opportunities of tourism positively and some will passively accept the change, but others may be more vocal or try to resist new developments.

Different types of tourism will have different forms and functions, and how they are managed will also influence the degree to which they can contribute to development. Whichever resort or attraction is constructed (for example, beach resort, village tourism, ecotourism lodge or a combination), it will act as a magnet for other development. In the case of beach resorts, it may even lead to the development of an urban tourism resort area. Over time, satellite tourist areas may open, as illustrated in Figure 9.1. Excursion roads to these attractions will themselves act as conduits for development as entrepreneurs try to sell their products to tourists at various rest stops or craft villages along the way. The overall impact spatially may well result in regional imbalances as tourism growth occurs around the tourism areas. The employment structure will change, especially if tourism develops in a major way. This will then lead to migration, and demands from immigrants for things such as housing, healthcare and schooling will need to be met. The challenge for any government is also how to redistribute the imbalances to ensure that those who do not work in the tourism sector and do not live near the tourism area will still be able to benefit from the industry. The final point raised in the centre of Figure 9.1 is the global-local interaction. These interactions are part of the tourism development trade-offs, which are explored below.

Tourism development trade-offs

The development of tourism will invariably result in impacts in the destination, as outlined in Chapter 7. As part of the tourism development dilemma, it will also typically result in some form of a trade-off. Broad examples of potential trade-offs are highlighted in Figure 9.1, and are meant as a framework for generating discussion rather than trying to present a comprehensive list. They are listed under the categories of economic, environmental, sociocultural and political. There can be trade-offs within these categories and between these categories. The trade-offs may be the result of a conscious planning or policy decision (by a particular level of government, organization or business) or they may evolve over time as individuals, companies, organizations or governments respond to the evolving business environment. The trade-offs are not static, but dynamic, as the tourism industry evolves and responses are made (both positive and negative) to the industry. Trade-offs have important geographical dimensions, including space and time (Adams 2005). In terms of space, tourism could be developed in a sustainable manner in one location and at the expense of the environment in another. Adams (2005) notes the importance of both political and ecological boundaries. In terms of time, sustainability implies balancing resources for future generations with the needs of the present generation, and the future is hard to predict. Unintentional trade-offs can also occur as a planning decision related to one area or activity in a destination may have wider unanticipated economic, sociocultural or political consequences. Finally, the trade-offs will be regarded differently by different groups at different time periods. Some may be willing to support a trade-off while others will be against the trade-off.

Under the economic category, for example, tourism may bring economic growth for those able to participate in the tourism industry, while, at the same time, it may result in economic leakages. If large-scale mass tourism is developed with multinational corporations, there will definitely be economic growth through increases in the number of tourists, development fees and taxes. However, if the multinational imports supplies and management personnel while repatriating profits, there will be a reduced multiplier effect. This may be a trade-off worth making if the developing country is not able to effectively enter the global tourism market without outside help. Liu and Wall (2006) argue that many, and perhaps most, communities in the developing world need an outside catalyst to stimulate interest in tourism and external expertise to take full advantage of their potential opportunities. Page and Connell (2006: 348) outline the economic costs and benefits of tourism. The economic benefits include the balance of payments, income and employment while the economic costs include inflation, opportunity costs, dependency, seasonality and leakage. Income and employment can also be a cost if the better-paid positions do not go to locals and if the income generated does not go to those who need it most (leaked out of the economy). As an example of a sociocultural trade-off, a community-based tourism project might promote empowerment in a local community; at the same time, however, it also may lead to the commodification of their culture as local crafts, products or performances are modified to appeal to tourists' needs and tastes.

Trade-offs can also occur between a number of the main categories in Figure 9.1. Economic growth may be pursued at the cost of the environment and the loss of local control. Tourism pursued as a path to economic growth is having environmental impacts in the form of climate change. Increased visitors arriving by aeroplane generate additional greenhouse gas emissions, which in turn contribute to a range of environmental changes such as rises in sea levels, extreme weather and droughts. Alternatively, ecotourism may be developed as a sustainable form of tourism; however, due to the restrictions in visitor numbers typically required, it may come at the cost of economic growth. It could be argued that enforcing strict and costly environmental policies may reduce the overall economic impact of tourism; however, environmental protection cannot be put off indefinitely. An example of a sociocultural and economic trade-off could be accepting or tolerating more tourists for financial gain, yet a traditional sacred ceremony can become secular entertainment for tourists (see Plate 9.3). These are just a few of the multitude of trade-offs that can occur.

The question as to who decides on the nature of the trade-offs can largely be answered by examining the power relations that exist surrounding the tourism development process. At the national level, the state may establish the regulatory environment, which will form the basis for a particular type of trade-off (for example, pursue a policy of limiting all-inclusive resorts so tourists will spend more

Plate 9.3 *Indonesia, Kuta Beach Bali: traditional Balinese ceremony on the beach continues in the presence of tourists*

Source: Photo by D. Telfer

money outside the resort – the trade-off being tour operators may be scared away), or conversely a powerful multinational may be able to dictate terms to local governments setting up a different type of trade-off (for example, demand to be able to import significant quantities of food – the trade-off being there may be more tourists as a result of the multinational, although there will be significant leakages). Power also rests with international funding agencies, which could dictate a certain type of tourism development is built, or with the international money market, which can influence exchange rates and thereby dictate whether a particular country is attractive to invest in.

It is important to recognize the tourism industry is very dynamic; the interactions within and between the categories happen simultaneously and there is not always an evident simple cause-and-effect relationship between the interactions in tourism. The best-laid plans may not always work and things may change in the future. There are multiple forces and actors interacting at the same time, as illustrated in Figure 9.1, and tourism will certainly not be the only industry in the destination, although it could be the largest. Within the framework of sustainability, the strategy will hopefully minimize the level of negative trade-offs. One final trade-off considered here is that of opportunity cost. The decision to put money into tourism results in that same money not being available for investment in another area of the economy, an area that could generate more development than tourism, depending on the situation.

Difficulties in implementing the ideals of sustainability

The trade-offs in tourism discussed in Figure 9.1 often imply that one sector or element may benefit at the expense of another. Decision-making means that competing needs and values must be considered so that the best conclusion can be reached in the numbing complexity of demands and circumstances (Wright 2008). Wright (2008) refers to sustainable development as an ideal, a goal towards which all human societies need to be moving even if it is yet to be completely achieved anywhere. Sustainable solutions come at the intersection of the concerns of sociologists (socially desirable), economists (economically feasible) and ecologists (ecologically feasible) (Wright 2008). The idea of sustainable solutions raises the question as to whether or not it is possible to move beyond the trade-offs in Figure 9.1 so that all three areas (environment, society and economy) benefit. In the case of tourism, for example, without an attractive environment, tourists will not come and no economic benefits will occur for the destination. If locals are not able to participate in the planning, implementation and benefits of tourism, and are only the object of commodification, then there may be resentment of the locals towards the tourists. While the end goal in this hypothetical example may be to move in the direction of having all three areas (environment, society and economy) benefit, it is important to recognize that tourism is a highly competitive global industry, and corporations, governments, various organizations and individuals all have values and ideologies guiding behaviour and outcomes. Wright (2008), for example, states that when governments linked to anti-environmental special interests are in power, the environment, as well as sustainability, is only of secondary concern. For now, as Adams (2005: 425) points out, the 'pursuit of sustainability demands choices about the distribution of costs and benefits in space and time'. The challenge then becomes how to make all forms of tourism more sustainable.

The weaknesses of sustainable development have been explored in Chapter 2 and a set of key questions are presented in Table 9.2. Challenges in implementation also relate to the policy process. Developing countries face global environmental agendas while at the same time facing their own unique environmental problems and development needs (Newell 2005). An examination of environmental policy in developing countries illustrates both differences and common challenges in terms of reconciling development needs with long-term environmental goals (Newell 2005). Countries have responded differently to the challenge of reconciling these issues as a result of variations in political systems, the nature of their economies and the level of civil engagement. Also, countries in the developing world have differing abilities to implement policies. Newell (2005) cites the example of India, which has some of the most impressive environmental legislation in the world. However, the lack of training resources and corruption among local pollution officials act as constraints to the effective implementation of this legislation. Conversely, in other countries, such as Kenya, which have a strong economic and developmental interest to ensure compliance with policies, then extra steps are taken. Due to the significance of revenues from wildlife safaris in Kenya, officials have gone to controversial lengths in tackling the problem of the illegal poaching of rhinos and elephants for their ivory, as well as 'banning tribal groups from culling animals for food even on their own ancestral land' (Newell 2005: 228). Part of sustainable development is local involvement, and while countries such as India and Mexico have strong traditions of active civil society engagement, these options are limited in places such as China or Singapore (Newell 2005). Newell (2005) also notes the differences to which there has been an overall 'greening' of industries, and the drivers of corporate environmental responsibility such as government incentives, consumer and investor pressure, and civil society watchdogs are currently undeveloped in many parts of the developing world. While there are often vast differences between countries, there have been similar problems in enforcement at the national level. Similar enforcement problems relate to constraints arising from economic relationships of trade, aid and debt and conflicts between often Northern-determined environmental priorities and other issues that appear to be more pressing at the local levels (Newell 2005).

Sustainable development has been increasingly linked to poverty alleviation and the Millennium Development Goals. The UN is currently in the process of developing the post-2015 development agenda, which will still clearly have links to the Millennium Development Goals. In a related example of the challenges of implementation, Hawkins and Mann (2007: 360) comment that:

> though the Millennium Development Goal imperative is driving individual country development strategies, these indicators are often not relevant to the political agendas of individual government leaders or ministers. The Minister of Tourism in any given country now is being asked to deliver a range of outcomes, including economic growth that can be empirically related to a poverty reduction strategy.

They go on to refer to Ravallion (2004), who comments that economic growth does not necessarily lead to poverty reduction.

The various examples above speak to the difficulties in implementing goals or ideas versus the realities in planning and managing tourism in developing countries. Wright (2008) argues, however, that there is increasing recognition among a wide variety of groups and individuals in different settings that 'business as usual' is not sustainable, and new ways of thinking and conducting business will be required. The challenges in changing the way businesses, governments, organizations and individuals both think and act are very real, yet there are a number of initiatives, as outlined in Chapters 2 and 5, that are taking tourism in a more sustainable direction linked to poverty alleviation.

Conclusion

Development is a highly contested term with a variety of different perspectives both on theory and practice. Development theory has changed as various paradigms have risen and fallen (though not completely disappeared), influencing how development is practised. Notions of development have changed from primarily a focus on economic growth to a more holistic notion of development incorporating economy, environment and society. Increasingly, development has been linked to sustainable development and poverty reduction, as illustrated in the Millennium Development Goals and the post-2015 development era. The emerging paradigm of global development calls for new forms of global governance to deal with evolving challenges, such as climate change. For developing countries, tourism continues to be one of the favoured development strategies, and there have been efforts made to make tourism more sustainable or have it directly relate to poverty reduction (see Plate 9.4). In considering using tourism as an agent of development, it is also necessary to frame the analysis with the broader issues of development as they relate to power and control. In this age of globalization, there is a wide range of actors involved directly or indirectly in the tourism development process, including international agencies, states, the private sector, citizens groups, NGOs and tourists, to name a few. All of these actors have their own values and goals that influence their actions, and they each have a certain amount of resources, technology, power and control that they may be able to utilize to influence the outcomes of development. Underhill (2006: 19) argues that the exercise of power in the international political economy takes place in a setting characterized by the complex interdependence among states and their societies and the economic structures at domestic and international levels. There are debates over the influence the free market, the state or civil society should have on the tourism development process. This book has examined the role that tourism can play in the overall development process and, as Addison (2005: 219) suggests:

> getting development policy right has the potential to lift millions out of poverty and misery. Making the right policy choices is not just a technical matter. It requires careful political judgement on how to promote economic and social change in ways that stand the best chance of succeeding.

The decision to pursue tourism leads to the tourism development dilemma, which has been the focus of this book. Developing

Plate 9.4 *Cambodia, Udong: sign promoting the benefits of tourism*

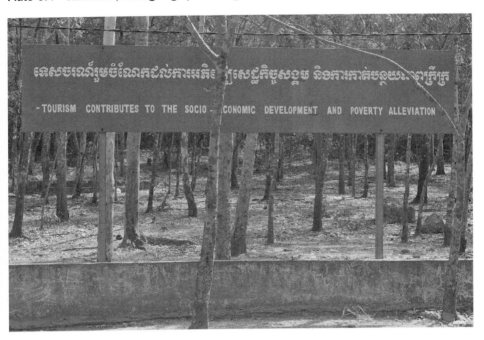

-TOURISM CONTRIBUTES TO THE SOCIO-ECONOMIC DEVELOPMENT AND POVERTY ALLEVIATION

Source: Photo by R. Sharpley

countries face significant challenges; tourism not only generates benefits, it also generates costs and trade-offs are made. These trade-offs not only occur in the destination, but also across national boundaries as well as across time. This chapter began by looking at the development imperative and the sustainability imperative. Sustainability has come to the forefront of the development debate and it is argued that all forms of tourism need to be more sustainable. It is presented in this chapter as an imperative involving not only environmental concerns, but also social and economic considerations. Local communities need to have greater participation in the planning and development process. It is also important to understand communities in terms of vulnerability and resilience to tourism development. In the context of Figure 9.1, the aim of sustainability would be to minimize the negative trade-offs within the tourism development dilemma. Although there is a great deal of debate surrounding the concept, efforts on a number of different fronts are being made to try to make tourism more sustainable.

With the shift towards globalization, there are concerns in the area of sustainability as multinational corporations are seeking out favourable destinations with good investment incentives and low production costs. The lower production costs may revolve around less stringent regulations on things such as the environment or labour laws. The globalization debate in tourism has also raised issues of dependency and loss of control. There may be a reduction in the power of the state if a country wanting to enter the global tourism market is pressured into opening their borders to multinationals. Globalization not only has economic, political and environmental consequences, but also cultural implications for developing countries. As seen in the aftermath of the 2008 global economic crisis, opening a destination economy to the global economy can have serious repercussions if a crisis occurs.

The government in a destination establishes the policy framework for tourism in their country as well as interacting with global corporations and so it still plays an important role. Many international agencies stress good governance. An open policy on tourism with numerous incentives will attract developers. However, caution needs to be taken when the priorities of corporation do not match those of the state. These competing interests in tourism development can be illustrated through the continuum for advice on tourism planning developed by Burns (1999b). At one end is 'Tourism First', where the industry is the focus of planning. At the other end of the continuum is 'Development First', where planning is done according to national developmental needs. A stronger planning focus on 'Development First' may facilitate a more direct link to poverty reduction in the context of the Millennium Development Goals. If a country adopts a policy to support large-scale projects with foreign investment, it also needs to make sure it supports small locally controlled projects with things such as microcredits, where the informal sector may have a greater role to play. Communities respond to the opportunities of tourism and they need to be heard to effectively involve more people in the benefits of tourism. A number of interesting programmes, such as fair trade in tourism and the pro-poor tourism approach, have been raised to specifically target poverty. However, there are significant challenges that need to be overcome. It is also important that development not be planned in isolation from the rest of the economy as it needs to be integrated into multiple sectors so that there is potential that more benefits may accrue.

While tourism development is often centred on governments, planners, corporations, communities and NGOs, it is important not to forget the role of tourists themselves in the development process. Decisions are made as to what markets are pursued. Different types of tourists will exhibit different types of behaviour and will have different spending patterns. Some countries have opted for large-scale mass tourism while others focus on smaller numbers of higher-paying tourists, while others have pursued a more mixed approach attracting a diversity of markets. How a country chooses to market and brand itself will influence the overall type of development along with the type of tourists that come to the destination.

All tourism developments have impacts, and different communities and environments may be more resilient and better able to handle these impacts than others. These impacts need not always be negative as communities and individuals respond to the opportunities presented by tourism (Wall and Mathieson 2006). For many developing countries, there is limited choice in terms of development options, and they may, in fact, have a real competitive advantage on a global scale to offer tourism services. In fact, as Lickorish (1991: 162) argues, tourism may prove more 'cost effective in its use of scarce foreign investment for creating additional prosperity than alternatives in primary or manufacturing industry'.

Tourism should not be planned and developed in isolation, as there are a number of interrelated challenges that have come to the forefront since the first edition of this book. Principal among these is climate change, with destinations facing changes in weather patterns ranging from extreme storms or drought to rising seawaters. Potential regulations to offset climate change will have implications for tourism in developing countries. The long-term sustainability of resources such as oil, water and food, and related shortages or price increases in all three, may cause visitor numbers to decrease or the duration of their stay to shorten. Other challenges include instability in the global economy and potential future economic crises, as well as the instability of political institutions linked to ineffective governance, terrorism and failing states. Conflict has the potential to spread, disrupting entire regions and negatively impacting tourism. Finally, Chapter 8 examined demographic changes, which present opportunities of new wealth in emerging destinations but also potential challenges as populations in major tourism generating regions are ageing. New issues will continue to evolve that may impact tourism and development, and, as Smil (2008: 251) states, 'no one knows which threats and concerns will soon be forgotten and which will become tragic realities'.

Tremendous steps have been taken towards achieving the UN Millennium Development Goals, yet there is much to be done in the post-2015 era as the new set of sustainable development goals are being implemented. The 2014 Human Development Report *Sustaining Human Progress: Reducing Vulnerabilities and Building Resilience* (UNDP 2014a) also shows overall global trends in human development as being positive, yet people of all ages are facing threats and challenges to their well-being through natural or human-induced crises and disasters. Concerns in the Report include South Asia having over 800 million poor and over 270 million near poor, totalling over 71 per cent of its population. Extensive income inequality also exists in locations such as Latin America and the Caribbean despite recent advances in these locations. A further UNDP (2014b) report on inequality in developing countries found that a significant majority of households, representing over 75 per cent of the population, are living in societies where income is more unequally distributed than it was during the 1990s. The report argues that inequality cannot be confronted effectively unless the inextricable links between inequality of opportunities and inequality of outcomes are taken into account. These diverse challenges lead to questions such as how tourism can best contribute towards reducing inequality of income and opportunity for those in developing countries and thereby contribute to the broader notions of development.

The purpose of this second edition of this book has been to further examine the role of tourism as an agent of development in developing countries and consider what role it can play in addressing the challenges of development not only highlighted in this chapter, but throughout the book. Tourism is but one agent of development being actively pursued by many countries, and what seems to work in one developing destination may not work in another. The relationship between tourism and development is multifaceted as theories, values, actors, power, strategies, politics, policies, plans, communities and environments all interact in a dynamic global system, a system that is undergoing climatic change. Exploring the interaction of these concepts and issues will help better understand the nature of the tourism development dilemma for developing countries.

Discussion questions

1   To what extent does tourism promote development and increase the standard of living for the local population?
2   Does sustainability represent a way forward to guide tourism to promote development or does it act as a barrier to development?

3   How can development be measured with respect to tourism?
4   What processes or goals should guide tourism development over the next 25 years?
5   What are the most pressing issues facing tourism development?

Further reading

Scheyvens, E. (2011) *Tourism and Poverty*, Abingdon: Routledge.

This book provides an overview of poverty and tourism before examining whether tourism entrenches poverty, if poverty attracts tourists and whether tourism reduces poverty from the perspectives of industry, government and development agencies.

Holden, A. (2013) *Tourism, Poverty and Development*, London: Routledge.

This book also examines the relationship between poverty and tourism development, as well as having chapters on the geography of poverty, global political economy and structural causes of poverty, critiquing tourism and poverty reduction, and concludes with a way forward.

Huybers, T. (ed.) (2007) *Tourism in Developing Countries*, Cheltenham: Edward Elgar.

This edited volume has 34 articles dating from 1974 to 2004 that explore a range of issues for tourism in developing countries. The book investigates the positive and negative aspects (economic, social, cultural and environmental) associated with tourism development.

Harrison, D. (ed.) (2001) *Tourism and the Less Developed World: Issues and Case Studies*, New York: CABI.

This book has an introductory section on issues related to tourism development in developing countries before moving on to 13 chapters on a variety of topics/case studies.

Websites

www2.unwto.org
This web page is for the UN World Tourism Organization with links to the various programmes that it offers.

www.un.org/millenniumgoals/poverty.shtml
This website is for the UN Millennium Development Goals and Beyond. It also provides updates on progress of each of the goals.

www.undp.org/content/undp/en/home/mdgoverview/post-2015-development-agenda.html
This website displays the 17 new Sustainable Development Goals adopted on 25 September 2015 at the United Nations Sustainable Development Summit. The goals are part of the 2030 Agenda for Sustainable Development (also see Box 2.1).

References

ABTA (2013) *Travel Trends Report 2013*, Association of British Travel Agents, available at: http://67d8396e010decf37f33-5facf23e658215b1771a91c2df41e9 fe.r14.cf3.rackcdn.com/publications/Travel_trends_report_2013.pdf (accessed 22 February 2015).

Accor (2015) *Brand Portfolio*, available at: www.accor.com/en/brands/brand-portfolio.html (accessed 10 April 2015).

Adam, M. and Urquhart, C. (2007) 'IT capacity building in developing countries: a model of the Maldivian tourism sector', *Information Technology for Development*, 13(4): 315–35.

Adams, W. (2005) 'Sustainable development?', in R. Johnston, P. Taylor and M. Watts (eds), *Geographies of Global Change Remapping the World*, Oxford: Blackwell, pp. 412–26.

Addison, T. (2005) 'Development', in P. Burnell and V. Randall (eds), *Politics in the Developing World*, Oxford: Oxford University Press, pp. 205–30.

Aguiló, E., Alegre, J. and Sard, M. (2005) 'The persistence of the *sun and sand* tourism model', *Tourism Management*, 26(2005): 219–31.

Air Transat (2015) *About Air Transat*, available at: www.airtransat.ca/en-CA/Corporate-information/About-Air-Transat (accessed 10 April 2015).

Airbnb (2015) *Hospitality, Havana Style: Airbnb Opens Its Doors in Cuba*, available at: www.airbnb.ca/press/news/hospitality-havana-style-airbnb-opens-its-doors-in-cuba (accessed 10 April 2015).

Airbus (2015) *A380*, available at: www.airbus.com/aircraftfamilies/passengeraircraft/a380family/ (accessed 10 April 2015).

Akis, S., Peristianis, N. and Warner, J. (1996) 'Resident attitudes to tourism development: the case of Cyprus', *Tourism Management*, 17(7): 481–94.

Al Sadik, A. and Servén, L. (2012) 'The global crisis: old and new lessons for macroeconomic policy', in A. Al Sadik and I. Elbadwi (eds), *The Global Economic Crisis and Consequences for Development Strategy in Dubai*, New York: Palgrave Macmillan, pp. 65–86.

Allen, J. (1995) 'Global worlds' in J. Allen and D. Massey (eds), *Geographical Worlds*, Oxford: Oxford University Press, pp. 105–44.

Andersen, V. (1991) *Alternative Economic Indicators*, London: Routledge.

Andreassen, B. and Marks, M. (eds) (2006) *Development as a Human Right*, Cambridge, MA: Harvard School of Public Health.

Ap, J. and Crompton, J. (1993) 'Residents' strategies for responding to tourism impacts', *Journal of Travel Research*, 31(3): 47–50.

Arai, S. (1996) 'Benefits of citizen participation in a healthy community initiative: linking community development and empowerment', *Journal of Applied Recreation Research*, 21: 25–44.

Archabald, K. and Naughton-Treves, L. (2001) 'Tourism revenue-sharing around national parks in Western Uganda: early efforts to identify and reward communities', *Environmental Conservation*, 28(2): 135–40.

Aref, F. (2011) 'Barriers to community capacity building for tourism development in communities in Shiraz, Iran', *Journal of Sustainable Tourism*, 19(3): 347–59.

Arellano, A. (2011) 'Tourism in poor regions and social inclusion: the porters of the Inca Trail to Machu Picchu', *World Leisure*, 53(2): 104–19.

Arnstein, S. (1969) 'A ladder of citizen participation', *American Institute of Planners Journal*, 35(4): 216–24.

Ashley, C., Boyd, C. and Goodwin, H. (2000) 'Pro-poor tourism: putting poverty at the heart of the tourism agenda', *Natural Resource Perspectives*, 51 (March), available at: www.odi.org.uk/nrp/51.htm (accessed 19 October 2006).

Ashley, C., Goodwin, H., McNab, D., Scott, M. and Chaves, L. (2006) *Making Tourism Count for the Local Economy in the Caribbean*, available at: www.propoortourism.org.uk/caribbean/caribbean-brief-whole.pdf (accessed 10 December 2006).

Associated Press (2006) 'New law worries India's young workers', *The Standard*, 11 October, B8.

ATAG (2014) *Facts and Figures. Air Transport Action Group*, available at: www.atag.org/facts-and-figures.html (accessed 12 January 2015).

Atlantica Hotels (2015) *Atlantic Hotels*, available at: www.atlanticahotels.com.br/nossos-hoteis/hoteis-abertos (accessed 23 April 2015).

Azarya, V. (2004) 'Globalisation and international tourism in developing countries: marginality as a commercial commodity', *Current Sociology*, 52(6): 949–67.

Azcárate, M., Baptista, I. and Rubio, F. (2014) 'Enclosures within enclosures and hurricane reconstruction in Cancún Mexico', *City and Society*, 26(1): 96–119.

Bah, A. and Goodwin, H. (2003) *Improving Access to the Informal Sector to Tourism in the Gambia*, PPT Working Paper No. 15, London: Pro-Poor Tourism Partnership.

Baker, S. (2006) *Sustainable Development*, London: Routledge.

Balint, P. and Mashinya, J. (2008) 'CAMPFIRE during Zimbabwe's national crisis: local impacts and broader implications for community-based wildlife management', *Society & Natural Resources: An International Journal*, 21(9): 783–96.

Banki, M. and Ismail, H. (2015) 'Understanding the characteristics of family owned tourism micro businesses in mountain destinations in developing countries: evidence from Nigeria', *Tourism Management Perspectives*, 13: 18–32.

Barkin, D. and Bouchez, C. (2002) 'NGO-community collaboration for eco-tourism: a strategy for sustainable regional development', *Current Issues in Tourism*, 5(3/4): 245–53.

Barratt Brown, M. (1993) *Fair Trade: Reform and Realities in the International Trading System*, London: Zed Books.

Basen, I. (2015) 'Sharing economy: it's really an old-style rental economy', *CBC News*, 22 February, available at: www.cbc.ca/news/sharing-economy-it-s-really-an-old-style-rental-economy-1.2965000 (accessed 10 April 2015).

Bastin, R. (1984) 'Small island tourism: development or dependency?', *Development Policy Review*, 2(1): 79–90.

Baud-Bovy, M. and Lawson, F. (1998) *Tourism and Recreation Handbook of Planning and Design*, Oxford: Architectural Press.

Bauer, T. and McKercher, B. (2003) *Sex and Tourism: Journeys of Romance, Love and Lust*, New York: Haworth Hospitality Press.

Bayart, J-F. (1993) *The State in Africa: The Politics of the Belly*, Longman: London.

BBC (2015) 'Hong Kong arrests after protest against mainland tourists', *BBC News*, 2 March, available at: www.bbc.com/news/world-asia-china-31689188 (accessed 6 May 2015).

Beaumont, N. (2011) 'The third criterion of ecotourism: are ecotourists more concerned about sustainability than other tourists?', *Journal of Ecotourism*, 10(2): 135–48.

Becken, S. (2008) 'Developing indicators for managing tourism in the face of peak oil', *Tourism Management*, 29(4): 695–705.

Becken, S. (2011) 'A critical review of tourism and oil', *Annals of Tourism Research*, 38(2): 359–79.

Becken, S. (2015) *Tourism and Oil: Preparing for the Challenge*, Bristol: Channel View Publications.

Becken, S., Rajan, R., Moore, S., Watt, M. and McLennan, C. (2013) *White Paper on Tourism and Water*, Griffith University, Queensland: EarthCheck Institute.

Beeton, S. (2006) *Community Development Through Tourism*, Collingwood: Landlinks.

Belk, R. (2014) 'You are what you can access: sharing and collaborative consumption', *Journal of Business Research*, 67(8): 1595–600.

Bell, C. and Newby, H. (1976) 'Communion, communalism, class and community action: the sources of new urban politics', in D. Herbert and R. Johnston (eds), *Social Areas in Cities, Vol. 2*, Chichester: Wiley.

Beresford, M., Nguon, S., Roy, R., Sau, S. and Namazie, C. (2004) *The Macroeconomics of Poverty Reduction in Cambodia*, Phnom Penh, Cambodia: United Nations Development Programme.

Berno, T. and Bricker, K. (2001) 'Sustainable tourism development: the long road from theory to practice', *International Journal of Economic Development*, 3(3): 1–18.

Beyer, M. (2014) *Tourism Planning in Development Cooperation: A Handbook*, Bonn: Deutsche Gesellschaft für Internationale Zusammenarbeit (GIZ) on behalf of Federal Ministry of Economic Cooperation and Development (BMZ), available at: www.giz.de/fachexpertise/downloads/giz2014-en-tourism-handbook.pdf (accessed 30 May 2015).

Bianchi, R. (2015) 'Towards a new political economy of global tourism revisited', in R. Sharpley and D. J. Telfer (eds), *Tourism and Development Concepts and Issues*, 2nd edn, Bristol: Channel View Publications, pp. 287–331.

Bigano, A., Hamilton, J., Lau, M., Tol, R. and Zhou, Y. (2004) *A Global Database of Domestic and International Tourist Numbers at National and Subnational Level*, available at: www.uni-hamburg.de/Wiss/FB/15/Sustainability/tourism-data.pdf.

Bintan Resorts (2006) *Investing in Bintan Resorts*, available at: www.bintan-resort.com (accessed 11 August 2006).

Bjorklund, E. M. and Philbrick, A. K. (1972) 'Spatial configurations of mental processes', unpublished paper, Department of Geography, University of Western Ontario, London, Ontario.

Black, M. and King, J. (2009) *The Atlas of Water: Mapping the World's Most Critical Resource*, London: Earthscan.

Blackstock, K. (2005) 'A critical look at community based tourism', *Community Development Journal*, 40(1): 39–49.

Blewitt, J. (2008) *Understanding Sustainable Development*, London: Earthscan.

Bloom, D., Canning, D. and Fink, G. (2010) 'Implications of population ageing for economic growth', *Oxford Review of Economic Policy*, 26(4): 583–612.

Bloom, D., Boersch-Supan, A., McGee, P. and Seike, A. (2011) *Population Aging: Facts, Challenges, and Responses*, PGDA Working Paper No. 71, available at: www.hsph.harvard.edu/program-on-the-global-demography-of-aging/Working Papers/2011/PGDA_WP_71.pdf (accessed 1 May 2015).

Boissevan, J. (1996) 'Introduction', in J. Boissevan (ed.), *Coping with Tourists: European Reactions to Mass Tourism*, Oxford: Berghan Books, pp. 1–26.

Bookman, M. (2006) *Tourists, Migrants and Refugees: Population Movements in Third World Development*, London: Lynne Rienner.

Borger, J. (2006) 'Half of global car exhaust produced by US vehicles', *The Guardian*, 29 June, available at: www.theguardian.com/environment/2006/jun/29/travelandtransport.usnews (accessed 15 March 2015).

Boulding, K. (1992) 'The economics of the coming spaceship earth', in A. Markandya and J. Richardson (eds), *The Earthscan Reader in Environmental Economics*, London: Earthscan, pp. 27–35.

Bramwell, B. and Lane, B. (1993) 'Sustainable tourism: an evolving global approach', *Journal of Sustainable Tourism*, 1(1): 1–5.

Bramwell, B. and Lane, B. (2000) *Tourism Collaboration and Partnership Politics, Practice and Sustainability*, Clevedon Hall: Channel View Publications.

Bramwell, B. and Lane, B. (2014) 'Critical research on the governance of tourism and sustainability', in B. Bramwell and B. Lane (eds), *Tourism Governance: Critical Perspectives on Governance and Sustainability*, London: Routledge, pp. 1–11.

Bratcher, E. (2014) 'The world's largest cruise ships', *Daily News*, 15 January 2015, available at: www.nydailynews.com/life-style/world-10-largest-cruise-ships-article-1.1581110 (accessed 14 April 2015).

Brewer, J. and Gibson, S. (eds) (2014) *Necessity Entrepreneurs: Microenterprise Education and Economic Development*, Cheltenham: Edward Elgar Publishing.

Briggs, H. (2015) 'Climate change: Paris "last chance" for action', BBC News Science and Environment, 22 April 2015, available at: www.bbc.com/news/science-environment-32399909 (accessed 22 April 2015).

Brinkerhoff, D. W. and Ingle, M. D. (1989) 'Integrating blueprint and process: a structured flexibility approach to development management', *Public Administration and Development*, 9: 487–503.

Britton, S. (1982) 'The political economy of tourism in the Third World', *Annals of Tourism Research*, 9: 331–358.

Britton, S. (1991) 'Tourism, capital and place: towards a critical geography of tourism', *Environment and Planning D: Society and Space*, 9: 451–78.

Brohman, J. (1996) *Popular Development: Rethinking the Theory and Practice of Development*, Oxford: Blackwell.

Brown, D. (2006) 'Gap year projects slammed as out-dated', *Eturbonews*, available at: www.travelwirenews.com/cgi-script/csArticles/articles/00009432-p.htm (accessed 18 August 2006).

Brown, F. (1998) *Tourism Reassessed: Blight or Blessing?* Oxford: Butterworth-Heinemann.

Brown, T. (1999) 'Antecedents of culturally significant tourist behaviour', *Annals of Tourism Research*, 26(3): 676–700.

Brunet, S., Bauer, J., De Lacy, T. and Tshering, K. (2001) 'Tourism development in Bhutan: tensions between tradition and modernity', *Journal of Sustainable Tourism*, 9(3): 243–63.

Bryman, A. (2004) *The Disneyization of Society*, London: Sage.

BTDC (n.d.) *Lombok Island The Mandalika Resorts Area*, available at: www.lombokdream.com/wp-content/uploads/2014/05/Mandalika-Resort-Lombok.pdf (accessed 24 April 2015).

Buckley, R. (2003) *Case Studies in Ecotourism*, Wallingford: CABI.

Buckley, R. (2013) 'Tourism and the sustainability of human societies', *Tourism Recreation Research*, 38(2): 226–31.

Bugsgang, M. (2015) 'Digital tourism', *Tourism: The Journal for the Tourism Industry*, 161(Spring): 20.

Buhalis, D. (2000) 'Marketing the competitive destination of the future', *Tourism Management*, 21(1): 97–116.

Buhalis, D. and Jun, S. (2011) *E-Tourism, Contemporary Tourism Reviews*, available at: www.goodfellowpublishers.com/free_files/fileEtourism.pdf (accessed 13 April 2015).

Buhalis, D. and Ujma, D. (2006) 'Intermediaries: travel agencies and tour operators', in C. Costa and D. Buhalis (eds), *Tourism Management Dynamics, Trends, Management and Tools*, London: Elsevier Butterworth Heinemann, pp. 172–80.

Burns, P. (1999a) 'Editorial – tourism NGOs', *Tourism Recreation Research*, 24(2): 3–6.

Burns, P. (1999b) 'Paradoxes in planning, tourism elitism or brutalism', *Annals of Tourism Research*, 26(2): 329–48.

Burns, P. and Holden, A. (1995) *Tourism: A New Perspective*, Hemel Hempstead: Prentice Hall.

Burns, P. and Novelli, M. (2006) *Tourism and Politics: Global Frameworks and Local Realities*, Oxford: Elsevier.

Burns, P. and Novelli, M. (2008) *Tourism Development, Growth, Myths, and Inequalities*, Wallingford: CABI.

Butcher, J. (2002) *The Moralisation of Tourism: Sun, Sand . . . and Saving the World?*, London: Routledge.

Butler, R. W. (1975) 'Tourism as an agent of social change', in F. Helleiner (ed.), *Tourism as a Factor in National and Regional Development*, Occasional Paper No. 4, Department of Geography, Trent University, Peterborough, Ontario, pp. 85–90.

Butler, R. (1980) 'The concept of a tourist area cycle of evolution', *Canadian Geographer*, 24: 5–12.

Butler, R. (1990) 'Alternative tourism: pious hope or trojan horse?', *Journal of Travel Research*, 28(3): 40–5.

Butler, R. (1993) 'Tourism – an evolutionary perspective', in J. Nelson, R. Butler and G. Wall (eds), *Tourism and Sustainable Development: Monitoring, Planning, Managing*, University of Waterloo: Department of Geography, pp. 27–43.

Butler, R. (2013) 'Sustainable tourism: the undefinable and unachievable pursued by the unrealistic?', *Tourism Recreation Research*, 38(2): 221–6.

Calgaro, E., Lloyd, K. and Dominey-Howes, D. (2014) 'From vulnerability to transformation: a framework for assessing the vulnerability and resilience of tourism destinations', *Journal of Sustainable Tourism*, 22(3): 341–360.

Campbell, F. (1999) 'Whispers and waste', *Our Planet*, 10(3) ('Small Islands'), available at: www.ourplanet.com/imgversn/103/07_whisp.htm (accessed 26 March 2007)

Campbell, L. and Smith, C. (2006) 'What makes them pay? Values of volunteer tourists working for sea turtle conservation', *Environmental Management*, 38(1): 84–98.

Cao, X. (2015) 'Challenges and potential improvements in the policy and regulatory framework for sustainable tourism planning in China: the case of Shanxi Province', *Journal of Sustainable Tourism*, 23(3): 455–76.

Caramel, L. (2014) 'Besieged by the rising tide of climate change, Kirabati buys land in Fiji' *The Guardian*, 1 July 2014, available at: www.theguardian.com/environment/2014/jul/01/kiribati-climate-change-fiji-vanua-levu (accessed 11 May 2015).

Carmichael, B. (2000) 'A matrix model for resident attitudes and behaviours in a rapidly changing tourist area', *Tourism Management*, 21(6): 601–11.

Carson, R. (1962) *Silent Spring*, Boston, MA: Houghton Mifflin.

Cater, C. and Garrod, B. (2015) *Encyclopaedia of Sustainable Tourism*, Wallingford: CABI.

Cater, E. (1993) 'Ecotourism in the Third World: problems for sustainable tourism development', *Tourism Management*, 14(2): 85–93.

Cater, E. (1994) 'Ecotourism in the Third World – problems and prospects for sustainability', in E. Cater and G. Lowman (eds), *Ecotourism a Sustainable Option?*, Chichester: John Wiley & Sons, pp. 69–86.

Cater, E. (2004) 'Ecotourism: theory and practice', in A. Lew, C. M. Hall and A. Williams (eds), *A Companion to Tourism*, Oxford: Blackwell, pp. 484–97.

Cater, E. (2006) 'Ecotourism as a Western construct', *Journal of Ecotourism*, 5(1/2): 23–39.

Cerviño, J. and Cubillo, M. (2005) 'Hotel and tourism development in Cuba', *Cornell Hotel and Restaurant Administration Quarterly*, 46(2): 223–46.

Chakravarty, S. and Irazábal, C. (2011) 'Golden geese or white elephants? The paradoxes of world heritage sites and community based tourism development', *Community Development*, 42(3): 359–76.

Chambers, D. and Airey, D. (2001) 'Tourism policy in Jamaica: a tale of two governments', *Current Issues in Tourism*, 4(2–4): 94–120.

Chambers, R. (1997) *Whose Reality Counts? Putting the First Last*, London: Intermediate Technology Publications.

Chandler, P. (1999) 'Fair Trade in Tourism', paper given at the *Achieving Fairly Traded Tourism Conference*, University of North London, 9 June 1999.

Chauvet, L., Collier, P. and Hoeffler, A. (2011) 'The cost of failing states and the limits to sovereignty', in W. Naudé, A. Santos-Paulino and M. McGillivray (eds), *Fragile States: Causes, Costs, and Responses*, Oxford: Oxford University Press, pp. 91–110.

Chheang, V. (2008) 'The political economy of tourism in Cambodia', *Asia Pacific Journal of Tourism Research*, 13(3): 281–97.

Choi, H. and Sirakaya, E. (2006) 'Sustainability indicators for managing community tourism', *Tourism Management*, 27(6): 1274–89.

Chok, S., Macbeth, J. and Warren, C. (2007) 'Tourism as a tool for poverty alleviation: a critical analysis of "pro-poor tourism" and implications for sustainability', *Current Issues in Tourism*, 10(2/3): 144–65.

Chossudovsky, M. (2008) 'The global crisis: food, water and fuel. Three fundamental necessities of life in jeopardy', Global Research, 5 June, available at: www.globalresearch.ca/the-global-crisis-food-water-and-fuel-three-funda-mental-necessities-of-life-in-jeopardy/9191 (accessed 13 May 2015).

Christie, I., Fernandes, E., Messerli, H. and Twining-Ward, L. (2013) *Tourism in Africa: Harnessing Tourism for Growth and Improved Livelihoods*, Washington, DC: World Bank, available at: www.wds.worldbank.org/external/default/WDSContentServer/WDSP/IB/2013/09/30/000442464_201309301244 58/Rendered/PDF/814680WP0P13260Box0379837B00PUBLIC0.pdf (accessed 23 April 2015).

Clancy, M. (1999) Tourism and development: evidence from Mexico, *Annals of Tourism Research*, 26(1): 1–20.

Cleverdon, R. (2001) 'Introduction: fair trade in tourism – applications and experience', *International Journal of Tourism Research*, 3: 347–9.

Cleverdon, R. and Kalisch, A. (2000) 'Fair trade in tourism', *International Journal of Tourism Research*, 2: 171–87.

Coccossis, H. and Mexa, A. (2004) *The Challenge of Tourism Carrying Capacity Assessment: Theory and Practice*, Wallingford: CABI.

Cohen, E. (1972) 'Towards a sociology of international tourism', *Social Research*, 39(1): 64–82.

Cohen, E. (1979) 'A phenomenonology of tourist experiences', *Sociology*, 13: 179–201.

Cohen, E. (1982) 'Marginal paradises: bungalow tourism on the islands of southern Thailand', *Annals of Tourism Research*, 9(2): 189–228.

Cohen, E. (2007) ' "Authenticity" in tourism studies: "Après la Lutte" ', *Tourism Recreation Research*, 32(2): 75–82.

Cohen, E. (2012) 'Globalisation, global crises and tourism', *Tourism Recreation Research*, 37(2): 103–11.

Colantonio, A. and Potter, R. (2006) 'The rise of urban tourism in Havana since 1989', *Geography*, 91(1): 23–33.

Coldwell, W. (2014) 'Nepal slashes cost of climbing Everest', *The Guardian*, 14 February, available at: www.theguardian.com/world/2014/feb/14/nepal-slashes-cost-climbing-mount-everest (accessed 9 April 2015).

Coles, T. and Church, A. (2007) 'Tourism, politics and the forgotten entanglements of power', in A. Church and T. Coles (eds), *Tourism, Power and Space*, London: Routledge, pp. 1–42.

Constantinescu, I., Dennis, A., Matto, A. and Ruta, M. (2015) 'What lies behind the global trade slowdown? Global economic prospects January 2015', *World Bank*, available at: www.worldbank.org/content/dam/Worldbank/GEP/GEP2015a/pdfs/GEP2015a_chapter4_report_trade.pdf (accessed 10 April 2015).

Cooper, M. (2003) 'The real Cancun: behind globalization's glitz', *The Nation*, 22 September.

Cooper, C., Fletcher, J., Fyall, A., Gilbert, D. and Wanhill, S. (2005) *Tourism: Principles and Practice*. Harlow: Pearson Education.

Cooperative Bank (2005) *The Ethical Consumerism Report 2005*, Manchester: The Cooperative Bank.

Cornet, C. (2015) 'Tourism development and resistance in China', *Annals of Tourism Reseearch*, 52(May): 29–43.

Costa, C. (2006) 'Tourism planning, development and territory', in D. Buhalis and C. Costa (eds), *Tourism Management Dynamics Trends, Management and Tools*, Oxford: Elsevier Butterworth Heineman, pp. 236–43.

Costa, C. and Buhalis, D. (2006) 'Introduction', in C. Costa and D. Buhalis (eds), *Tourism Management Dynamics, Trends, Management and Tools*, London: Elsevier Butterworth Heinemann, pp. 1–5.

Cowe, R. and Williams, S. (2000) *Who Is the Ethical Consumer?*, Manchester: Co-Operative Bank.

Cowen, M. and Shenton, R. (1996) *Doctrines of Development*, London: Routledge.

Croall, J. (1995) *Preserve or Destroy: Tourism and the Environment*, London: Calouste Gulbenkian Foundation.

Crompton, J. (1979) 'Motivations for pleasure vacation', *Annals of Tourism Research*, 6(4): 408–24.

Cronin, L. (1990) 'A strategy for tourism and sustainable developments', *World Leisure and Recreation*, 32(3): 12–18.

Cruise Market Watch (2015) *2015 World Wide Market Share*, available at: www.cruisemarketwatch.com/market-share/ (accessed 11 April 2015).

CST (2015) Sustainable Tourism CST: Certification for Sustainable Tourism in Costa Rica, available at: www.turismo-sostenible.co.cr/en (accessed 29 April 2015).

Cukier, J. (2002) 'Tourism employment issues in developing countries: examples from Indonesia', in R. Sharpely and D. J. Telfer (eds), *Tourism and Development: Concepts and Issues*, Clevedon: Channel View Publications, pp. 165–201.

Cukier, J. and Wall, G. (1994) 'Informal tourism employment: vendors in Bali, Indonesia', *Tourism Management*, 15(6): 464–7.

D'Amico, B. (2005) *A Touch of Africa: Part II – On to the Amazon*, Bloomington, IN: Authorhouse.

Daher, R. (2005) 'Urban regeneration/heritage tourism endeavours: the case of Salt, Jordan – local actors, international donors and the state', *International Journal of Heritage Studies*, 11(4): 289–308.

Dann, G. (1981) 'Tourist motivation: an appraisal', *Annals of Tourism Research*, 8(2): 187–219.

Dar, R. and Rashid, I. (2014) 'Sustainability of winter tourism in a changing climate over Kashmir Himalaya', *Environmental Monitoring & Assessment*, 186: 2549–62.

Darma Putra, I. and Hitchcock, M. (2006) 'The Bali bombs and the tourism development cycle', *Progress in Development Studies*, 6(2): 157–66.

Dasgupta, P. and Weale, M (1992) 'On measuring the quality of life', *World Development*, 20(1): 119–31.

Davidson, L. and Sahli, M. (2015) 'Foreign direct investment in tourism, poverty alleviations and sustainable development: a review of the Gambian hotel sector', *Journal of Sustainable Tourism*, 23(2): 167–87.

Davis, H. D. and Simmons, J. A. (1982) 'World Bank experience with tourism projects', *Tourism Management*, 3(4): 212–17.

Davison, N. (2015) 'The ultimate package holiday: Chinese billionaire takes 6,400 staff to France', *Daily Telegraph*, 10 May. Available at: www.telegraph.co.uk/news/worldnews/asia/china/11595649/The-ultimate-package-holiday-Chinese-billionaire-takes-6400-staff-to-France.html (accessed 12 May 2015).

de Araujo, L. and Bramwell, B. (1999) 'Stakeholder assessment and collaborative tourism planning: the case of Brazil's Costa Dourada Project', *Journal of Sustainable Tourism*, 7(3/4): 356–78.

de Grosbois, D. (2012) 'Corporate social responsibility reporting by the global hotel industry: commitment, initiatives, and performance', *International Journal of Hospitality Management*, 31(3): 896–905.

de Kadt, E. (1979) *Tourism: Passport to Development?*, New York: Oxford University Press.

de la Dehesa, G. (2006) *Winners and Losers in Globalisation*, Oxford: Blackwell Publishing.

de Rivero, O. (2001) *The Myth of Development: Non-Viable Economies of the 21st Century*, London: Zed Books.

Deery, M., Jago, L. and Fredline, L. (2012) 'Rethinking social impacts of tourism research: a new research agenda', *Tourism Management*, 33(1): 64–73.

Dela Santa, E. (2015) 'The evolution of Philippine tourism policy implementation from 1973 to 2009', *Tourism Planning and Development*, 12(2): 155–75.

Dezan Shira & Associates (2013) 'Foreign direct investment in Indian tourism', *India Briefing*, 19 April 2013, available at: www.india-briefing.com/news/foreign-direct-investment-indian-tourism-6153.html/ (accessed 23 April 2015).

Di John, J. (2010) 'The concept, causes and consequences of failed states: a critical review of the literature and agenda for research with specific reference to sub-Saharan Africa', *European Journal of Development Research*, 22(1): 10–30.

Diamantis, D. (1999) 'Green strategies for tourism worldwide', *Travel & Tourism Analyst*, 4: 89–112.

Diamantis, D. (2004) 'Ecotourism management: an overview', in D. Diamantis (ed.), *Ecotourism: Management and Assessment*, London: Thomson, pp. 1–26.

Diamond, J. (1977) 'Tourism's role in economic development: the case re-examined, *Economic Development and Cultural Change*, 25(3): 539–53.

Digence, J. (2003) 'Pilgrimage at contested sites', *Annals of Tourism Research*, 30(1): 143–59.

Dogan, H. (1989) 'Forms of adjustment: sociocultural impacts of tourism', *Annals of Tourism Research*, 16(2): 216–36.

Dollar, D. (2005) 'Globalization, poverty and inequality', in M. Weinstein (ed.), *Globalization: What's New?*, New York: Columbia University Press, pp. 96–128.

Dolnicar, S., Crouch G. and Long, P. (2008) 'Environment-friendly tourists: what do we really know about them?', *Journal of Sustainable Tourism*, 16(2): 197–210.

Dorji, T. (2001) 'Sustainability of tourism in Bhutan', *Journal of Bhutan Studies*, 3(1): 84–104.

Dowling, R. (1992) 'Tourism and environmental integration: the journey from idealism to realism', in C. Cooper and A. Lockwood (eds), *Progress in Tourism, Recreation and Hospitality Management, Vol. 4*, London: Bellhaven Press, pp. 33–46.

Doxey, G. (1975) 'A causation theory of visitor-resident irritants: methodology and research inferences', *Proceedings of the Travel Research Association*, 6th Annual Conference, San Diego, pp. 195–8.

Doxey, G. (1976) 'When enough's enough: the natives are restless in Old Niagara', *Heritage Canada*, 2(2): 26–7.

Dredge, D. and Jenkins, J. (2011a) 'New spaces of tourism planning and policy', in D. Dredge and J. Jenkins (eds), *Stories of Practice: Tourism Policy and Planning*, Farnham, Surrey: Ashgate, pp. 1–12.

Dredge, D. and Jenkins, J. (2011b) 'Conclusion', in D. Dredge and J. Jenkins (eds), *Stories of Practice: Tourism Policy and Planning*, Farnham, Surrey: Ashgate, pp. 359–68.

Dredge, D., Jenkins, J. and Whitford, M. (2011) 'Tourism planning and policy: historical development and contemporary challenges', in D. Dredge and J. Jenkins (eds), *Stories of Practice: Tourism Policy and Planning*, Farnham, Surrey: Ashgate, pp. 13–36.

Dresner, S. (2002) *The Principles of Sustainability*, London: Earthscan.

Dubai Attractions (2015) *Exploring New Tourism Related Development in Dubai*, available at: www.dubaiattractions.com/dubai-development.html#sthash.H3M FQg3N.dpbs (accessed 15 February 2015).

Duffy, R. (2000) 'Shadow players: ecotourism development, corruption and state politics in Belize', *Third World Quarterly*, 21(3): 549–65.

Duffy, R. (2002) *A Trip Too Far*, London: Earthscan.

Duffy, R. (2006) 'The politics of ecotourism and the developing world', *Journal of Ecotourism*, 5(1/2): 1–6.

Dwyer, G. (2008) *Climate Wars*, Toronto: Random House Canada.

Easterly, W. (2005) 'The rich have markets, the poor the bureaucrats' in M. Weinstein (ed.), *Globalization: What's New?*, New York: Columbia University Press, pp. 170–95.

Eber, S. (1992) *Beyond The Green Horizon: Principles for Sustainable Tourism*, Godalming: WWF.

EC (1993) *Taking Account of Environment in Tourism Development*, DG XXIII Tourism Unit, Luxembourg: Commission of the European Communities.

Echtner, C. and Prasad, P. (2003) 'The context of Third World tourism marketing', *Annals of Tourism Research*, 30(3): 660–82.

Economy Watch (2010) *Types of Financial Markets*, available at: www.economy watch.com/market/market-types/financial-market-types.html (accessed 14 May 2015).

Elijah-Mensah, A. (2009) 'Motivation and age: an empirical study of women-owners of tourism ventures in Ghana', *Journal of Travel and Tourism Research*, 9(2): 139–53.

Elliott, J. (1997) *Tourism and Public Policy*, London: Routledge.

Elliott, J. (2013) *An Introduction to Sustainable Development*, 4th edn, Abingdon: Routledge.

Eshun, G. and Tagoe-Darko, E. (2015) 'Ecotourism development in Ghana: a post colonial analysis', *Development Southern Africa*, 32(3): 392–406.

ETB (1991) *The Green Light: A Guide to Sustainable Tourism*, London: English Tourist Board.

ETN (2008) 'Haiti's tourism dreams deferred by riots', *ETN Global Travel Industry News*, 2 June, available at: www.eturbonews.com/2769/haitis-tourism-dreams-deferred-riots (accessed 13 May 2015).

eTN Global Travel Industry News (2015) *Pegasus Expands Its UK Network, Launches London Gatwick-Istanbul Route*, available at: www.eturbonews.com/57447/pegasus-expands-its-uk-network-launches-london-gatwick-istanbul- (accessed 10 April 2015).

European Commission (2009) *Economic Crisis in Europe: Causes, Consequences and Responses*, Brussels: European Commission, available at: http://ec.europa.eu/economy_finance/publications/publication15887_en.pdf (accessed 14 May 2015).

Evans, G. (2005) 'Mundo Maya: from Cancún to city of culture: world heritage in post-colonial Mesoamerica', in D. Harrison and M. Hitchcock (eds), *The Politics of World Heritage Negotiating Tourism and Conservation*, Clevedon: Channel View Publications, pp. 35–49.

Evans, G. and Cleverdon, R. (2000) 'Fair trade in tourism: community development or marketing tool', in D. Hall and G. Richards (eds), *Tourism and Sustainable Community Development*, London: Routledge, pp. 137–53.

FAO (2015) *The State of Food Insecurity in the World 2015*, Food and Agriculture Organization of the UN, available at: www.fao.org/3/a-i4646e/index.html (accessed 14 May 2015).

FAO, IFAD and WFP (2014) *The State of Food Insecurity in the World 2014: Strengthening the Enabling Environment for Food and Security*, Rome: FAO, available at: www.fao.org/3/a-i4037e.pdf (accessed 18 May 2015).

Farver, J. (1984) 'Tourism and employment in the Gambia', *Annals of Tourism Research*, 11(2): 249–65.

Faulkner, B. (2001) 'Towards a framework for tourism disaster management', *Tourism Management*, 22(2): 135–47.

Fennell, D. (1999) *Ecotourism: An Introduction*. London: Routledge.

Fennell, D. (2008) *Ecotourism*, 3rd edn, Abingdon: Routledge.

Figueroa, A. (2015) 'Cancun travel continues to grow', *Travel Agent*, 23 February 2015, available at: www.travelagentcentral.com/cancun/cancun-travel-continues-grow-49905 (accessed 11 April 2015).

Fletcher, J., Fyall, A., Gilbert, D. and Wanhill, S. (2013) *Tourism: Principles and Practice*, 5th edn, Harlow: Pearson Education.

Flora, J. (1998) 'Social capital and communities of place', *Rural Sociology*, 63(4): 481–506.

FoE (2014) *Friends of the Earth News Release*, available at: www.foe.org/news/news-releases/2013-10-cruise-ships-flushed-more-than-1-billion-gallons-of-sewage-last-year#sthash.Sot7Q1Dd.dpuf (accessed 4 March 2015).

Fonatur (2006) *About Fonatur*, available at: www.fonatur.gob.mx (accessed 21 September 2006).

Foo, J., McGuiggan, R. and Yiannakis, A. (2004) 'Roles tourists play: an Australian perspective', *Annals of Tourism Research*, 31(2): 408–27.

Forsyth, T. (1995) 'Business attitudes to sustainable tourism: self-regulation in the UK outgoing tourism industry', *Journal of Sustainable Tourism*, 3(4): 210–31.

Fox News Latino (2013) *422 Hotels to Be Built in Brazil Ahead of 2016 Olympics*, available at: http://latino.foxnews.com/latino/money/2013/11/22/422-hotels-to-be-built-in-brazil-ahead-2016-olympics/ (accessed 26 April 2015).

Freeman, M. (2005) 'Putting law in its place: an interdisciplinary evaluation of national amnesty laws', in S. Meckled-Garcia and B. Cali (eds), *The Legalization of Human Rights: Multidisciplinary Perspectives on Human Rights and Human Rights Law*, New York: Routledge, pp. 45–59.

Freitag, T. (1984) 'Enclave tourism development: for whom the benefits roll?', *Annals of Tourism Research*, 21(3): 538–54.

Frenzel, F. (2014) 'Slum tourism and urban regeneration: touring inner Johannesburg', *Urban Forum*, 25(4): 431–47.

FTT (2015) *Get Certified, Fair Trade Tourism*, available at: www.fairtrade.travel/get-certified (accessed 7 May 2015).

Funke, N. and Solomon, H. (2002) *The Shadow State in Africa: A Discussion*, DPMF Occasional Paper, No. 5, available at: www.dpmf.org/Publications/Occassional%20Papers/occasionalpaper5.pdf (accessed 5 June 2012).

fvw (2015a) *TUI, Cook Dominate European Tourism Market*, available at: www.fvw.com/european-tour-operators-ranking-tui-cook-dominate-european-tourism-market/393/126948/11245 (accessed 10 April 2015).

fvw (2015b) *TUI Closer Airline and Cruise Cooperation*, available at: www.fvw.com/index.cfm?cid=11245&pk=140392&event=showarticle (accessed 11 April 2015).

Galtung, J. (1986) 'Towards a new economics: on the theory and practice of self-reliance', in P. Ekins (ed.), *The Living Economy: A New Economy in the Making*, London: Routledge, pp. 97–109.

Garrod, B. and Fennell, D. (2004) 'An analysis of whalewatching codes of conduct', *Annals of Tourism Research*, 31(2): 334–52.

Gartner, W. (2004) 'Factors affecting small firms in tourism: a Ghanaian perspective', in R. Thomas (ed.), *Small Firms in Tourism: International Perspectives*, Oxford: Elsevier, pp. 35–70.

Gee, E. (2002) 'Misconceptions and misapprehensions about population ageing', *International Journal of Epidemiology*, 31: 750–3.

Gerritsen, P. (2014) 'Working with indigenous women on multifunctionality and sustainable rural tourism development in Western Mexico', *Journal of Rural & Community Development*, 9(3): 243–57.

Getz, D. (1987) 'Tourism planning and research: traditions, models and futures', paper presented at the Australian Travel Research Workshop, Bunbury, Western Australia, 5–6 November.

Ghani, A. and Lockhart, C. (2008) *Fixing Failed States*, Oxford: Oxford University Press.

Ghimire, K. (2001) 'The growth of national and regional tourism in developing countries: an overview', in K. Ghimire (ed.), *The Native Tourist: Mass Tourism within Developing Countries*, London: Earthscan, pp. 1–29.

Giampiccoli, A., Jugmohan, S. and Mtapuri, O. (2014) 'International cooperation, community-based tourism and capacity building: results from a Mpondoland village in South Africa', *Mediterranean Journal of Social Science*, 5(2): 657–67.

Giddings, B., Hopwood, B. and O'Brien, G. (2002) 'Environment, economy and society: fitting them together into sustainable development', *Sustainable Development*, 10(4): 187–96.

GIEWS (2015) 'Countries requiring external assistance for food, March 2015', *Food and Agricultural Organisation of the United Nations*, available at: www.fao.org/Giews/english/hotspots/index.htm (accessed 13 May 2015).

Glaesser, D. (2006) *Crisis Management in the Tourism Industry*, Oxford: Butterworth-Heinemann.

Global Business Guide Indonesia (2013) *Indonesia's Tourism Industry and the Creative Economy*, available at: www.gbgindonesia.com/en/services/article/2012/indonesia_s_tourism_industry_and_the_creative_economy.php (accessed 24 April 2015).

Go, F. and Pine, R. (1995) *Globalisation Strategy in the Hotel Industry*, London: Routledge.

Godfrey, K. (1996) 'Towards sustainability? Tourism in the republic of Cyprus', in L. Harrison and W. Husbands (eds), *Practising Responsible Tourism: International Case Studies in Tourism Planning, Policy and Development*, Chichester: John Wiley & Sons, pp. 58–79.

GOL (2015) *Profile*, available at: www.voegol.com.br/en-zw/a-gol/quem-somos/perfil/paginas/default.aspx (accessed 13 April 2015).

Gold, J. (1980) *An Introduction to Behavioural Geography*, Oxford: Oxford University Press.

Goldsworthy, D. (1988) 'Thinking politically about development', *Development and Change*, 19(3): 505–30.

Goodland, R. (1992) 'The case that the world has reached its limits', in R. Goodland, H. Daly, S. Serafy and B. von Droste (eds), *Environmentally Sustainable Economic Development: Building on Brundtland*, Paris: UNESCO, pp. 15–27.

Goodwin, H. (2011) *Taking Responsibility for Tourism*, Oxford: Goodfellow Publishers.

Gössling, S., Hall, C.M. and Scott, D. (2015) *Tourism and Water*, Bristol: Channel View Publications.

Gössling, S., Peeters, P., Ceron, J., Dubois, G., Patterson, T. and Richardson, R. (2005) 'The eco-efficiency of tourism', *Ecological Economics*, 54(4): 417–34.

Gössling, S., Peeters, P., Hall, C.M., Ceron, J-P., Dubois, G., Lehmann, L.V. and Scott, D. (2012) 'Tourism and water use: supply, demand, and security. An international review', *Tourism Management*, 33(1): 1–15.

Gössling, S. and Schultz, U. (2005) 'Tourism-related migration in Zanzibar, Tanzania', *Tourism Geographies*, 7(1): 43–62.

Goudie, A. and Viles, H. (1997) *The Earth Transformed: An Introduction to Human Impacts on the Environment*, Oxford: Blackwell.

Goulet, D. (1992) 'Participation in development: new avenues', *World Development*, 17(2): 165–78.

Government of Nepal (2013) *Nepal Tourism Statistics*, Ministry of Culture, Tourism & Civil Aviation, available at: www.tourism.gov.np/uploaded/TourrismStat2012.pdf (accessed 9 April 2015).

Graham, A. (2006) 'Transport and transit: air, land and sea', in C. Costa and D. Buhalis (eds), *Tourism Management Dynamics, Trends, Management and Tools*, London: Elsevier Butterworth Heinemann, pp. 181–90.

Gray, H. (1970) *International Travel – International Trade*, Lexington, KY: DC Heath.

Grihault, N. (2011) *Country Report: Cambodia*, London: Mintel.

GSTC (2013) *Global Sustainable Tourism Council Criteria for Destinations*, available at: www.gstcouncil.org/gstc-criteria/criteria-for-destinations.html (accessed 23 April 2015).

GSTC (2015) *Welcome to the Global Sustainable Tourism Council*, available at: www.gstcouncil.org/about/learn-about-gstc.html (accessed 30 January 2015).

Gulf Times (2014) *Qatar Tourism Grows Further in 2013*, available at: www.gulf-times.com/Mobile/Qatar/178/details/386831/Qatar-tourism-grows-further-in-2013 (accessed 10 January 2015).

Gunn, C. and Var, T. (2002) *Tourism Planning: Basics, Concepts, Cases*, 4th edn, New York: Routledge.

Gurung, D. and Seeland, K. (2011) 'Ecotourism benefits and livelihood improvement for sustainable development in the nature conservation areas of Bhutan', *Sustainable Development*, 19(5): 348–58.

Guttentag, D. (2009) 'The possible negative impacts of volunteer tourism', *International Journal of Tourism Research*, 11(6): 537–51.

Hall, C. M. (1994) *Tourism and Politics: Policy, Power and Place*, Chichester: John Wiley & Sons.

Hall, C. M. (2002) 'Local initiatives for local regional development: the role of food, wine and tourism', in E. Arola, J. Kärkkäinen and M. Siltari (eds), *Tourism and Well Being: The 2nd Tourism Industry and Education Symposium*, 16–18 May, Jyväsklä, Finland Symposium Proceedings, Jyväsklä Polytechnic, pp. 47–63.

Hall, C. M. (2005) *Tourism: Rethinking the Social Science of Mobility*, Harlow: Pearson Prentice Hall.

Hall, C. M. (2007) 'Editorial, pro-poor tourism: do "tourism exchanges benefit primarily the countries of the south?"', *Current Issues in Tourism*, 10(2/3): 111–18.

Hall, C. M. (2008) *Tourism Planning: Policies, Processes and Relationships*, 2nd edn, Harlow: Prentice Hall.

Hall, C. M. and Jenkins, J. (1998) 'The policy dimensions of rural tourism and recreation', in R. Butler, C. M. Hall and J. Jenkins (eds), *Tourism and Recreation in Rural Areas*, Chichester: Wiley, pp. 19–42.

Hall, C. M. and Jenkins, J. (2004) 'Tourism and public policy', in A. Lew, C. M. Hall and A. M. Williams (eds), *A Companion to Tourism*, Oxford: Blackwell, pp. 525–40.

Hall, C. M. and Page, S. (2006) *The Geography of Tourism and Recreation*, 3rd edn, London: Routledge.

Hall, C. M. and Lew, A. (2009) *Understanding and Managing Tourism Impacts: An Integrated Approach*, Abingdon: Routledge.

Hall, C. M., Scott, D. and Gössling, S. (2013) 'The primacy of climate change for sustainable international tourism', *Sustainable Development*, 21(2): 112–21.

Hall, C. M., Scott, D. and Gössling, S. (2015) 'Tourism, climate change and development', in R. Sharpley and D. Telfer (eds), *Tourism and Development: Concepts and Issues*, 2nd edn, Bristol: Channel View Publications, pp. 332–57.

Hall, D. (2000) 'Identity, community and sustainability prospects for rural tourism in Albania', in G. Richards and D. Hall,(eds), *Tourism and Sustainable Community Development*, London: Routledge, pp. 48–59.

Hambira, W. and Saarinen, J. (2015) 'Policy-makers' perceptions of the tourism-climate change nexus: policy needs and constraints in Botswana', *Development Southern Africa*, 32(3): 350–62.

Hannam, K., Sheller, M. and Urry, J. (2006) 'Editorial: mobilities, immobilities and moorings', *Mobilities*, 1(1): 1–22.

Hard Rock Café (2015) *Hard Rock Locations*, available at: www.hardrock.com/locations.aspx (accessed 9 April 2015).

Hardin, G. (1968) 'The tragedy of the commons', *Science*, 162: 1243–8.

Harrigan, J. and Mosley, P. (1991) 'Evaluating the impact of World Bank structural adjustment lending', *Journal of Development Studies*, 27(3): 63–94.

Harrison, D. (1988) *The Sociology of Modernisation and Development*, London: Routledge.

Harrison, D. (ed.) (2001) *Tourism and the Less Developed World: Issues and Case Studies*, New York: CABI.

Harrison, D. (2008) 'Pro-poor tourism: a critique', *Third World Quarterly*, 29(5): 851–68.

Harrison, D. and Price, M. (1996) 'Fragile environments, fragile communities? An introduction', in M. Price and V. Smith (eds) *People and Tourism in Fragile Environments*, Chichester: John Wiley & Sons, pp. 1–18.

Harrison, D. and Schipani, S. (2007) 'Lao tourism and poverty alleviation: community-based tourism and the private sector', *Current Issues in Tourism*, 10(2/3): 194–230.

Harvey, D. (1989) *The Condition of Postmodernity*, Oxford: Basil Blackwell.

Hashimoto (2004) Personal communication, Human Rights and Tourism Lecture, Brock University.

Hashimoto, A. (2015) 'Tourism and sociocultural development issues', in R. Sharpley and D. J. Telfer (eds), *Tourism and Development Concepts and Issues*, Bristol: Channel View Publications, pp. 205–36.

Hashimoto, A. and Telfer, D. J. (2007) 'Geographic representations embedded within souvenirs in Niagara: the case of geographically displaced authenticity', *Tourism Geographies*, 9(2): 191–217.

Hasler, H. and Ott, J. (2008) 'Diving down the reefs? Intensive diving tourism threatens the reefs of the northern Red Sea', *Marine Pollution Bulletin*, 56(10): 1788–94.

Hatton, M. (1999) *Community-Based Tourism in the Asia-Pacific*, Toronto: Canadian Tourism Commission, Asia-Pacific Economic Cooperation and Canadian International Development Agency.

Hawkins, D. and Mann, S. (2007) 'The World Bank's role in tourism development', *Annals of Tourism Research*, 34(2): 348–63.

Hazbun, W. (2004) 'Globalisation, reterritorialisation and the political economy of tourism development in the Middle East', *Geopolitics*, 9(2): 310–41.

Hecho En GeoCuba (1997) Las Terrazas Complejo Turistico Tourist Map.

Hedlund, T. (2011) 'The impact of values, environmental concern, and willingness to accept economic sacrifices to protect the environment on tourists' intentions to buy ecologically sustainable tourism alternatives', *Tourism and Hospitality Research*, 11(4): 278–88.

Held, D. (2010) *Cosmopolitanism Ideals and Realities*, Malden, MA: Polity Press.

Helman, G. and Ratner, R. (1993) 'Saving failed states', *Foreign Policy*, 89: 3–20.

Henderson, J. (2003) 'The politics of tourism in Myanmar', *Current Issues in Tourism*, 6(2): 97–118.

Henderson, J. (2014) 'Global Gulf cities and tourism: a review of Abu Dhabi, Doha and Dubai', *Tourism Recreation Research*, 39(1): 107–14.

Herrera, A. (2013) 'Heritage tourism, identity and development in Peru', *International Journal of Historical Archaeology*, 17: 275–95.

Hettne, B. (2009) *Thinking about Development*, London: Zed Books.

Hiernaux-Nicolas, D. (1999) 'Cancún bliss', in D. Judd and S. Fainstein (eds), *The Tourist City*, New Haven, CT: Yale University Press, pp. 124–39.

Hinch, T. and Butler, R. (1996) 'Indigenous tourism: a common ground for discussion', in R. Butler and T. Hinch (eds), *Tourism and Indigenous Peoples*, London: International Thomson Business Press, pp. 3–19.

Hinch, T. and Butler, R. (2007) 'Introduction: revisiting common ground', in T. Hinch and R. Butler (eds), *Tourism and Indigenous Peoples: Issues and Implications*, Oxford: Butterworth-Heinemann, pp. 1–15.

Hitchcock, M. and Darma Putra, I. (2005) 'The Bali bombings: tourism crisis management and conflict avoidance', *Current Issues in Tourism*, 8(1): 62–76.

Hjalager, A. (2007) 'Stages in the economic globalization of tourism', *Annals of Tourism Research*, 34(2): 437–57.

Hof, A. and Schmidt, T. (2011) 'Urban and tourist land use patterns and water consumption: evidence from Mallorca, Balearic Islands', *Land Use Policy*, 28(4): 792–804.

Høivik, T. and Heiberg, T. (1980) 'Centre-periphery tourism and self-reliance', *International Social Science Journal*, 32(1): 69–98.

Holden, A. (2000) *Environment and Tourism*. London: Routledge.

Holden, A. (2007) *Environment and Tourism*, 2nd edn, Abingdon: Routledge.

Holden, A. (2013) *Tourism Poverty and Development*, London: Routledge.

Holladay, P. and Powell, R. (2013) 'Resident perceptions of social-ecological resilience and the sustainability of community based tourism development in the Commonwealth of Dominica', *Journal of Sustainable Tourism*, 21(8): 118–211.

Holloway, J.C. and Humphreys, C. (2012) *The Business of Tourism*, 9th edn, Harlow: Pearson Education.

Holt, D. (1995) 'How consumers consume: a typology of consumption practices', *Journal of Consumer Research*, 22(June): 1–16.

Hoogvelt, A. (1997) *Globalisation and the Postcolonial World*, London: Macmillan Press.

Horan, J. (2002) 'Indigenous wealth and development: micro-credit schemes in Tonga', *Asia Pacific Viewpoint*, 43(2): 205–21.

Howie, F. (2003) *Managing the Tourist Destination*, London: Thomson.

Høyer, K. (2000) 'Sustainable tourism or sustainable mobility? The Norwegian case', *Journal of Sustainable Tourism*, 8(2): 147–60.

Hsu, C. and Huang, S. (2008) 'Travel motivation: a critical review of the concept's development', in A. Woodside and D. Martin (eds), *Tourism Management: Analysis, Behaviour and Strategy*, Wallingford: CABI, pp. 14–27.

Hughes, G. (1995) 'Authenticity in tourism', *Annals of Tourism Research*, 22(4): 781–803.

Hunter, C. (1995) 'On the need to re-conceptualise sustainable tourism development', *Journal of Sustainable Tourism*, 3(3): 155–65.

Hunter, C. and Green, H. (1995) *Tourism and the Environment: A Sustainable Relationship?*, London: Routledge.

Huybers, T. (ed.) (2007) *Tourism in Developing Countries*, Cheltenham: Edward Elgar.

IBRD/World Bank (2007) *A Decade of Measuring the Quality of Governance*, Washington, DC: International Bank for Reconstruction & Development/World Bank, available at: http://info.worldbank.org/governance/wgi/index.aspx#doc (accessed 28 April 2015).

IDB (2015) *BR-L1211: Tourism Development Program – Rio Grande do Norte State*, Inter-American Development Bank, available at: www.iadb.org/en/projects/project-description-title,1303.html?id=BR-L1211 (accessed 25 April 2015).

IEP (2014) *Global Terrorism Index 2014: Measuring and Understanding the Impact of Terrorism*, Institute for Economic and Peace, available at: http://economicsandpeace.org/research/iep-indices-data/global-peace-index (accessed 5 April 2015).

IFTO (1994) *Planning for Sustainable Tourism: The ECOMOST Project*, Lewes: International Federation of Tour Operators.

Iliau, R. (1997) 'The chaining role of women in Tonga', unpublished MA dissertation, University of Auckland.

IMF (2006) *Poverty Reduction Strategy Papers*, available at: www.imf.org/external/np/prsp/prsp.aspx (accessed 1 August 2006).

IMF (2008) *Haiti: Poverty Reduction Strategy Paper*, Washington, DC: International Monetary Fund.

IMF (2009) *Initial Lessons of the Crisis, 6 Feb*, available at: www.imf.org/external/np/pp/eng/2009/020609.pdf (accessed 14 May 2015).

India eNews (2006) *Bengal to Woo Foreign Investment in Tourism*, available at: www.indiaenews.com (accessed 17 August 2006).

Inskeep, E. (1991) *Tourism Planning: An Integrated and Sustainable Development Approach*, New York: Van Nostrand Reinhold.

Inskeep, I. and Kallenberger, M. (1992) *An Integrated Approach to Resort Development: Six Case Studies*, Madrid: World Tourism Organization.

Institution of Mechanical Engineers (2015) Mitigation, Adaptation and Geo-Engineering, available at: www.imeche.org/knowledge/themes/environment/climate-change/mitigation-adaptation-geo-engineering (accessed 12 May 2015).

InterContinental (2015) *IHG Hotel & Room World Stats*, available at: www.ihgplc.com/files/pdf/factsheets/factsheet_worldstats.pdf (accessed 23 April 2015).

International Energy Statistics (2015) *Total Oil Supply*, available at: www.eia.gov/cfapps/ipdbproject/iedindex3.cfm?tid=5&pid=53&aid=1&cid=ww,&syid=2000&eyid=2014&unit=TBPD (accessed 10 May 2015).

Internet World Stats (2015) *Internet Growth Statistics*, available at: www.internetworldstats.com/emarketing.htm (accessed 13 April 2015).

Ioannides, D. and Debbage, K. (1998) 'Neo-Fordism and flexible specialisation', in D. Ioannides and K. Debbage (eds), *The Economic Geography of the Tourist Industry*, London: Routledge, pp. 99–122.

Iorio, M. and Corsale, A. (2014) 'Community-based tourism and networking: Viscri, Romania', *Journal of Sustainable Tourism*, 22(2): 234–55.

IPCC (2014) *Climate Change 2014 Impacts, Adaptation, and Vulnerability, Summary for Policy Makers*, Working Group II Contribution to the Fifth Assessment Report of the Intergovernmental Panel on Climate Change, available at: www.ipcc.ch/pdf/assessment-report/ar5/wg2/ar5_wgII_spm_en.pdf (accessed 12 April 2015).

Issa, J. and Jayawardena, C. (2003) 'The "all-inclusive" concept in the Caribbean', *International Journal of Contemporary Hospitality Management*, 15(3): 167–71.

IUCN (1980) *World Conservation Strategy: Living Resources Conservation for Sustainable Development*, Gland, Switzerland: World Conservation Union.

IUCN (1991) *Caring for the Earth: A Strategy for Sustainable Living*, Gland, Switzerland: World Conservation Union.

Jackson, G. and Morpeth, N. (1999) 'Local Agenda 21 and community participation in tourism policy and planning: future or fallacy', *Current Issues in Tourism*, 2(1): 1–38.

Jafari, J. (1989) 'Sociocultural dimensions of tourism: an English language literature review', in J. Bystranowski (ed.), *Tourism as a Factor of Change: A Sociocultural Study*, Vienna: Vienna Centre, pp. 17–60.

Jamal, T. and Dredge, D. (2015) 'Tourism and community development issues', in R. Sharpley and D. Telfer (eds), *Tourism and Development: Concepts and Issues*, 2nd edn, Bristol: Channel View Publications, pp. 178–204.

Jamal, T. and Getz, D. (1995) 'Collaboration theory and community tourism planning', *Annals of Tourism Research*, 22(1): 186–204.

Jamal, T. and Jamrozy, U. (2006) 'Collaborative networks and partnerships for integrated destination management', in D. Buhalis and C. Costa (eds), *Tourism Management Dynamics: Trends, Management and Tools*, Oxford: Elsevier Butterworth Heinemann, pp. 164–172.

Jamal, T., Borgers, M. and Stronza, A. (2006) 'The institutionalisation of ecotourism: certification, cultural equity and praxis', *Journal of Ecotourism*, 5(3): 145–75.

James, P. (2006) *Globalism, Nationalism and Tribalism: Bringing Theory Back In*, London: Sage.

Jayoti, D. and DiRienzo, C. (2010) 'Tourism competitiveness and corruption: a cross country analysis', *Tourism Economics*, 16(3): 477–92.

JCP Inc. (1987) *Nusa Tenggara, Tourism Development Plan for Lombok*, Tokyo: JCP Inc. Planners, Architects & Consulting Engineers.

Jenkins, C. (1980) 'Tourism policies in developing countries: a critique', *International Journal of Tourism Management*, 1(1): 22–9.

Jenkins, C. (1991) 'Development strategies', in L. Lickorish, A. Jefferson, J. Bodlender and C. Jenkins (eds), *Developing Tourist Destinations*, London: Longman, pp. 59–118.

JLL Hotels and Hospitality Group (2015) *Hotel Investment Outlook 2015*, available at: www.jll.com/Research/JLL%20Hotel%20Investment%20Outlook%202015.pdf?ee1160d5-9c6b-4761-b690-eddc19c4a7bd (accessed 11 April 2015).

Johansson, Y. and Diamantis, D. (2004) 'Ecotourism in Thailand and Kenya: a private sector perspective', in D. Diamantis (ed.), *Ecotourism Management and Assessment*, London: Thomson, pp. 298–312.

Johnson, L. (2014) 'How the biggest 5 online travel agencies are expanding beyond mobile bookings', *The Mobile Marketer*, available at: www.mobilemarketer. com/cms/news/strategy/17579.html (accessed 11 April 2015).

Jones, S. (2005) 'Community-based ecotourism: the significance of social capital', *Annals of Tourism Research*, 32(2): 303–24.

Jordan, E., Vogt, C., and DeShon, R. (2015) 'A stress and coping framework for understanding resident responses to tourism development', *Tourism Management*, 48(June): 500–12.

Kastarlak, B. and Barber, B. (2011) *Fundamentals of Planning and Developing Tourism*, Harlow: Prentice Hall.

Kates, R., Parris, T. and Leiserowitz, A. (2005) 'What is sustainable development? Goals, indicators, values and practice', *Environment*, 47(3): 8–21.

Kepp, M. (2005) 'Beach blanket Brazil: global hotel chains see tourism on the rise in Brazil and are spending big bucks now', *Latin Trade*, March, available at: http://findarticles.com/p/articles/mi_m0BEK/is_3_13/ai_n13619929 (accessed 28 September 2006).

Kerala Tourism Watch (n.d.) 'Fisherwomen protest against Word Tourism Day celebrations', *Kerala Tourism Watch Rhetoric of Responsible Tourism: Irresponsible Practices?*, available at: www.keralatourismwatch.org/node/ 20 (accessed 9 April 2015).

Kerrigan, F., Shivanandan, J. and Hede, A. (2012) 'National branding: a critical appraisal of Incredible India', *Journal of Macromarketing*, 32(3): 319–27.

Kimball, A. (2006) *Risky Trade Infections: Disease in the Era of Global Trade*, Aldershot: Ashgate.

King, R. and Dinkoksung, S. (2014) 'Ban Pa-Ao, pro-poor tourism and uneven development', *Tourism Geographies*, 16(4): 687–703.

Kiribati Climate Change (2015) *Kiribati Climate Change*, available at: www. climate.gov.ki/category/effects/ (accessed 11 May 2015).

Kiribati National Tourism Office Action Plan (n.d.) *Kiribati National Tourism Action Plan 2009–2014*, available at: www.globalislands.net/userfiles/ Kiribati9.pdf (accessed 11 May 2015).

Kiribati Tourism (2015) *Things to Do*, available at: www.kiribatitourism.gov. ki/index.php/thingstodo/thingstodooverview (accessed 11 May 2015).

Knowles, T., Diamantis, D. and El-Mourabi, J. (2001) *The Globalisation of Tourism: A Strategic Perspective*, London: Continuum.

Knox, P., Agnew, J. and McCarthy, L. (2003) *The Geography of the World Economy*, 4th edn, London: Arnold.

Koch, E. and Massyn, P. (2001) 'South Africa's domestic tourism sector: promises and problems', in K. Ghimire (ed.), *The Native Tourist: Mass Tourism within Developing Countries*, London: Earthscan, pp. 142–71.

Komaladat, S. (2009) 'Health tourism destination in Thailand: a case study of Raksawarin Hot Spring', *International Journal of Leisure and Tourism Marketing*, 30(1): 238.

Kousis, M. (2000) 'Tourism and the environment: a social movements perspective', *Annals of Tourism Research*, 27(2): 468–89.

Koutra, C. and Edwards, J. (2012) 'Capacity building through socially responsible tourism development: a Ghanaian case study', *Journal of Travel Research*, 51(6): 779–92.

Krippendorf, J. (1986) 'Tourism in the system of industrial society', *Annals of Tourism Research*, 13(4): 517–32.

Krippendorf, J. (1987) *The Holiday Makers*, Oxford: Heinemann.

Kruger, E. and Douglas, A. (2015) 'Constraints to consumption of South Africa's national parks among the emerging domestic tourism market', *Development Southern Africa*, 32(3): 303–19.

Kumar, A. (2010) *Half of India's Population Lives Below the Poverty Line*, available at: www.wsws.org/en/articles/2010/08/indi-a02.html (accessed 8 January 2015).

Kusluvan, S. and Karamustafa, K. (2001) 'Multinational hotel development in developing countries: an exploratory analysis of critical policy issues', *International Journal of Tourism Research*, 3(2001): 179–97.

Lai, K., Li, Y. and Feng, X. (2006) 'Gap between tourism planning and implementation: a case of China', *Tourism Management*, 27(2006): 1171–80.

Lane, B. (1990) 'Sustaining host areas, holiday makers and operators alike', *Conference Proceedings, Sustainable Tourism Development Conference*, Queen Margaret College, Edinburgh, November 1990.

Lane, J. (2005) *Globalization and Politics*, Aldershot: Ashgate.

Larsen, R., Calgaro, E. and Thomalla, F. (2011) 'Governing resilience building in Thailand's tourism-dependent coastal communities: conceptualising stakeholder agency in social-ecological systems', *Global Environmental Change*, 21(2): 481–91.

Last Frontiers (2015) *Code of Ethics*, available at: www.lastfrontiers.com/about-us/responsible-tourism/last-frontiers-code-of-ethics-for-travellers (accessed 15 March 2015).

Lea, J. (1988) *Tourism and Development in the Third World*, London: Routledge.

Lehmann, L. V. (2009) 'The relationship between tourism and water in dry land regions', *Proceedings of the Environmental Research Event*, Noosa Heads, Queensland, pp. 1–8, available at: http://espace.library.uq.edu.au/view/UQ:17 9617/Lehmann_-_ERE2009.pdf (accessed 12 May 2015).

Leigh, J. (2011) 'New tourism in a new society arises from "peak oil" ', *Tourismos*, 6(1): 165–91.

Leinbach, T. and Bowen, J. (2004) 'Airspaces: air transport, technology and society', in S. Brunn, S. Cutter and J. Harrington (eds), *Geography and Technology*, London: Kluwer Academic Publications, pp. 285–313.

Leiper, N. (1979) 'The framework of tourism: towards a definition of tourism, tourist, and the tourist industry', *Annals of Tourism Research*, 6(4): 390–407.

Leiper, N. (1990) *Tourism Systems: An Interdisciplinary Perspective*, Palmerston North, New Zealand: Massey University.

Leisen, B. (2001) 'Image segmentation: the case of a tourism destination', *Journal of Services Marketing*, 15(1): 49–66.

Lett, J. (1989) 'Epilogue to touristic studies in anthropological perspective', in V. Smith (ed.), *Hosts and Guests: The Anthropology of Tourism*, 2nd edn, Philadelphia, PA: University of Pennsylvania Press, pp. 265–79.

Leung, D., Law, R., van Hoof, H. and Buhalis, D. (2013) 'Social media in tourism and hospitality: a literature review', *Journal of Travel & Touirsm Marketing*, 30(1/2): 3–23.

Leung, Y-F., Spenceley, A., Hvenegaard, G. and Buckley, R. (2014) *Tourism and Visitor Management in Protected Areas: Guidelines for Sustainability*, IUCN, available at: http://iucn.oscar.ncsu.edu/mediawiki/images/3/3a/Sustainable_ Tourism_BPG_Full_Review_Copy_for_WPC14_v2.pdf (accessed 18 February 2015).

Lew, A. (2104) 'Scale, change and resilience in community tourism planning', *Tourism Geographies*, 16(1): 14–22.

Li, Y. (2004) 'Exploring community tourism in China: the case of Nanshan Cultural Tourism Zone', *Journal of Sustainable Tourism*, 12(3): 175–93.

Lichrou, M., O'Malley, L. and Patterson, M. (2010) 'Narrative of a tourism destination: local particularities and their implications for place marketing and branding', *Place Branding and Public Diplomacy*, 6(2): 134–44.

Lickorish, L. (1991) 'International agencies', in L. Lickorish, A. Jefferson, J. Bodlender and C. Jenkins (eds), *Developing Tourism Destinations Policies and Perspectives*, Harlow: Longman, pp. 147–65.

Lipscomb, A. (1998) 'Village-based tourism in the Solomon Islands impediments and impacts', in E. Laws, B. Faulkner and G. Moscardo (eds), *Embracing and Managing Change in Tourism*, London: Routledge, pp. 185–201.

Liu, A. and Wall, G. (2006) 'Planning tourism employment: a developing country perspective', *Tourism Management*, 27(2006): 159–70.

Liu, C. and Lee, T. (2015) 'Promoting entrepreneurial orientation through the accumulation of social capital', *International Journal of Hospitality Management*, 46(April): 138–50.

Lloyd, K. (2004) 'Tourism and transitional geographies: mismatched expectations of tourism investment in Vietnam', *Asia Pacific Viewpoint*, 45(2): 197–215.

Logar, I. and van der Bergh, J. (2013) 'The impact of peak oil on tourism in Spain: an input–output analysis of price, demand and economy-wide effects', *Energy*, 54(1): 155–66.

Long, V. (1993) 'Techniques for socially sustainable tourism development: lessons from Mexico', in J. Nelson, R. Butler and G. Wall (eds), *Tourism and Sustainable Development: Monitoring, Planning, Managing*, Waterloo: Heritage Resources Centre Joint Publication Number 1, University of Waterloo, pp. 201–19.

Ludwig, D., Hilborn, R. and Walters, C. (1993) 'Uncertainty, resource exploitation, and conservation: lessons from history', *Science*, 269(5104): 17, 36.

Lupoli, C., Morse, W., Bailey, C. and Schelhas, J. (2014) 'Assessing the impacts of international volunteer tourism in host communities: a new approach to organizing and prioritizing indicators', *Journal of Sustainable Tourism*, 22(6): 898–921.

Lury, C. (2011) *Consumer Culture*, 2nd edn, Cambridge: Polity Press.

Mabogunje, A. (1980) *The Development Process: A Spatial Perspective*, London: Hutchinson.

MacCannell, D. (1989) *The Tourist: A New Theory of the Leisure Class*, 2nd edn, New York: Shocken Books.

McCool, S. and Lime, D. (2001) 'Tourism carrying capacity: tempting fantasy or useful reality?', *Journal of Sustainable Tourism*, 9(5): 372–88.

McCormick, J. (1995) *The Global Environmental Movement*, Chichester: John Wiley & Sons.

McElroy, J. and de Albuquerque, K. (2002) 'Problems for managing sustainable tourism in small islands', in Y. Apostlolpoulos and D. Gayle (eds), *Island Tourism and Sustainable Development: Caribbean, Pacific and Mediterranean Experiences*, Westport, CT: Praeger, pp. 15–34.

McGehee, N. and Anderek, K. (2004) 'Factors predicting rural residents' support for tourism', *Journal of Travel Research*, 43(2): 131–40.

McGillivray, M. (2008) 'What is development?', in D. Kingsbury, J. McKay, J. Hunt, M. McGillivray and M. Clarke (eds), *International Development: Issues and Challenges*, Houndmills: Palgrave Macmillan, pp. 21–50.

McGuire, F., Uysal, M. and McDonald, C. (1988) 'Attracting the older traveller', *Tourism Management*, 9(2): 161–4.

Mackay, R. and Palmer, S. (2015) 'Tourism, world heritage and local communities: an ethical framework in practice at Angkor', in T. Ireland and J. Schofield (eds), *The Ethics of Cultural Heritage*, London: Springer, pp. 165–83.

MacKenzie, M. (2006) 'Temples Doomed by Tourism', *The Independent*, available at: http://travel.independent.co.uk/news_and_advice/article1090291.ece.

McKercher, B. (1993) 'Some fundamental truths about tourism: understanding tourism's social and environmental impacts', *Journal of Sustainable Tourism*, 1(1): 6–16.

McKercher, B. and Chon, K. (2004) 'The over-reaction to SARS and the collapse of Asian tourism', *Annals of Tourism Research*, 31(3); 716–19.

MacLellan, R., Dieke, P. and Thopo, B. (2000) 'Mountain tourism and public policy in Nepal', in P. Godde, M. Price and F. Zimmerman (eds), *Tourism and Development in Mountain Regions*, Wallingford: CABI, pp. 173–97.

Macleod, D. (2004) *Tourism, Globalisation and Cultural Change: An Island Community Perspective*, Clevedon: Channel View Publications.

McMichael, P. (2004) *Development and Social Change: A Global Perspective*, 3rd edn, London: Pine Forge Press.

Macnaught, T. (1982) 'Mass tourism and the dilemmas of modernization in Pacific island communities', *Annals of Tourism Research*, 9(3): 359–81.

Madrigal, R. and Kahle, L. (1994) 'Predicting vacation activity preferences on the basis of value-system segmentation', *Journal of Travel Research*, 32(3): 22–8.

Mair, J. (2011) 'Exploring air travellers' voluntary carbon-offsetting behaviour', *Journal of Sustainable Tourism*, 19(2): 215–30.

Malta Tourism Authority (2015) *Hotels Height Limitation Adjustment Policy*, available at: www.mta.com.mt/hotel-height-policy (accessed 24 April 2015).

Mann, M. (1986) *The Sources of Social Power, Volume 1: A History of Power from the Beginning to A.D. 1760*, Cambridge: Polity Press.

Mann, M. (2000) *The Community Tourism Guide*, London: Earthscan.

Mansbach, R. and Rafferty, K. (2008) *Introduction to Global Politics*, London: Routledge.

Mao, N., Delacy, T. and Grunfeld, H. (2013) 'Local livelihoods and the tourism value chain: a case study in Siem Reap-Angkor Region, Cambodia', *International Journal of Environmental and Rural Development*, 4(2): 120–6.

Martell, L. (1994) *Ecology and Society*, Cambridge: Polity Press.

Mason, P. (2008) *Tourism Impacts, Planning and Management*, 2nd edn, Abingdon: Routledge.

Mason, P. and Mowforth, M. (1995) *Codes of Conduct in Tourism*, Occasional Papers in Geography No. 1, University of Plymouth, Department of Geographical Sciences.

Mason, P. and Mowforth, M. (1996) 'Codes of conduct in tourism', *Progress in Tourism and Hospitality Research*, 2(2): 151–64.

Mathieson, A. and Wall, G. (1982) *Tourism: Economic, Physical and Social Impacts*, Harlow: Longman.

Matthews, H. and Richter, L. (1991) 'Political science and tourism', *Annals of Tourism Research*, 18(1): 120–35.

Mazar, N. and Zhong, C. (2010) 'Do green products make us better people?', *Psychological Science*, 21(4): 494–8.

Mbaiwa, J. (2005) 'Enclave tourism and its socio-economic impacts in the Okavango Delta, Botswana', *Tourism Management*, 26(2): 157–72.

Mehmetoglu, M. (2007) 'Typologising nature-based tourists by activity: theoretical and practical implications', *Tourism Management*, 28(3): 651–60.

Mellor, W. (2014) 'Medical tourists flock to Thailand spurring post-coup economy', *Bloomberg*, 18 November, available at: www.bloomberg.com/news/articles/2014-11-18/medical-tourists-flock-to-thailand-spurring-post-coup-economy (accessed 19 May 2015).

Mexico Premiere (2012) *Mexico Surpasses Targets for Tourist Arrivals and Private Investment in Tourism Projects*, 28 November, available at: www.mexicopremiere.com/2012/11/ (accessed 23 April 2015).

Mieczkowski, Z. (1995) *Environmental Issues of Tourism and Recreation*, Lanham, MD: University Press of America.

MIGA (2006) *MIGA: Supporting Tourism and Hospitality Investments, Tourism and Hospitality Brief*, available at: www.miga.org/documents/touris0.6pdf (accessed 22 September 2006).

Mihalič, T. (2015) 'Tourism and economic development issues', in R. Sharpley and D. Telfer (eds), *Tourism and Development: Concepts and Issues*, 2nd edn, Bristol: Channel View Publications, pp. 77–117.

Miller, G., Rathouse, K., Scarles, C., Holmes, K. and Tribe, J. (2010) 'Public understanding of sustainable tourism', *Annals of Tourism Research*, 37(3): 627–45.

Milne, S. and Ewing, G. (2004) 'Community participation in Caribbean tourism problems and prospects', in D. Duval (ed.), *Tourism in the Caribbean: Trends, Development, Prospects*, London: Routledge, pp. 205–17.

Milner, H. (1991) 'The assumption of anarchy in international relations: a critique', *Review of International Studies*, 17(1): 67.

Minghetti, V. and Buhalis, D. (2011) *Digital Divide in Tourism: Modelling the Effects of the Digital Gap on Tourists, Businesses and Destinations*, ENTER 2010 Conference, available at: http://virgo.unive.it/ciset/website/it/allegati/ENTER_2010_Minghetti_Buhalis.pdf (accessed 15 April 2015).

Miossec, J. M. (1976) 'Elements pour une théorie de l'espace touristique' Les Cahiers du tourisme C-336, CHET, Aix-in-Province: Mintel.

Mintel (1994) *The Green Consumer I: The Green Conscience*, London: Mintel International.

Mishan, E. (1969) *The Costs of Economic Growth*, Harmondsworth: Penguin.

Mitchell, B. (2002) *Resource and Environmental Management*, 2nd edn, London: Routledge.

Mitchell, D. (2000) *Cultural Geography: A Critical Introduction*, Oxford: Blackwell.

Mitchell, J. (2012) 'Value chain approaches to assessing the impact of tourism on low-income households in developing countries', *Journal of Sustainable Tourism*, 20(3): 45–75.

Mitchell, J. and Ashley, C. (2010) *Tourism and Poverty Reduction: Pathways to Prosperity*, London: Earthscan.

Mohan, G. and Mohan, J. (2002) 'Placing social capital', *Progress in Human Geography*, 26(2): 191–210.

Momsen, J. H. (2004) *Gender and Development*, London: Routledge.

Morakabati, Y., Beavis, J. and Fletcher, J. (2014) 'Without oil: tourism and economic diversification, a battle of perceptions', *Tourism Planning and Development*, 11(4): 415–34.

Morgan, N., Pritchard, A. and Pride, R. (2011) 'Tourism places, brands and reputation management', in N. Morgan, A. Pritchard and R. Pride (eds), *Destinations Brands Managing Place Reputation*, 3rd edn, Oxford: Butterworth-Heinemann, pp. 3–12.

Moscardo, G. (2008) 'Community capacity building: an emerging challenge for tourism development', in G. Moscardo (ed.), *Building Community Capacity for Tourism Development*, Wallingford: CABI, pp. 1–15.

Mosley, P. and Toye, J. (1988) 'The design of Structural Adjustment Programmes', *Development Policy Review*, 6(4): 395–413.

MoT (2014) *Tourism Statistics Report December 2014*, Cambodia: Ministry of Tourism, available at: www.nagacorp.com/eng/ir/tourism/tourism_statistics_201412.pdf (accessed 28 April 2015).

Mouzelis N. (1994) 'The state in late development: historical and comparative perspectives', in D. Booth (ed.), *Rethinking Social Development: Theory and Practice*, Harlow: Addison Wesley Longman, pp. 126–51.

Mowforth, M. and Munt, I. (1998) *Tourism and Sustainability: New Tourism in the Third World*, London: Routledge.

Mowforth, M. and Munt, I. (2003) *Tourism and Sustainability Development and New Tourism in the Third World*, 2nd edn, London: Routledge.

Mowforth, M. and Munt, I. (2009) *Tourism and Sustainability: Development, Globalisation and New Tourism in the Third World*, 3rd edn, London: Routledge.

Mowl, G. (2002) 'Tourism and the environment', in R. Sharpley (ed.), *The Tourism Business: An Introduction*, Sunderland: Business Education Publishers, pp. 219–42.

Muangasame, K. and McKercher, B. (2015) 'The challenge of implementing sustainable tourism policy: a 360-degree assessment of Thailand's 7 Greens sustainable tourism policy', *Journal of Sustainable Tourism*, 23(4): 497–516.

Munanura, I. and Backman, K. (2012) 'Stakeholder collaboration as a tool for tourism planning – a developing country's perspective', *Journal of Tourism*, 13(1): 23–39.

Mundt, J. (2011) *Tourism and Sustainable Development: Reconsidering a Concept of Vague Policies*, Berlin: Erich Schmidt Verlag.

Murphy, P. (1985) *Tourism: A Community Approach*, London: Routledge.

Murugesan, S. (2007) 'Understanding Web 2.0', *IT Professional*, 9(4): 34–41.

Mycoo, M. (2014) 'Sustainable tourism, climate change and seal level rise adaptation policies in Barbados', *Natural Resources Forum*, 38(1): 47–57.

Nash, D. (1989) 'Tourism as a form of imperialism', in V. Smith (ed.), *Hosts and Guests: The Anthropology of Tourism*, 2nd edn, Philadelphia, PA: University of Pennsylvania Press, pp. 37–52.

National Bureau of Statistics (2013) *Domestic Tourism*, available at: www.quandl.com/STATCHINA-National-Bureau-of-Statistics-China/Q1819-Domestic-Tourism (accessed 24 July 2013).

Nepal, S. (2000) 'Tourism in protected areas: the Nepalese Himalaya', *Annals of Tourism Research*, 27(3): 661–81.

Neumayer, E. (2004) 'The impact of political violence on tourism: dynamic cross-national estimation', *Journal of Conflict Resolution*, 48(2): 259–81.

Newell, P. (2005) 'Environment', in P. Burnell and V. Randall (eds), *Politics in the Developing World*, Oxford: Oxford University Press, pp. 221–36.

Newsome, D., Moore, S. and Dowling, R. (2000) *Natural Area Tourism: Ecology, Impacts and Management*, Clevedon: Channel View Publications.

Ngwira, P. and Mbaiwa, J. (2013) 'Community based natural resource management, tourism and poverty alleviation in Southern Africa: what works and what doesn't work', *Chinese Business Review*, 12(12): 789–806.

Nkyi, E. and Hashimoto, A. (2015) 'Human rights issues in tourism development', in R. Sharpley and D. J. Telfer (eds), *Tourism and Development: Concepts and Issues*, 2nd edn, Bristol: Channel View Publications, pp. 378–99.

Noronha, L., Siqueira, A., Sreekesh, S., Qureshy, L. and Kazi, S. (2002) 'Goa: tourism migrations and ecosystem transformations', *Ambio*, 31(4): 295–302.

Nozick, M. (1993) 'Five principles of sustainable community development', in E. Shragge (ed.), *Community Economic Development: In Search of Empowerment and Alteration*, Montreal: Black Rose Books, pp. 18–43.

Nunkoo, R., Smith, S. and Ramkissoon, M. (2013) 'Resident attitudes to tourism: a longitudinal study of 140 articles from 1984 to 2010', *Journal of Sustainable Tourism*, 21(1): 5–25.

Nyaupane, G. and Timothy, D. (2010) 'Power, regionalism and tourism policy in Bhutan', *Annals of Tourism Research*, 37(4): 969–88.

O'Connell, J. and Aurélise, B. (2015) 'Dynamic packaging spells the end of European charter airlines', *Journal of Vacation Marketing*, 21(2): 175–89.

O'Connor, P., Buhalis, D. and Frew, A. (2001) 'The transformation of tourism distribution channels through information technology', in D. Buhalis and E. Laws (eds), *Tourism Distribution Channels Practices, Issues and Transformations*, London: Continuum, pp. 315–31.

O'Grady, R. (1980) *Third World Stopover*, Geneva: World Council of Churches.

O'Reilly, K. (2003) 'When is a tourist? The articulation of tourism and migration in Spain's Costa del Sol', *Tourist Studies*, 3(3): 301–17.

OECD (1981) *The Impact of Tourism on the Environment*, Paris: Organisation for Economic Co-Operation and Development.

Olesen, A. (2006) 'Train route opens travel to remote Tibet' *Baltimore Sun*, available at: www.baltimoresun.com (accessed 14 August 2006).

Oneworld (2015) *Overview*, available at: www.oneworld.com/member-airlines/overview (accessed 9 April 2015).

Opperman, M. and Chon, K. (1997) *Tourism in Developing Countries*, London: International Thomson Business Press.

Ottaway, M. (2005) 'Civil society', in P. Burnell and V. Randall (eds), *Politics in the Developing World*, Oxford: Oxford University Press, pp. 120–35.

Page, S. and Connell, J. (2006) *Tourism: A Modern Synthesis*, 2nd edn, London: Thomson.

Page, S. and Connell, J. (2014) *Tourism: A Modern Synthesis*, 4th edn, Andover: Cengage Learning EMEA.

Pal, M. (1994) 'Constraints facing the small-scale informal sectors in developing economies', *Ecodecision*, (Fall): 79–81.

Palma, G. (1995) 'Underdevelopment and Marxism: from Marx to the theories of imperialism and dependency', in R. Ayers (ed.), *Development Studies: An Introduction through Selected Readings*, Dartford: Greenwich University Press, pp. 161–210.

Papatheodorou, A. (2006) 'Liberalisation and deregulation for tourism: implications for competition', in C. Costa and D. Buhalis (eds), *Tourism Management Dynamics, Trends, Management and Tools*, London: Elsevier Butterworth Heinemann, pp. 68–77.

Parinello, G. (1993) 'Motivation and anticipation in post-industrial tourism', *Annals of Tourism Research*, 20(2): 233–49.

Park, D., Lee, K., Chio, H. and Yoon, Y. (2012) 'Factors influencing social capital in rural tourism communities in South Korea', *Tourism Management*, 33(6): 1511–20.

Patel, R. (2012) *Stuffed and Starved: The Hidden Battle for the World Food System*, Brooklyn, NY: Melville House Publishing.

Patullo, P. (1996) *Last Resorts: The Cost of Tourism in the Caribbean*, Kingston, Jamaica: Ian Randle Publishers.

Payne, G., Moore, S., Griffis, S. and Autry, C. (2011) 'Multilevel challenges and opportunities in social capital research', *Journal of Management*, 37(2): 491–520.

Pearce, D. (1989) *Tourist Development*, 2nd edn, Harlow: Longmans.

Pearce, D., Markandya, A. and Barbier, E. (1989) *Blueprint for a Green Economy*, London: Earthscan.

Pearce, P. (1992) 'Fundamentals of tourist motivation', in D. Pearce and R. Butler (eds), *Tourism Research: Critiques and Challenges*, London: Routledge, pp. 113–34.

Pearce, P. (2005) *Tourist Behaviour Themes and Conceptual Schemes*, Clevedon: Channel View Publications.

Peet, R. and Hartwick, E. (1999) *Theories of Development*, London: Guildford Press.

Peet, R. and Hartwick, E. (2015) *Theories of Development: Contentions, Arguments, Alternatives*, New York: Guildford Press.

Pegas, F., Weaver, D. and Castley, G. (2015) 'Domestic tourism and sustainability in an emerging economy: Brazil's littoral pleasure periphery', *Journal of Sustainable Tourism*, 23(5): 748–69.

Perkins, J. (2004) *Confessions of an Economic Hit Man*, London: Penguin Books.

Peterson, R. (1979) 'Revitalizing the culture concept', *Annual Review of Sociology*, 5: 137–66.

Phillips, P. and Moutinho, L. (2014) 'Critial review of strategic planning research in hospitality and tourism', *Annals of Tourism Research*, 48: 96–120.

Pigram, J. (1990) 'Sustainable tourism – policy considerations', *Journal of Tourism Studies*, 1(2): 2–9.

Pike, S. (2002) 'Destination image analysis – a review of 142 papers from 1973 to 2000', *Tourism Management*, 23(4): 541–9.

Pillay, M. and Rogerson, C. (2013) 'Agriculture-tourism linkages and pro-poor impacts: the accommodation sector of urban coastal KwaZulu-Natal, South Africa', *Applied Geography*, 36: 49–58.

Pires, E. (2014) 'Vertical farming: fighting poverty and food insecurity in Kibera, Kenya', *Innovate Development*, 30 March, available at: http://innovatedevel opment.org/2014/03/30/vertical-farming-fighting-poverty-and-food-insecurity-in-kibera-kenya/ (accessed 13 May 2015).

Pizam, A. and Mansfeld, Y. (eds) (1996) *Tourism, Crime and International Security Issues*, Chichester: John Wiley & Sons.

Plog, S. (1977) 'Why destinations rise and fall in popularity', in E. Kelly (ed.), *Domestic and International Tourism*, Wellesley, MA: Institute of Certified Travel Agents.

Plummer, R. and Fitzgibbon, J. (2004) 'Some observations on the terminology in co-operative environmental management', *Journal of Environmental Management*, 70(1): 63–72.

Poon, A. (1989) 'Competitive strategies for a "new tourism"', in C. Cooper (ed.), *Progress in Tourism, Recreation and Hospitality Management, Vol. 1*, London: Belhaven Press.

Poon, A. (1993) *Tourism, Technology and Competitive Strategies*, Wallingford: CAB International.

Porter, M. (1998) *On Competition: A Harvard Business Review Book*, Boston, MA: Harvard Business School Publishing.

Porter, M., Stern, S. and Artavia Loría, R. (2013) *Social Progress Index 2013*, Washington, DC: Social Progress Imperative.

Potter, D. (2000) 'Democratisation, "good governance" and development', in T. Allen and A. Thomas (eds), *Poverty and Development into the 21st Century*, Oxford: Oxford University Press, pp. 365–82.

Potter, R. (1995) 'Urbanisation and development in the Caribbean', *Geography*, 80: 334–41.

Potter, R. (2002) 'Theories, strategies and ideologies of development', in V. Desai and R. B. Potter (eds), *The Companion to Development Studies*, New York: Oxford University Press, pp. 61–5.

Potter, R., Binns, T., Elliot, J. A. and Smith, D. (1999) *Geographies of Development*, London: Prentice Hall.

Powell, R. and Ham, S. (2008) 'Can ecotourism interpretation really lead to pro-conservation knowledge, attitudes and behaviour? Evidence from the Galapagos Islands', *Journal of Sustainable Tourism*, 16(4): 467–89.

PPT (2009) Pro-poor tourism info sheets, sheet No. 9: Tourism in Poverty Reduction Strategy Papers (PRSPs), www.propoortourism.org.uk (accessed 8 August 2006), available at: www.iztzg.hr/UserFiles/Pdf/sustainable/Pro-poor-tourism-info-sheets.pdf

Prasad, E., Rogoff, K., Wei, S. and Kose, A. (2003) *Effects of Financial Globalisation on Developing Countries, Some Empirical Evidence*, International Monetary Fund Occasional Paper 220, 9 September, International Monetary Fund, available at: www.imf.org/external/pubs/nft/op/220/index.htm (accessed 8 May 2007).

Preston, P. (1996) *Development Theory: An Introduction*, Oxford: Blackwell.

Preston-Whyte, R. and Watson, H. (2005) 'Nature tourism and climate change in Southern Africa', in C. M. Hall and J. Higham (eds), *Tourism, Recreation and Climate Change*, Clevedon: Channel View Publications, pp. 130–42.

Proops, J. and Wilkinson, D. (2000) 'Sustainability, knowledge, ethics and the law', in M. Redclift (ed.), *Sustainability, Life Chances and Livelihoods*, London: Routledge, pp. 17–34.

Putnam, R. (1993) 'The prosperous community: social capital and public life', *American Prospect*, 13(1): 35–42.

Qatar Tourism Authority (2015) *Qatar National Tourism Sector Strategy 2030/ Q&A*, available at: http://corporate.qatartourism.gov.qa/Portals/0/QAENG LISHFINAL.pdf (accessed 8 January 2015).

Ragazzi, M., Catellani, R., Rada, E., Torretta, V. and Salaazar-Valenzuela, X. (2014) 'Management of municipal solid waste in on the Galapagos Islands', *Sustainability*, 6(12): 9080–95.

Rahnema, M. and Bawtree, V. (eds) (1997) *The Post-Development Reader*, London: Zed Books.

Rain, D. and Brooker-Gross, S. (2004) 'A world on demand: geography of the 24-hour global TV news', in S. Brunn, S. Cutter and J. Harrington (eds), *Geography and Technology*, London: Kluwer Academic, pp. 315–37.

Rakodi, C. (1995) 'Poverty lines or household strategies?', *Habitat International*, 19(4): 407–26.

Rastogi, A., Thapliyal, S. and Hickey, G. (2014) 'Community action and tiger conservation: assessing the role of social capital', *Society and Natural Resources*, 27: 1271–87.

Ravallion, M. (2004) *Pro-Poor Growth: A Primer*, Washington, DC: World Bank.

Ray, C. (1998) 'Culture, intellectual property and territorial rural development', *Sociologia Ruralis*, 38: 3–20.

Redclift, M. (1987) *Sustainable Development: Exploring the Contradictions*, London: Routledge.

Redclift, M. (2000) 'Introduction', in M. Redclift (ed.), *Sustainability, Life Chances and Livelihoods*, London: Routledge, pp. 1–13.

Reid, D. (1995) *Sustainable Development: An Introductory Guide*, London: Earthscan.

Reid, D. (2003) *Tourism, Globalization and Development: Responsible Tourism Planning*, London: Pluto Press.

Reisinger, Y. and Turner, L. (2003) *Cross-Cultural Behaviour in Tourism Concepts and Analysis*, Oxford: Butterwoth-Heinemann.

Reno, W. (2000) 'Clandestine economies, violence and states in Africa', *Journal of International Affairs*, 53(2): 433–60.

Responsible Travel (2015) *About the Responsible Tourism Awards*, available at: www.responsibletravel.com/awards/about/ (accessed 29 January 2015).

Richards, G. and Hall, D. (2000a) 'The community: a sustainable concept in tourism development?', in D. Hall and G. Richards (eds), *Tourism and Sustainable Community Development*, London: Routledge, pp. 1–14.

Richards, G. and Hall, D. (2000b) 'Conclusions' in D. Hall and G. Richards (eds), *Tourism and Sustainable Community Development*, London: Routledge, pp. 297–306.

Ritchie, B. (2009) *Crisis and Disaster Management for Tourism*, Clevedon: Channel View Publications.

Ritzer, G. (2010) *The McDonaldization of Society 6*, Thousand Oaks, CA: Pine Forge Press.

Robbins, P. (2001) *Greening the Corporation: Management Strategies and the Environmental Challenge*, London: Earthscan.

Roberts, S. (2002) 'Global regulation and trans-state organisation', in R. Johnston, P. Taylor and M. Watts (eds), *Geographies of Global Change Remapping the World*, Oxford: Blackwell, pp. 143–57.

Robins, K. (1997) 'What in the world is going on?', in P. du Gay (ed.), *Production of Culture/Cultures of Production*, London: Sage, pp. 11–66.

Robinson, L. (2004) 'Squaring the circle? Some thoughts on the idea of sustainable development', *Ecological Economics*, 48(4): 369–84.

Rodrik, D. (2005) 'Feasible globalisations', in M. Weinstein (ed.), *Globalization: What's New?*, New York: Columbia University Press, pp. 196–213.

Rogerson, C. (2006) 'Pro-poor local economic development in South Africa: the role of pro-poor tourism', *Local Environment*, 11(1): 37–60.

Rogerson, C. (2012) 'Strengthening agriculture-tourism linkages in the developing world: opportunities, barriers and current initiatives', *African Journal of Agricultural Research*, 7(4): 616–23.

Rokeach, M. (1973) *The Open and Closed Mind*, New York: Basic Books.

Rostow, W. (1967) *The Stages of Economic Growth: A Non-Communist Manifesto*, 2nd edn, Cambridge: Cambridge University Press.

Rothman, J., Erlich, M. and Tropman, J. E. (1995) *Strategies of Community Intervention*, 5th edn, Itasca, IL: F. E. Peacock.

Routledge, P. (2001) ' "Selling the rain", resisting the sale: resistant identities and the conflict over tourism in Goa', *Social and Cultural Geography*, 2(2): 221–40.

RWSSD (2002) *Report of the World Summit on Sustainable Development*, New York: United Nations, available at: www.un.org/summit/html/documents/summit_docs.html (accessed 2 February 2015).

Ryan, C. (1991a) *Recreational Tourism: A Social Science Perspective*, London: Routledge.

Ryan, C. (1991b) *Tourism, Terrorism and Violence: The Risks of Wider World Travel*, London: Research Institute for the Study of Conflict and Terrorism.

Ryan, C. (1997) 'The chase of a dream, the end of a play', in C. Ryan (ed.), *Tourist Experience: A New Introduction*, London: Cassell, pp. 1–24.

Ryan, C. (2005) 'Introduction: tourist-host nexus – research considerations', in C. Ryan and M. Aicken (eds), *Indigenous Tourism: The Commodification and Management of Culture*, London: Elsevier, pp. 1–11.

Ryan, C. and Hall, C. M. (2001) *Sex Tourism: Marginal People and Liminalities*, London: Routledge.

S. R. (2014) 'Why China is creating a new "World Bank" ', *The Economist*, 11 November 2014, available at: www.economist.com/blogs/economist-explains/2014/11/economist-explains-6 (accessed 14 April 2015).

Saarinen, J. and Lenao, M. (2014) 'Integrating tourism to rural development and planning in the developing world', *Development Southern Africa*, 31(3): 363–72.

Sachs, J. (2005) *The End of Poverty Economic Possibilities for Our Time*, New York: Penguin Books.

Sachs, W. (1996) 'Introduction', in W. Sachs (ed.), *The Development Dictionary: A Guide to Knowledge and Power*, London: Zed Books, pp. 2–5.

Said, E. (1978) *Orientalism*, London: Routledge.

Salem, N. (1994) 'Water rights', *Tourism in Focus (Tourism Concern)*, 17: 4–5.

Salerno, F., Viviano, G., Manfredi, E., Caroli, P., Thakuri, S. and Tartari, G. (2013) 'Multiple carrying capacities from a management-oriented perspective to operationalize sustainable tourism in protected areas', *Journal of Environmental Management*, 128: 116–25.

Sandals Foundation (2015) *Sandals Foundation*, available at: www.sandalsfoundation.org (accessed 15 May 2015).

Saul, J. R. (2005) *The Collapse of Globalism and the Reinvention of the World*, Toronto: Viking Canada.

Schaeffer, R. (2009) *Understanding Globalization: The Social Consequences of Political, Economic and Environmental Change*, Plymouth: Rowman & Littlefield.

Schemo, J. (1995) 'Galapagos Island journal: homo sapiens at war on Darwin's peaceful island', *New York Times*, available at: www.nytimes.com/1995/11/28/world/galapagos-islands-journal-homo-sapiens-at-war-on-darwin-s-peaceful-isles.html (accessed 15 August 2006).

Scheyvens, R. (2002) *Tourism for Development Empowering Communities*, London: Prentice Hall.

Scheyvens, R. (2003) 'Local involvement in managing tourism', in S. Singh, D. Timothy and R. Dowling (eds), *Tourism in Destination Communities*, Wallingford: CABI, pp. 229–52.

Scheyvens, R. (2011) *Tourism and Poverty*, Abingdon: Routledge.

Scheyvens, R. (2015) 'Tourism and poverty reduction', in R. Sharpley and D. Telfer (eds), *Tourism and Development: Concepts and Issues*, 2nd edn, Bristol: Channel View Publications, pp. 118–39.

Schilcher, D. (2007) 'Growth versus equity: the continuum of pro-poor tourism and neoliberal governance', *Current Issues in Tourism*, 10(2/3): 166–93.

Schindler, J. and Zittel, W. (2008) *Crude Oil: The Supply Outlook*, Berlin: Energy Watch Group.

Schmidt, H. (1989) 'What makes development?', *Development and Cooperation*, 6: 19–26.

Scholte, J. (2005) *Globalisation: A Critical Introduction*, New York: Palgrave.

Schumacher, E. (1974) *Small Is Beautiful: A Study of Economics as if People Mattered*, London: Abacus.

Schuurman, F. (1993) *Beyond the Impasse: New Directions in Development Theory*, London: Zed Books.

Schuurman, F. (1996) 'Introduction: development theory in the 1990s', in F. Schuurman (ed.), *Beyond the Impasse: New Direction in Development Theory*. London: Zed Books, pp. 1–48.

Seckelman, A. (2002) 'Domestic tourism – a chance for regional development in Turkey?', *Tourism Management*, 23(1): 85–92.

Seers, D. (1969) 'The meaning of development', *International Development Review*, 11(4): 2–6.

Sen, A. (1999) *Development as Freedom*, New York: Anchor Books.

Shah, K. (2000) *Tourism, the Poor and Other Stakeholders: Asian Experience*, ODI Fair-Trade Tourism Paper, London: ODI.

Sharma, R. (2013) *Breakout Nations in the Pursuit of the Next Economic Miracles*, London: W. W. Norton.

Shackley, M. (1996) *Wildlife Tourism*, London: International Thomson Business Press.

Sharpley, R. (2000) 'Tourism and sustainable development: exploring the theoretical divide', *Journal of Sustainable Tourism*, 8(1): 1–19.

Sharpley, R. (2001) 'Tourism in Cyprus: challenges and opportunities', *Tourism Geographies*, 3(1): 64–86.

Sharpley, R. (2002) *The Tourism Business: An Introduction*, Sunderland: Business Education.

Sharpley, R. (2005) 'The tsunami and tourism: a comment', *Current Issues in Tourism*, 8(4): 344–9.

Sharpley, R. (2006) 'Ecotourism: a consumption perspective', *Journal of Ecotourism*, 5(1/2): 7–22.

Sharpley, R. (2008) *Tourism, Tourists and Society*, 4th edn, Huntingdon: Elm Publications.

Sharpley, R. (2009a) 'Tourism, religion and spirituality', in T. Jamal and M. Robinson (eds), *Sage Handbook of Tourism Studies*, London: Sage, pp. 237–53.

Sharpley, R. (2009b) 'Tourism and development challenges in the least developed countries: the case of the Gambia', *Current Issues in Tourism*, 12(4): 337–58.

Sharpley, R. (2009c) *Tourism Development and the Environment: Beyond Sustainability*, London: Earthscan.

Sharpley (2009/2010) *The Myth of Sustainable Tourism*, CSD Working Paper Series 2009/2010 – No. 4, Centre for Sustainable Development, University of Central Lancashire, available at: www.researchgate.net/profile/Lisa_Ruhanen/publication/50996204_The_myth_of_sustainable_tourism/links/540d205a0cf2f2b29a3826cc.pdf (accessed 19 May 2015).

Sharpley, R. (2012) 'Responsible tourism: whose responsibility?', in A. Holden and D. Fennel (eds), *Handbook of Tourism and the Environment*, Abingdon: Routledge, pp. 382–91.

Sharpley, R. (2014) 'Host perceptions of tourism: a review of the research', *Tourism Management*, 42(1): 37–49.

Sharpley, R. and Knight, M. (2009) 'Tourism and the state in Cuba: from the past to the future', *International Journal of Tourism Research*, 11(3): 241–54.

Sharpley, R. and Naidoo, P. (2010) 'Tourism and poverty reduction: the case of Mauritius', *Tourism and Hospitality Planning and Development*, 7(2): 145–62.

Sharpley, R. and Telfer, D. (eds) (2015) *Tourism and Development: Concepts and Issues*, 2nd edn, Bristol: Channel View Publications.

Sharpley, R. and Ussi, M. (2014) Tourism and governance in Small Island Developing States (SIDS): the case of Zanzibar, *International Journal of Tourism Research*, 16(1): 87–96.

Sharpley, R., Sharpley, J. and Adams, J. (1996) 'Travel advice or trade embargo: the impacts and implications of official travel advice', *Tourism Management*, 17(1): 1–7.

Shaw, B. and Shaw, G. (1999) ' "Sun, sand and sales": enclave tourism and local entrepreneurship in Indonesia', *Current Issues in Tourism*, 2(1): 68–81.

Sherwood, S. (2006) 'Is Qatar the next Dubai?', *New York Times*, available at: www.nytimes.com/2006/06/04/travel/04qatar.html?pagewanted=all&_r=0 (accessed 8 January 2015).

Shoo, R. and Songorowa, A. (2013) 'Contribution of ecotourism to nature conservation and improvement of livelihoods around Amani nature reserve, Tanzania', *Journal of Ecotourism*, 12(2): 75–89.

Silver, I. (1993) 'Marketing authenticity in Third World countries', *Annals of Tourism Research*, 20(2): 302–18.

Simão, J. and Partidário, M. (2012) 'How does tourism planning contribute to sustainable development?', *Sustainable Development*, 20(6): 372–85.

Simpson, B. (1993) 'Tourism and tradition: from healing to heritage', *Annals of Tourism Research*, 20(2): 164–81.

Simpson, K. (2004) ' "Doing development": the gap year volunteer-tourists and a popular practice of development', *Journal of International Development*, 16(7): 681–92.

Sin, H. (2009) 'Volunteer tourism: "Involve me and I will learn"?', Annals of Tourism Research, 36(3): 480–501.

Sinclair-Maragh, G. and Gursoy, D. (2015) 'Imperialism and tourism: the case of developing island countries', *Annals of Tourism Research*, 50: 143–58.

Singh, S. (2001) 'Indian tourism: policy, performance and pitfalls', in D. Harrison (ed.), *Tourism and the Less Developed World: Issues and Case Studies*, Wallingford: CABI, pp. 137–49.

Singh, S. (2009a) 'Domestic tourism: searching for an Asianic Perspective', in S. Singh (ed.), *Domestic Tourism in Asia: Diversity and Divergence*, London: Earthscan, pp. 1–25.

Singh, S. (ed.) (2009b) *Domestic Tourism in Asia: Diversity and Divergence*, London: Earthscan.

Singh, S., Timothy, D. and Dowling, R. (2003) 'Tourism and destination communities', in S. Singh, D. Timothy and R. Dowling (eds), *Tourism in Destination Communities*, Wallingford: CABI, pp. 3–18.

Skidmore, J. (2008) 'Britons: more mean than green', *Telegraph Travel*, 14 June, T4.

Sklair, L. (1995) *Sociology of the Global System*. Baltimore, MD: Johns Hopkins University Press.

SkyTeam (2015) *SkyTeam Member Airlines*, available at: www.skyteam.com/About-us/Our-members/ (accessed 9 April 2015).

Smil, V. (2008) *Global Catastrophes and Trends: The Next Fifty Years*, London: MIT Press.

Smith, M. (2007) 'Cultural tourism in a changing world', *Tourism: The Journal for the Tourism Industry*, 1(1): 18–19.

Smith, S. (1994) 'The tourism product', *Annals of Tourism Research*, 21(3): 582–95.

Smith, V. (ed.) (1977) *Hosts and Guests: The Anthropology of Tourism*, 1st edn, Philadelphia, PA: University of Pennsylvania Press.

Smith, V. (ed.) (1989) *Hosts and Guests: The Anthropology of Tourism*, 2nd edn, Philadelphia, PA: University of Pennsylvania Press.

Smith, V. (1996) 'Indigenous tourism: the four Hs', in R. Butler and T. Hinch (eds), *Tourism and Indigenous Peoples*, London: International Thomson Business Press, pp. 283–307.

Smith, V. and Eadington, W. (eds) (1992) *Tourism Alternatives: Potentials and Problems in the Development of Tourism*, Philadelphia, PA: University of Pennsylvania Press.

Snyman, S. (2014) 'The impact of ecotourism employment on rural household incomes and social welfare in six southern African countries', *Tourism and Hospitality Research*, 14(1/2): 37–52.

Sofreavia in association with P.T. Asana Wirasta Setia and P.T. Desigras (1993) *Feasibility Study for Airport Development in Lombok, Master Plan Executive Summary.*

Solomon, M. (1994) *Consumer Behaviour: Buying, Having, Being*, 2nd edn, Needham Heights, MA: Allyn & Bacon.

Sönmez, S. (1998) 'Tourism, terrorism and political instability', *Annals of Tourism Research*, 25(2): 416–56.

Sönmez, S. and Graefe, A. (1998) 'Influence of terrorism risk on foreign tourism decisions', *Annals of Tourism Research*, 25(1): 112–44.

Southgate, C. and Sharpley, R. (2015) 'Tourism, development and the environment', in R. Sharpley and D. Telfer (eds), *Tourism and Development: Concepts and Issues*, 2nd edn, Bristol: Channel View Publications, pp. 250–84.

Speakman, M. and Sharpley, R. (2012) 'A chaos theory perspective on destination crisis management: evidence from Mexico', *Journal of Destination Marketing and Management*, 1(1): 67–77.

Spenceley, A. (2004) 'Responsible nature-based tourism planning in South Africa and the commercialisation of Kruger National Park', in D. Diamantis (ed.), *Ecotourism Management and Assessment*, London: Thomson, pp. 267–80.

SpiceJet (2015) *SpiceJet Company Profile*, available at: www.spicejet.com/Media Kit.aspx (accessed 8 April 2015).

Spring Airlines (2015) *About Us*, available at: http://help.ch.com/en/Static/Spring Airlines.shtml (accessed 10 April 2015).

Star Alliance (2015) *Travel the World with the Star Alliance*, available at: www.staralliance.com/en/about/member_airlines/ (accessed 9 April 2015).

Starmer-Smith, C. (2004) 'Eco-friendly tourism on the rise', *Daily Telegraph Travel*, 6 November, p. 4.

Steck, B. and Wood, K. (2010) *Tourism: More Value for Zanzibar: Value Chain Analysis: Summary Report*, SNV – Netherlands Development Organisation, VSO Voluntary Overseas, The Hague, available at: www.snvworld.org/files/publications/tourism_-_more_value_for_zanzibar.pdf (accessed 16 April 2015).

Steer, A. and Wade-Gery, W. (1993) 'Sustainable development: theory and practice for a sustainable future', *Sustainable Development*, 1(3): 23–35.

Stephen, C. (2015) 'Fear of Tunisia's democracy led Isis to launch an attack on its tourist economy', *The Guardian*, 22 March, available at: www.theguardian.com/world/2015/mar/22/tunisia-terror-attack-tourists (accessed 9 April 2015).

Stevens, R. (2002) *Sustainable Tourism in National Parks and Protected Areas: An Overview*, Scottish Natural Heritage Commissioned Report F01NC04, Inverness: Scottish Natural Heritage.

Stoddart, H. and Rogerson, C. (2004) 'Volunteer tourism: the case of Habitat for Humanity South Africa', *GeoJournal*, 60(3): 311–18.

Stone, A. (2014) 'Tourism falling off in Africa, far beyond the Ebola zone', *National Geographic*, 31 October, available at: http://news.nationalgeographic.com/news/2014/11/141101-ebola-travel-africa-tourism-world-health-medicine/ (accessed 11 April 2015).

Stone, M. (2015) 'Community-based ecotourism: a collaborative partnership perspective', *Journal of Ecotourism*, 18 March 2015, available at: www.tandfonline.com/doi/abs/10.1080/14724049.2015.1023309# (accessed 18 April 2015).

Strauß, S. (2015) 'Alliances across ideologies: networking with NGOs in a tourism dispute in Northern Bali', *The Asia Pacific Journal of Anthropology*, 16(2): 123–40.

Streeten, P. (1977) 'The basic features of a basic needs approach to development', *International Development Review*, 3: 8–16.

Stubbs, R. and Underhill, G. (eds) (2006) *Political Economy and the Changing Global Order*, 3rd edn, Don Mills, Ontario: Oxford University Press.

Swarbooke, J. and Horner, S. (2007) *Consumer Behaviour in Tourism*, 2nd edn, Oxford: Butterworth-Heinemann.

Tani, C. (2015) *The Chinese Travel Market in 2015: A Breakdown of Tour Operators*, available at: www.rezdy.com/blog/chinese-travel-2015-tour-opera tors/ (accessed 11 April 2015).

Taylor, I. (2014) 'Analysis: holidays tipped to top 100 million', *Travel Weekly*, 27 February, available at: www.travelweekly.co.uk/articles/2014/02/27/47 112/analysis+holidays+tipped+to+top+100+million.html (accessed 4 February 2015).

Taylor, P., Watts, M. and Johnston, R. (2002) 'Geography/globalisation', in R. Johnston, P. Taylor and M. Watts (eds), *Geographies of Global Change Remapping the World*, Oxford: Blackwell, pp. 1–17.

TCB (2014) *Bhutan Tourism Monitor, Annual Report 2013*, Tourism Council of Bhutan, available at: www.abto.org.bt/wp-content/uploads/2014/06/BTM-2013.pdf (accessed 2 May 2015).

Tearfund (2000) *Tourism – An Ethical Issue*, Market Research Report, Teddington: Tearfund.

Telfer, D. (1996) 'Food purchases in a five-star hotel: a case study of the Aquila Prambanan Hotel, Yogyakarta, Indonesia', *Tourism Economics*, 2(4): 321–38.

Telfer, D. J. (2000) 'Agritourism – a path to community development? The case of Bangunkerto, Indonesia', in D. Hall and G. Richards (eds), *Tourism and Sustainable Community Development*, London: Routledge, pp. 242–57.

Telfer, D. J. (2001) 'Tourism and community development in a biosphere: Sierra Del Rosario, Cuba', unpublished paper.

Telfer, D. J (2002a) 'Tourism and regional development issues', in R. Sharpley and D. J Telfer (eds), *Tourism and Development: Concepts and Issues*, Clevedon: Channel View Publications, pp. 112–48.

Telfer, D. J. (2002b) 'The evolution of tourism and development theory', in R. Sharpley and D. J. Telfer (eds), *Tourism and Development: Concepts and Issues*, Clevedon: Channel View Publications, pp. 35–78.

Telfer, D. J. (2013) 'The Brundtland Report (Our Common Future) and tourism', in A. Holden and D. Fennel (eds), *The Routledge Handbook of Tourism and Environment*, London: Routledge, pp. 213–27.

Telfer, D. J (2015a) 'The evolution of development theory and tourism', in R. Sharpley and D. Telfer (eds), *Tourism and Development: Concepts and Issues*, 2nd edn, Bristol: Channel View Publications, pp. 31–79.

Telfer, D. J. (2015b) 'Tourism and regional development issues', in R. Sharpely and D. J. Telfer (eds), *Tourism and Development Concepts and Issues*, 2nd edn, Bristol: Channel View Publications, pp. 140–77.

Telfer, D. J. and Hashimoto, A. (2006) 'Recourse management: social, cultural, physical environment and the optimization of impacts', in D. Buhalis and C. Costa (eds), *Tourism Management Dynamics: Trends, Management and Tools*, Oxford: Elsevier, pp. 145–54.

Telfer, D. J and Hashimoto, A. (2015) 'Tourism, development and international studies', in R. Sharpley and D. J. Telfer (eds), *Tourism and Development: Concepts and Issues*, 2nd edn, Bristol: Channel View Publications, pp. 400–27.

Telfer, D. J. and Wall, G. (1996) 'Linkages between tourism and food production', *Annals of Tourism Research*, 23(3): 635–53.

Telfer, D. J. and Wall, G. (2000) 'Strengthening backward economic linkages: local food purchasing by three Indonesian hotels', *Tourism Geographies*, 2(4): 421–47.

Teo, P. (2002) 'Striking a balance for sustainable tourism: implications of the discourse on globalisation', *Journal of Sustainable Tourism*, 10(6): 459–74.

The Economist (2007) 'Globalisation's offspring', *The Economist*, 383(8523): 11.

The Economist (2009) *Political Instability Index: Aux barricades!*, available at: www.economist.com/node/13349331 (accessed 6 May 2015).

The Economist (2013) 'Social progress beyond GDP', *The Economist*, 18 April 2013, available at: www.economist.com/blogs/feastandfamine/2013/04/social-progress (accessed 10 April 2015).

Thomas, A. (2000) 'Poverty and the "end of development"', in T. Allen and A. Thomas (eds), *Poverty and Development into the 21st Century*, Oxford: Oxford University Press, pp. 3–22.

Thurau, B., Seekamp, E., Carver, A. and Lee, J. (2015) 'Should cruise ports market ecotourism? A comparative analysis of passenger spending expectations within the Panama Canal Watershed', *International Journal of Tourism Research*, 17(1): 45–54.

Timothy, D. J. (1998) 'Cooperative tourism planning in a developing destination', *Journal of Sustainable Tourism*, 6(1): 52–68.

Timothy, D. (1999) 'Participatory planning a view of tourism in Indonesia', *Annals of Tourism Research*, 26(2): 371–91.

Timothy, D. J. (2000) 'Tourism planning in Southeast Asia: bringing down borders through cooperation', in K. Chon (ed.), *Tourism in Southeast Asia: A New Direction*, New York: The Haworth Hospitality Press, pp. 21–38.

Timothy, D. J. (2004) 'Tourism and supranationalism in the Caribbean', in D. Duval (ed.), *Tourism in the Caribbean: Trends, Development, Prospects*, London: Routledge, pp. 119–35.

Timothy, D. J. (2005) *Shopping Tourism, Retailing and Leisure*, Clevedon: Channel View Publications.

Timothy, D. J. and Boyd, S. (2003) *Heritage Tourism*, London: Prentice Hall.

Todaro, M. (1997) *Economic Development*, 6th edn, Harlow: Addison-Wesley.

Todaro, M. and Smith, S. (2011) *Economic Development*, 11th edn, Harlow: Pearson Education.

Tomazos, K. and Butler, R. (2010) 'The volunteer tourist as "hero"', *Current Issues in Tourism*, 13(4): 363–80.

Torres, R. (2002) 'Cancun's tourism development from a Fordist spectrum of analysis', *Tourist Studies*, 2(1): 87–116.

Torres, M. and Anderson, M. (2004) *Fragile States: Defining Difficult Environments for Poverty Reduction*, DfIDL PRDE Working Paper 1, available at: www.ineesite.org/uploads/files/resources/doc_1_FS-Diff_environ_for_pov_reduc.pdf2012 (accessed 6 May 2015).

Torres, R. and Momsen, J. H. (2004) 'Challenges and potential of linking tourism and agriculture to achieve pro-poor tourism objectives', *Progress in Development Studies*, 4(4): 294–318.

Tosun, C. (2000) 'Limits to community participation in the tourism development process in developing countries', *Tourism Management*, 21(6): 613–33.

Tourism Concern (2013) *Save Bimini*, available at: http://tourismconcern.org.uk/save-bimini/ (accessed 24 April 2015).

Tourism Concern (2014a) *Displacement and Land Rights in Sri Lanka*, available at: http://tourismconcern.org.uk/displacement-and-land-rights-in-sri-lanka/ (accessed 6 May 2015).

Tourism Concern (2014b) *Burma, The Issue*, available at: http://tourismconcern.org.uk/burma/ (accessed 18 May 2015).

Tourism Concern (n.d.) *Behind the Smile: The Tsunami of Tourism*, London: Tourism Concern.

Tran, L. and Walter, P. (2014) 'Ecotourism, gender and development in northern Vietnam', *Annals of Tourism Research*, 44(9): 116–30.

Tribe, J., Font, X., Griffiths, N., Vickery, R. and Yale, K. (2000) *Environmental Management for Rural Tourism and Recreation*, London: Cassell.

Tsogo Sun (2015) *About Tsogo Sun*, available at: www.tsogosun.com/about-us/Pages/default.aspx (accessed 26 April 2015).

Tucker, H. and Boonabaana, B. (2012) 'A critical analysis of tourism, gender and poverty reduction', *Journal of Sustainable Tourism*, 20(3): 437–55.

Turkstat (2015) *Number of Trips of Domestic Visitors*, available at: www.turkstat.gov.tr/PreTablo.do?alt_id=1072 (accessed 5 February 2015).

Turner, L. and Ash, J. (1975) *The Golden Hordes: International Tourism and the Pleasure Periphery*, London: Constable.

TVNZ (2006) *Coup Costing Fiji Tourism Millions*, available at: http://tvnz.co.nz/view/page/411419/930768 (accessed 15 February 2007).

Uber (2015) *UN Women + Uber = A Vision for Equality*, available at: https://blog.uber.com/un-women (accessed 10 April 2015).

UN (1948) *The Universal Declaration of Human Rights*, United Nations, available at: www.un.org/en/documents/udhr/ (accessed 2 May 2015).

UN (1955) *Social Progress Through Community Development*, New York: United Nations.

UN (1999) *The World at Six Billion*, United Nations, Department of Economic and Social Affairs, available at: www.un.org/esa/population/publications/sixbillion/sixbilcover.pdf (accessed 10 May 2015).

UN (2003) *Poverty Alleviation through Sustainable Tourism Development*, New York: United Nations.

UN (2007) *Universal Declaration of Human Rights*, available at: www.un.org/en/events/humanrightsday/2007/hrphotos/declaration%20_eng.pdf (accessed 12 May 2015).

UN (2011) *Global Food Security*, available at: www.un-foodsecurity.org/background (accessed 12 May 2015).

UN (2013) *World Population Prospects: The 2012 Revision*, United Nations, Department of Economic and Social Affairs, available at: http://esa.un.org/wpp/documentation/pdf/wpp2012_highlights.pdf (accessed 2 May 2015).

UN (2014a) *The Millennium Development Goals Report 2014*, New York: United Nations.

UN (2014b) *The Road to Dignity by 2030: Ending Poverty, Transforming All Lives and Protecting the Planet*, Synthesis report of the Secretary General on the post-2015 sustainable development agenda, December, available at: www.un.

org/en/ga/search/view_doc.asp?symbol=A/69/700&referer=http://www.un.org/en/documents/&Lang=E (accessed 8 April 2015).

UN (2014c) *Report of the Open Working Group of the General Assembly on Sustainable Development Goals*, August, available at: www.un.org/ga/search/view_doc.asp?symbol=A/68/970&Lang=E (accessed 8 April 2015).

UN (2015) *We Can End Poverty: Millennium Development Goals and Beyond 2015*, available at: www.un.org/millenniumgoals/poverty.shtml (accessed 17 May 2015).

UN Climate Change Conference (2015) *COP21 Main Issues*, available at: www.cop21.gouv.fr/en/cop21-cmp11/cop21-main-issues (accessed 11 May 2015).

UN Framework Convention on Climate Change (2014) *First Steps to a Safer Future: Introducing the United Nations Framework Convention on Climate Change*, available at: http://unfccc.int/essential_background/convention/items/6036.php (accessed 11 May 2015).

UN Global Compact (2008) *Global Compact and World Tourism Organization Launch TOURpact.GC – a New Framework for CSR in Tourism*, available at: www.unglobalcompact.org/newsandevents/news_archives/2008_10_01.html, (accessed 24 April 2015).

UNCSD (2012a) *Report of the United Nations Conference on Sustainable Development, Rio de Janeiro*, New York: United Nations.

UNCSD (2012b) *Green Economy in the Context of Sustainable Development and Poverty Eradication*, Rio + 20 United Nations Conference on Sustainable Development, available at: www.uncsd2012.org/greeneconomy.html (accessed 6 May 2015).

UNCTAD (2001) *Tourism and Development in the Least Developed Countries*, Third UN Conference on the Least Developed Countries, Las Palmas, Canary Islands, available at: http://unctad.org/en/Docs/poldcm64.en.pdf (accessed 7 May 2015).

UNCTAD (2007) *FDI in Tourism: The Development Dimension*, New York: United Nations, available at: http://unctad.org/en/Docs/iteiia20075_en.pdf (accessed 23 April 2015).

Underhill, G. (2006) 'Conceptualising the changing global order', in R. Stubbs and G. Underhill (eds), *Political Economy and the Changing Global Order*, 3rd edn, Don Mills, Ontario: Oxford University Press, pp. 3–23.

UNDP (1992) *Tourism Sector Programming and Policy Development, Output 1: National Tourism Strategy*, New York: United Nations Development Programme.

UNDP (2010) *Human Development Report 2010*, New York: OUP.

UNDP (2014a) *Human Development Report Sustaining Human Progress: Reducing Vulnerabilities and Building Resilience*, New York: UNDP, available at: www.undp.org/content/dam/undp/library/corporate/HDR/2014HDR/HDR 2014-Summary-English.pdf (accessed 20 May 2015).

UNDP (2014b) *Humanity Divided: Confronting Inequality in Developing Countries*, New York: UNDP, available at: www.undp.org/content/undp/en/home/librarypage/poverty-reduction/humanity-divided-confronting-inequality-in-developing-countries.html (accessed 20 May 2015).

UNDP (2015) *Frequently Asked Questions – Human Development Index (HDI)*, available at: http://hdr.undp.org/en/faq-page/human-development-index-hdi#t 292n61 (accessed 28 January 2015).

UNEP (2002) *Tourism's Three Main Impact Areas*, available at: www.uneptie.org/pc/tourism/sust-tourism/env-3main.htm.

UNEP (2005) *Integrating Sustainability into Business: A Management Guide for Responsible Tour Operations*, Paris: United Nations Environment Programme.

UNEP (2009) *Sustainable Coastal Tourism: An Integrated Planning and Management Approach*, Paris: United Nations Environment Programme, available at: www.unep.org/pdf/DTIE_PDFS/DTIx1091xPA-SustainableCoastal Tourism-Planning.pdf (accessed 24 April 2015).

UNEP (2011) *Towards a Green Economy: Pathways to Sustainable Development and Poverty Reduction*, available at: www.unep.org/greeneconomy (accessed 6 May 2015).

UNEP/WTO (2005) *Making Tourism More Sustainable: A Guide for Policy Makers*, Paris/Madrid: United Nations Environment Programme/World Tourism Organization.

UNFCCC (1992) *United Nations Framework Convention on Climate Change*, New York: United Nations, available at: www.ipcc.ch/pdf/assessment-report/ar5/wg2/ar5_wgII_spm_en.pdf (accessed 11 May 2015).

UNHSP (2003) *The Challenge of Slums: Global Report on Human Settlements 2003*, UN Human Settlements Programme, London: Earthscan.

UNWTO (2006) *Technical Cooperation: An Effective Tool for Development Assistance*, available at: www.world-tourism.org/techcoop/eng/objectives.htm (accessed 24 August 2006).

UNWTO (2009a) *From Davos to Copenhagen and Beyond: Advancing Tourism's Response to Climate Change*, Madrid: UNWTO, available at: http://sdt.unwto.org/sites/all/files/docpdf/fromdavostocopenhagenbeyondunwtopaperelectronicversion.pdf (accessed 12 May 2015).

UNWTO (2009b) *Roadmap for Recovery, Tourism & Travel: A Primary Vehicle for Job Creation and Economic Recovery*, Report of the Secretary General on Sustainable Tourism in Challenging Times, Madrid: UNWTO, available at: www.unwto.org/conferences/ga/en/pdf/18_08.pdf (accessed 14 May 2015).

UNWTO (2011) *UNWTO Technical Product Portfolio*, Madrid: UNWTO, available at: http://dtxtq4w60xqpw.cloudfront.net/sites/all/files/pdf/unwto_technical_product_portfolio.pdf (accessed 29 April 2015).

UNWTO (2012) *Challenges and Opportunities for Tourism Development in Small Island Developing States*, Madrid: UNWTO, available at: www.e-unwto.org/content/m47h4q (accessed 6 May 2012).

UNWTO (2013a) *Tourism Highlights, 2013 Edition*, available at: http://dtxtq4w60xqpw.cloudfront.net/sites/all/files/pdf/unwto_highlights13_en_hr.pdf (accessed 12 February 2014).

UNWTO (2013b) *Sustainable Tourism for Development Guidebook*, Madrid: World Tourism Organization, available at: http://dtxtq4w60xqpw.cloudfront.net/sites/all/files/docpdf/devcoengfinal.pdf (accessed 4 February 2015).

UNWTO (2013c) *Key Outbound Tourism Markets in South-East Asia – Indonesia, Malaysia, Singapore's, Thailand and Vietnam*, Madrid: UNWTO, available at: http://pub.unwto.org/WebRoot/Store/Shops/Infoshop/516D/592B/0A56/0A3A/2264/C0A8/0164/E474/130415_KSE_asian_markets_excerpt.pdf (accessed 19 May 2013).

UNWTO (2014a) *UNWTO Tourism Highlights, 2014 Edition*, available at: http://mkt.unwto.org/en/barometer (accessed 24 January 2015).

UNWTO (2014b) *UNWTO Annual Report 2013*, Madrid: UNWTO.

UNWTO (2014c) *Tourism in Small Island Developing States (SIDS): Building a More Sustainable Future for the People of Islands*, Madrid: UNWTO, available at: http://dtxtq4w60xqpw.cloudfront.net/sites/all/files/pdf/unwto_tourism_in_sids_a4_w.tables.pdf (accessed 19 May 2015).

UNWTO (2015a) *Sustainable Development of Tourism*, available at: http://sdt.unwto.org/content/about-us-5 (accessed 3 February 2015).

UNWTO (2015b) *About Technical Cooperation and Services*, available at: http://cooperation.unwto.org/content/about-technical-cooperation-and-services (accessed 29 April 2015).

UNWTO (2015c) *Resilience of Tourism Development*, Madrid: UNWTO, available at: http://rcm.unwto.org/en/content/about-us-7 (accessed 24 April 2015).

UNWTO and ILO (2013) *Economic Crisis, International Tourism Decline and Its Impact on the Poor*, Madrid: UNWTO, available at: www.ilo.org/wcmsp5/groups/public/—ed_dialogue/--sector/documents/publication/wcms_214576.pdf (accessed 14 May 2015).

Uriely, N. (1997) 'Theories of modern and postmodern tourism', *Annals of Tourism Research*, 24(4): 982–5.

Uriely, N. (2009) 'Deconstructing tourist typologies: the case of backpacking', *International Journal of Culture, Tourism and Hospitality Research*, 3(4): 306–12.

Urry, J. (1990) *The Tourist Gaze*, London: Sage.

Urry, J. (1995) *Consuming Places*, London: Routledge.

Urry, J. (2012) *The Tourist Gaze*, 3rd edn, London: Sage.

Urry, J. and Larsen, J. (2012) *The Tourist Gaze 3.0*, London: Sage.

US-Mexico Chamber of Commerce (2011) *Issue Paper 3: Tourism Development, Medical Tourism, and Safe and Secure Tourism in Mexico*, available at: www.usmcoc.org/papers-current/3-Tourism-Development-Medical-Tourism-and-Safe-and-Secure-Tourism-in-Mexico.pdf (accessed 30 April 2015).

USAID (2014) *Food Assistance Fact Sheet – Kenya*, available at: www.usaid.gov/kenya/food-assistance (accessed 13 May 2105).

Van Uffelen (2013) 'The de-disasterisation of food crises: structural reproduction or change in policy development and response options? A case study from Ethiopia', in D. Hilhorst (ed.), *Disaster, Conflict and Society in Crises*, London: Routledge, pp. 58–75.

Van Wijk, J., Van der Duim, R. and Sumba, D. (2015) 'The emergence of institutional innovations in tourism: the evolution of the African Wildlife Foundation's tourism conservation enterprise', *Journal of Sustainable Tourism*, 23(1): 104–25.

Vargas, C. (2000) 'Community development and micro-enterprises: fostering sustainable development', *Sustainable Development*, 8(1): 11–26.

Vatikiotis, A., Kalamara, A., Skourtis, G., Assiouras, I. and Koniordus, M. (2011) 'Fair trade tourism: the case of South Africa', *International Journal of Management Cases*, 13(3): 476–83.

Visit Mexico Press (2006a) *Private Investment in Mexico's Tourism Sector Booming, Sectur Expects Year-End Total to Surpass US$12 Billion*, available at: www.visitmexicopress.com?press_release02.ap?pressID=179 (accessed 21 September 2006).

Visit Mexico Press (2006b) *Fonatur Outlines Accomplishments and Future Development Strategy at Tiangus*, available at: www.visitmexicopress.com? press_release02.ap?pressID=162 (accessed 21 September 2006).

Vörösmarty, C., Green, P., Salisbury, J. and Lammers, R. (2000) 'Global water resources: vulnerability from climate change and population growth', *Science*, 289: 284–8.

Wade, R. (2004) *Governing the Market*, 2nd edn, Princeton, NJ: Princeton University Press.

Wahab, S. and Cooper, C. (eds) (2001) *Tourism in the Age of Globalization*, London: Routledge.

Wall, G. (1993) 'Towards a tourism typology', in J. G. Nelson, R. W. Butler and G. Wall (eds), *Tourism and Sustainable Development: Monitoring, Planning, Managing*, Waterloo, Ontario: University of Waterloo; Heritage Resources Centre Joint Publication, pp. 45–58.

Wall, G. and Mathieson, A. (2006) *Tourism Change, Impacts and Opportunities*, Toronto: Pearson Prentice Hall.

Wall, G. and Xie, P. (2005) 'Authenticating ethnic tourism: Li Dancers' perspectives', *Asia Pacific Journal of Tourism Research*, 10(1): 1–21.

Wang, A. and Edwards, A. (2012) 'China wealth gap yawn wider as Beijing pours money into tourist town', *Financial Post*, 7 May 2012, available at: http://business.financialpost.com/news/economy/china-wealth-gap-yawns-wider-as-beijing-pours-money-into-tourist-town (accessed 13 May 2015).

Wang, Y. and Wall, G. (2005) 'Resorts and residents: stress and conservatism in a displaced community', *Tourism Analysis*, 10(1): 37–53.

Washington, H. (2015) *Demystifying Sustainability: Towards Real Solutions*, Abingdon: Routledge.

WCED (1987) *Our Common Future*, Oxford: Oxford University Press.

Wearing, S. (2001) *Volunteer Tourism: Experiences That Make a Difference*, Wallingford: CABI.

Wearing, S. and McGehee, N. (2013) 'Volunteer tourism: a review', *Tourism Management*, 38(1): 120–30.

Weaver, D. (2004) 'Manifestations of ecotourism in the Caribbean', in D. Duval (ed.), *Tourism in the Caribbean Trends, Development, Prospects*, London: Routledge, pp. 172–86.

Weaver, D. (2010) 'Geopolitical dimensions of sustainable tourism', *Tourism Recreation Research*, 35(1): 45–51.

Weaver, D. (2014) 'Asymmetrical dialectics of sustainable tourism: towards enlightened mass tourism', *Journal of Travel Research*, 53: 131–40.

Web, T. (2010) 'Few air travellers offset carbon emissions, study finds', *The Guardian*, 30 August, available at: www.theguardian.com/business/2010/aug/30/carbon-emissions-offset-civil-aviation-authority (accessed 8 April 2015).

Wei, F. (2013) *Compendium of Best Practices in Sustainable Tourism*, UN Department of Social and Economic Affairs, available at: https://sustainable development.un.org/content/documents/3322Compendium%20of%20Best%20Practices%20in%20Sustainable%20Tourism%20-%20Fen%20Wei%2001032014.pdf (accessed 15 December 2014).

Weinstein, M. (2005) 'Introduction', in M. Weinstein (ed.), *Globalization: What's New?*, New York: Columbia University Press, pp. 1–18.

WFP (2015) *What Causes Hunger?*, available at: www.wfp.org/hunger/causes (accessed 13 May 2015).

Wheeller, B. (1991) 'Tourism's troubled times: responsible tourism is not the answer', *Tourism Management*, 12(2): 91–6.

Wheeller, B. (1992) 'Eco or ego tourism: new wave tourism', *Insights*, Vol. III, London: English Tourist Board, D41–44.

Wiarda, H. J. (1983) 'Toward a nonethnocentric theory of development: alternative conceptions from the Third World', *Journal of Developing Areas*, 17, reprinted in C. K. Wilber (ed.) (1988), *The Political Economy of Development and Underdevelopment*, 4th edn, Toronto: McGraw-Hill, pp. 59–82.

Wilbanks, T. (2004) 'Geography and technology', in S. Brunn, S. Cutter and J. Harrington (eds), *Geography and Technology*, London: Kluwer Academic Publications, pp. 1–16.

Williams, A. and Hall, C. M. (2000) 'Tourism and migration: new relationships between production and consumption', *Tourism Geographies*, 2(1): 5–27.

Winter, T. (2008) 'Post-conflict heritage and tourism in Cambodia: the burden of Angkor', *International Journal of Heritage Studies*, 14(6): 525–39.

Wood, K. and House, S. (1991) *The Good Tourist: A Worldwide Guide for the Green Traveller*, London: Mandarin.

Wood, R. (1997) 'Tourism and the state: ethnic options and constructions of otherness', in M. Picard and R. Wood (eds), *Tourism, Ethnicity and the State in Asian and Pacific Countries*, Honolulu, HI: University of Hawaii Press, pp. 1–34.

Wood, R. (2004) 'Global currents: cruise ships in the Caribbean', in D. Duval (ed.), *Tourism in the Caribbean Trends, Development, Prospects*, London: Routledge, pp. 152–71.

World Bank (2005) *Classification of Economies*, available at: www.worldbank.org/data/aboutdata/errata03/Class.htm (accessed 16 April 2005).

World Bank (2012) *Gender Equality and Development*, available at: http://econ.worldbank.org/WBSITE/EXTERNAL/EXTDEC/EXTRESEARCH/EXTWDRS/EXTWDR2012/0,,contentMDK:22999750~menuPK:8154981~pagePK:64167689~piPK:64167673~theSitePK:7778063,00.html (accessed 7 May 2015).

World Bank (2015a) *Updated Income Classifications, Country Groups*, available at: http://data.worldbank.org/about/country-and-lending-group (accessed 8 January 2015).

World Bank (2015b) *Poverty and Equity, China*, available at: http://povertydata.worldbank.org/poverty/country/CHN (accessed 8 January 2015).

World Bank (2015c) *GDP per Capita*, available at: http://data.worldbank.org/indicator/NY.GDP.PCAP.CD (accessed 12 January 2015).

World Bank (2015d) *Food Security Overview*, available at: www.worldbank.org/en/topic/foodsecurity/overview#1 (accessed 12 May 2015).

World Bank (2015e) *What Is Governance?*, available at: http://web.worldbank.org/WBSITE/EXTERNAL/COUNTRIES/MENAEXT/EXTMNAREGTOPGOVERNANCE/0,,contentMDK:20513159~pagePK:34004173~piPK:34003707~theSitePK:497024,00.html (accessed 5 May 2015).

World Economic Forum (2015) *The Global Risks Report 2015*, available at: www.weforum.org/reports/global-risks-report-2015 (accessed 14 May 2015).

World Responsible Tourism Awards (2015) *Past Winners: 2008 Responsible Tourism Awards*, available at: www.responsibletravel.com/awards/winners/2008.htm (accessed 14 May 2015).

World Trade Organization (2001) *Doha WTO Ministerial Declaration*, available at: wto.org/english/thewto_e/minist_e/min01_e/mindecl_e.htm (accessed 31 July 2006).

Wright, H., Haq, S. and Reeves, J. (2015) *Impact of Climate Change on Least Developed Countries: Are the SDGs Possible?*, International Centre for Climate Change and Development Briefing, May 2015, available at: http://pubs.iied.org/pdfs/17298IIED.pdf (accessed 20 May 2015).

Wright, R. (2008) *Environmental Science toward a Sustainable Future*, 10th edn, instructors edn, Upper Saddle River, NJ: Pearson Prentice Hall.

WTO (1980) *Manila Declaration on World Tourism*, Madrid: World Tourism Organization.

WTO (1981) *The Social and Cultural Dimension of Tourism*, Madrid: World Tourism Organization.

WTO (1986) *Village Tourism Development Programme for Nusa Tenggara*, Madrid: World Tourism Organization, UNDP.

WTO (1993) *Sustainable Tourism Development: A Guide for Local Planners*, Madrid: World Tourism Organization.

WTO (1996) *Agenda 21 for the Travel and Tourism Industry: Towards Environmentally Sustainable Development*, Madrid: World Tourism Organization.

WTO (1998) *Tourism – 2020 Vision: Influences, Directional Flows and Key Influences*, Madrid: World Tourism Organization.

WTO (1999) *Sustainable Tourism Development: An Annotated Bibliography*, Madrid: World Tourism Organization.

WTO (2004a) *Indicators of Sustainable Development for Tourism Destinations: A Guidebook*, Madrid: World Tourism Organization.

WTO (2004b) *Tourism Highlights Edition 2004*, Madrid: World Tourism Organization.

WTO (2005) *World Tourism Barometer 3(2)*, Madrid: World Tourism Organization.

WTO (2006) 'Asian outbound tourism takes off', World Tourism News Release June 14, 2006, available at: www.world-tourism.org/newsroom/Release/2006/june/asianoutbound.html (accessed 28 October 2006).

WTO/UNSTAT (1994) *Recommendations on Tourism Statistics*, Madrid: World Tourism Organization.

WTO/WTTC (1996) *Agenda 21 for the Travel & Tourism Industry: Towards Environmentally Sustainable Development*, Madrid: World Tourism Organization/World Travel and Tourism Council.

WTTC (2003) *Blueprint for a New Tourism*, London: World Travel and Tourism Council.

WTTC (2004) *Country League Tables*, London: World Travel and Tourism Council.

WTTC (2012) *The Comparative Economic Impact of Travel and Tourism*, London: World Travel and Tourism Council.

WTTC (2014) *League Table Summary*, available at: www.wttc.org/focus/research-for-action/economic-impact-analysis/league-table-summary (accessed 8 January 2015).

WTTC (2015) *Tourism for Tomorrow Awards: World Travel and Tourism Council*, available at: www.wttc.org/tourism-for-tomorrow-awards/ (accessed 21 April 2015).

Xiao, C. (2013) 'Investors interested in Kenya', *China Daily USA*, available at: http://usa.chinadaily.com.cn/epaper/2013-08/21/content_16910850.htm (accessed 11 April 2015).

Xinhau (2014) 'China's investment in Australian tourism set to surge', *Capital News*, available at: www.capitalfm.co.ke/news/2014/07/chinas-investment-in-australian-tourism-set-to-surge/ (accessed 10 April 2015).

Yamamura, T. (2005) 'Donga Art in Lijinag, China: indigenous culture, local community and tourism', in C. Ryan and M. Aicken (eds), *Indigenous Tourism: The Commodification and Management of Culture*, London: Elsevier, pp. 181–99.

Yang, J., Ryan, C. and Zhang, L. (2014) 'Sustaining culture and seeking a just destination: governments, power and tension – a life cycle approach to analysing tourism development in an ethnic-inhabited scenic area in Xinjiang, China', *Journal of Sustainable Tourism*, 22(8): 1151–74.

Yeoman, I. (2012) *2050: Tomorrow's Tourism*, Bristol: Channel View Publications.

Yiannakis, A. and Gibson, H. (1992) 'Roles tourists play', *Annals of Tourism Research*, 19(3): 287–303.

Young, G. (1973) *Tourism: Blessing or Blight?* Harmondsworth: Penguin.

Zahara, A. and McGehee, N. (2013) 'Volunteer tourism: a host community capital perspective', *Annals of Tourism Research*, 42: 22–45.

Zaidi, A. (2008) *Features and Challenges of Population Ageing: The European Perspective*, Vienna: European Centre for Social Welfare Policy and Research.

Zhang, H., Chong, K. and Ap, J. (1999) 'An analysis of tourism development policy in modern China', *Tourism Management*, 20(4): 471–85.

Zhang, H., Pine, R. and Lam, T. (2005) *Tourism and Hotel Development in China*, New York: The Haworth Hospitality Press.

Zhao, W. (2009) 'The nature and roles of small tourism businesses in poverty alleviation: evidence from Guangxi, China', *Asia Pacific Journal of Tourism Research*, 14(2): 169–82.

Index